# Radiation Detection Systems

**Devices, Circuits, and Systems**
Series Editor - Krzysztof Iniewski

**Wireless Technologies**
Circuits, Systems, and Devices
*Krzysztof Iniewski*

**Circuits at the Nanoscale**
Communications, Imaging, and Sensing
*Krzysztof Iniewski*

**Internet Networks**
Wired, Wireless, and Optical Technologies
*Krzysztof Iniewski*

**Semiconductor Radiation Detection Systems**
*Krzysztof Iniewski*

**Electronics for Radiation Detection**
*Krzysztof Iniewski*

**Radiation Effects in Semiconductors**
*Krzysztof Iniewski*

**Electrical Solitons**
Theory, Design, and Applications
*David Ricketts and Donhee Ham*

**Semiconductors**
Integrated Circuit Design for Manufacturability
*Artur Balasinski*

**Integrated Microsystems**
Electronics, Photonics, and Biotechnology
*Krzysztof Iniewski*

**Nano-Semiconductors**
Devices and Technology
*Krzysztof Iniewski*

**Atomic Nanoscale Technology in the Nuclear Industry**
*Taeho Woo*

**Telecommunication Networks**
*Eugenio Iannone*

**Optical, Acoustic, Magnetic, and Mechanical Sensor Technologies**
*Krzysztof Iniewski*

**Biological and Medical Sensor Technologies**
*Krzysztof Iniewski*

**Graphene, Carbon Nanotubes, and Nanostuctures**
Techniques and Applications
*James E. Morris and Krzysztof Iniewski*

**Low Power Emerging Wireless Technologies**
*Reza Mahmoudi and Krzysztof Iniewski*

**High-Speed Photonics Interconnects**
*Lukas Chrostowski and Krzysztof Iniewski*

**Smart Sensors for Industrial Applications**
*Krzysztof Iniewski*

**MEMS: Fundamental Technology and Applications**
*Vikas Choudhary and Krzysztof Iniewski*

**Nanoelectronic Device Applications Handbook**
*James E. Morris and Krzysztof Iniewski*

**Novel Advances in Microsystems Technologies and Their Applications**
*Laurent A. Francis and Krzysztof Iniewski*

**Building Sensor Networks: From Design to Applications**
*Ioanis Nikolaidis and Krzysztof Iniewski*

**Embedded and Networking Systems**
Design, Software, and Implementation
*Gul N. Khan and Krzysztof Iniewski*

**Medical Imaging**
Technology and Applications
*Troy Farncombe and Krzysztof Iniewski*

**Nanoplasmonics**
Advanced Device Applications
*James W. M. Chon and Krzysztof Iniewski*

**Testing for Small-Delay Defects in Nanoscale CMOS Integrated Circuits**
*Sandeep K. Goel and Krishnendu Chakrabarty*

**Energy Harvesting with Functional Materials and Microsystems**
*Madhu Bhaskaran, Sharath Sriram, and Krzysztof Iniewski*

**Energy Harvesting with Functional Materials and Microsystems**
*Madhu Bhaskaran, Sharath Sriram, and Krzysztof Iniewski*

**Nanoscale Semiconductor Memories**
Technology and Applications
*Santosh K. Kurinec and Krzysztof Iniewski*

**Integrated Power Devices and TCAD Simulation**
*Yue Fu, Zhanming Li, Wai Tung Ng, and Johnny K.O. Sin*

**MIMO Power Line Communications**
Narrow and Broadband Standards, EMC, and Advanced Processing
*Lars Torsten Berger, Andreas Schwager, Pascal Pagani, and Daniel Schneider*

**Technologies for Smart Sensors and Sensor Fusion**
*Kevin Yallup and Krzysztof Iniewski*

**Smart Grids**
Clouds, Communications, Open Source, and Automation
*David Bakken*

**Microfluidics and Nanotechnology**
Biosensing to the Single Molecule Limit
*Eric Lagally*

**Nanopatterning and Nanoscale Devices for Biological Applications**
*Šeila Selimovic´*

**High-Speed Devices and Circuits with THz Applications**
*Jung Han Choi*

**Metallic Spintronic Devices**
*Xiaobin Wang*

**Mobile Point-of-Care Monitors and Diagnostic Device Design**
*Walter Karlen*

**VLSI**
Circuits for Emerging Applications
*Tomasz Wojcicki*

**Micro- and Nanoelectronics**
Emerging Device Challenges and Solutions
*Tomasz Brozek*

**Design of 3D Integrated Circuits and Systems**
*Rohit Sharma*

**Wireless Transceiver Circuits**
System Perspectives and Design Aspects
*Woogeun Rhee*

**Soft Errors**
From Particles to Circuits
*Jean-Luc Autran and Daniela Munteanu*

**Optical Fiber Sensors**
Advanced Techniques and Applications
*Ginu Rajan*

**Laser-Based Optical Detection of Explosives**
*Paul M. Pellegrino, Ellen L. Holthoff, and Mikella E. Farrell*

**Organic Solar Cells**
Materials, Devices, Interfaces, and Modeling
*Qiquan Qiao*

**Solid-State Radiation Detectors**
Technology and Applications
*Salah Awadalla*

**CMOS**
Front-End Electronics for Radiation Sensors
*Angelo Rivetti*

**Multisensor Data Fusion**
From Algorithm and Architecture Design to Applications
*Hassen Fourati*

**Electrostatic Discharge Protection**
Advances and Applications
*Juin J. Liou*

**Optical Imaging Devices**
New Technologies and Applications
*Ajit Khosla and Dongsoo Kim*

**Radiation Detectors for Medical Imaging**
*Jan S. Iwanczyk*

**Gallium Nitride (GaN)**
Physics, Devices, and Technology
*Farid Medjdoub*

**Mixed-Signal Circuits**
*Thomas Noulis*

**MRI**
Physics, Image Reconstruction, and Analysis
*Angshul Majumdar and Rabab Ward*

**Reconfigurable Logic**
Architecture, Tools, and Applications
*Pierre-Emmanuel Gaillardon*

**Ionizing Radiation Effects in Electronics**
From Memories to Imagers
*Marta Bagatin and Simone Gerardin*

**CMOS Time-Mode Circuits and Systems**
Fundamentals and Applications
*Fei Yuan*

**Tunable RF Components and Circuits**
Applications in Mobile Handsets
*Jeffrey L. Hilbert*

**Cell and Material Interface**
Advances in Tissue Engineering, Biosensor, Implant, and Imaging Technologies
*Nihal Engin Vrana*

**Nanomaterials**
A Guide to Fabrication and Applications
*Sivashankar Krishnamoorthy*

**Physical Design for 3D Integrated Circuits**
*Aida Todri-Sanial and Chuan Seng Tan*

**Wireless Medical Systems and Algorithms**
Design and Applications
*Pietro Salvo and Miguel Hernandez-Silveira*

**High Performance CMOS Range Imaging**
Device Technology and Systems Considerations
*Andreas Süss*

**Analog Electronics for Radiation Detection**
*Renato Turchetta*

**Power Management Integrated Circuits and Technologies**
*Mona M. Hella and Patrick Mercier*

**Circuits and Systems for Security and Privacy**
*Farhana Sheikh and Leonel Sousa*

**Multisensor Attitude Estimation**
Fundamental Concepts and Applications
*Hassen Fourati and Djamel Eddine Chouaib Belkhiat*

**Structural Health Monitoring of Composite Structures Using Fiber Optic Methods**
*Ginu Rajan and Gangadhara Prusty*

**Advances in Imaging and Sensing**
*Shuo Tang and Daryoosh Saeedkia*

**Semiconductor Devices in Harsh Conditions**
*Kirsten Weide-Zaage and Malgorzata Chrzanowska-Jeske*

**Introduction to Smart eHealth and eCare Technologies**
*Sari Merilampi, Krzysztof Iniewski, and Andrew Sirkka*

**Diagnostic Devices with Microfluidics**
*Francesco Piraino and Šeila Selimović*

**Magnetic Sensors and Devices**
Technologies and Applications
*Kirill Poletkin and Laurent A. Francis*

**Semiconductor Radiation Detectors**
Technology, and Applications
*Salim Reza*

**Noise Coupling in System-on-Chip**
*Thomas Noulis*

**High Frequency Communication and Sensing**
Traveling-Wave Techniques
*Ahmet Tekin and Ahmed Emira*

**3D Integration in VLSI Circuits**
Design, Architecture, and Implementation Technologies
*Katsuyuki Sakuma*

**IoT and Low-Power Wireless: Circuits, Architectures, and Techniques**
*Christopher Siu and Krzysztof Iniewski*

**Radio Frequency Integrated Circuit Design**
*Sebastian Magierowski*

**Low Power Semiconductor Devices and Processes for Emerging Applications in Communications, Computing, and Sensing**
*Sumeet Walia and Krzysztof Iniewski*

**Sensors for Diagnostics and Monitoring**
*Kevin Yallup and Laura Basiricò*

**Biomaterials and Immune Response**
Complications, Mechanisms and Immunomodulation
*Nihal Engin Vrana*

**High-Speed and Lower Power Technologies**
Electronics and Photonics
*Jung Han Choi and Krzysztof Iniewski*

**X-Ray Diffraction Imaging**
Technology and Applications
*Joel Greenberg and Krzysztof Iniewski*

**Compressed Sensing for Engineers**
*Angshul Majumdar*

**Low Power Circuits for Emerging Applications in Communications, Computing, and Sensing**
*Krzysztof Iniewski and Fei Yuan*

**Labs on Chip**
Principles, Design, and Technology
*Eugenio Iannone*

**Energy Efficient Computing & Electronics**
Devices to Systems
*Santosh K. Kurinec and Sumeet Walia*

**Spectral, Photon Counting Computed Tomography**
Technology and Applications
*Katsuyuki Taguchi, Ira Blevis and Krzysztof Iniewski*

**Flexible, Wearable, and Stretchable Electronics**
*Katsuyuki Sakuma*

**Radiation Detection Systems**
Sensor Materials, Systems, Technology, and Characterization Measurements
*Jan S. Iwanczyk and Krzysztof Iniewski*

**Radiation Detection Systems**
Medical Imaging, Industrial Testing and Security Applications
*Jan S. Iwanczyk and Krzysztof Iniewski*

For more information about this series, please visit: https://www.crcpress.com/Devices-Circuits-and-Systems/book-series/CRCDEVCIRSYS

# Radiation Detection Systems

## Sensor Materials, Systems, Technology, and Characterization Measurements

*2nd Edition*

**Edited by**

Jan S. Iwanczyk
Krzysztof Iniewski

CRC Press
Taylor & Francis Group
Boca Raton London New York

CRC Press is an imprint of the
Taylor & Francis Group, an **informa** business

CRC Press
Boca Raton and London
Second edition published 2022

by CRC Press
6000 Broken Sound Parkway NW, Suite 300, Boca Raton, FL 33487-2742

and by CRC Press
2 Park Square, Milton Park, Abingdon, Oxon, OX14 4RN



First edition published by CRC Press 2015

CRC Press is an imprint of Taylor & Francis Group, LLC



*Library of Congress Cataloging-in-Publication Data*

Names: Iwanczyk, Jan S., editor. | Iniewski, Krzysztof, 1960- editor.
Title: Radiation detection systems. Sensor materials, systems, technology and characterization measurements / edited by Jan Iwanczyk, Krzysztof Iniewski.
Description: Second edition. | Boca Raton : CRC Press, 2022. | Series: Devices, circuits, and systems | Includes bibliographical references and index.
Identifiers: LCCN 2021035713 (print) | ISBN 9780367707156 ( (hbk) | ISBN 9780367707170 (pbk) | ISBN 9781003147633 (ebk)
Subjects: LCSH: Radiation—Measurement—Instruments. | Radiography, Medical—Instruments. | Radiography—Equipment and supplies. | Photon detectors. | Semiconductor nuclear counters. | X-ray spectroscopy.
Classification: LCC QC795.5 .R383 2022 (print) | LCC QC795.5 (ebook) | DDC 539.7/7—dc23
LC record available at https://lccn.loc.gov/2021035713
LC ebook record available at https://lccn.loc.gov/2021035714

ISBN: 978-0-367-70715-6 (hbk)
ISBN: 978-0-367-70717-0 (pbk)
ISBN: 978-1-003-14763-3 (ebk)

DOI: 10.1201/9781003219446

Typeset in Palatino LT Std
by KnowledgeWorks Global Ltd.

# Contents

# *Preface*

The advances in semiconductor detectors, scintillators, photodetectors such as silicon photomultiplier (SiPM), and readout electronics in the past decades have led to significant progress in terms of performance and greater choice of the detection tools in many applications. This book presents the state-of-the-art in the design of detectors and integrated circuit design, in the context of medical imaging using ionizing radiation. It addresses exciting new opportunities in x-ray detection, computed tomography, (CT), bone dosimetry, and nuclear medicine (positron emission tomography, PET; single photon emission computed tomography, SPECT). In addition to medical imaging, the book explores other applications of radiation detection systems in security applications such as luggage scanning, dirty bomb detection, and border control.

The material in the book has been divided into two volumes. Volume I puts more emphasis on sensor materials, detector, and front end electronics technology and designs as well as system optimization for different applications. Also includes characterization measurements of the developed detection systems. Volume II is devoted to more specific applications of detection systems in medical imaging, industrial testing, and security applications. However, there is an unavoidable certain overlap in topics between both volumes.

A significant portion of a book describes new advances in development of detection systems based on CdZnTe (CZT) and CdTe detectors. The use of these detectors in fast growing medical and security applications is possible due to recent progress in material/detector technologies combined with the availability of application specific integrated circuits (ASIC) that provide a very compact low noise amplification and processing of the signal from individual detector pixels in imaging arrays. These new detectors have already been commercialized for use in surgical probes, gamma cameras, SPECT systems, and bone mineral density scanners. On the way there is a great effort to introduce this technology to the largest medical diagnostic imaging modality namely CT. Spectral x-ray photon counting possible with these new detectors will allow to reduce the radiation dose to the patient and to improve the imaging contrast but also can offer many other advantages. One of the most exciting possibilities is the future use of CT scanners not only as an anatomical modality but also as a functional modality that provides functional information. In security applications CZT and CdTe detection systems are used to detect hidden explosives, radioactive sources in luggage or radiological dispersal devices (dirty bombs) that are transported. There is a great effort to develop new imaging detectors for luggage scanning systems in airports for a direct x-ray transmission and for x-ray diffraction. CT systems for luggage scanning have similar requirements regarding photon counting detectors to that used in medical imaging. However, composition of luggage content is usually more complex than the patient body and for this reason it is advantageous to use in the readout more energy bins than that in medical applications. On the other hand, there are much less concerns about delivered radiation dose. In diffraction systems extremely high-count rate is not the major requirement because detectors are not placed in the direct x-ray beam. In the diffraction applications rather an excellent energy resolution and a large field of view of the detector arrays are the prerequisite.

Currently, SiPMs show very promising results in many fields. SiPMs used for reading the light from scintillators are starting to make a big impact on the design concepts for new nuclear medicine equipment for the gamma cameras used in SPECT and PET applications. These new designs allow for construction of more compact imagers with better performance that are not sensitive to magnetic fields, as are designs utilizing conventional photomultiplier tubes, allowing for the construction of SPECT and PET systems combined with magnetic resonance imaging (MRI) scanners in multimodality systems. Particularly fascinating is the renewed interest in Time-of-Flight (TOF) PET systems now becoming possible with development of combined very fast scintillators with novel SiPM structures making more feasible quest toward 10 ps the coincidence resolving time (CRT) to improve the spatial resolution and to enable the detection of abnormalities at the earliest possible stage.

Individual chapters of the book deal with variety of radiation detection systems beyond mentioned above systems based on CZT, CdTe, and SiPM technologies giving readers a broader view of radiation detection systems.

**Jan S. Iwanczyk, PhD,**
*Los Angeles, CA, USA*

*and*

**Krzysztof (Kris) Iniewski, PhD,**
*Vancouver, BC, Canada*
*February 19, 2021*

# Editors

**Dr. Jan S. Iwanczyk** is a consultant to universities and private companies since July 2017. He has served as a president and CEO of DxRay, Inc., Northridge, California, from 2005 to 2017. In 2017 DxRay, Inc., has been sold to OSI/Rapican Systems one of the three largest companies in the world that provides equipment for security in the airports. He previously was affiliated with several start-up private and publicly traded companies and centered on bringing novel scientific and medical technologies to the market. During the period from 1979 to 1989, Dr. Iwanczyk was associate professor at the University of Southern California, School of Medicine. He holds Master degree in Electronics and PhD degree in Physics. His multi-faceted experience combines operations, organizational development with strong scientific research and technical project management qualifications. Dr. Iwanczyk's technical expertise is in the field of x-ray and gamma ray imaging detectors and systems. In recent years he has been developing photon-counting, energy dispersive x-ray imaging detectors based on CdTe, CZT, and Si for medical and security applications. He is the author of over 200 scientific papers, 1 book, several book chapters, and 20 patents. He also lectures at major symposia worldwide as an invited speaker and has received numerous honors and awards including 2002 Merit Award, IEEE Nuclear and Plasma Sciences Society and the 2016 Scientist Award IEEE – RTSD for lifetime achievements.

**Dr. Krzysztof (Kris) Iniewski** is managing R&D activities at Redlen Technologies Inc., a detector company based in British Columbia, Canada. During his 15 years at Redlen he has managed development of highly integrated CZT detector products in medical imaging and security applications. Prior to Redlen, Kris held various management and academic positions at PMC-Sierra, University of Alberta, SFU, UBC, and University of Toronto.

Dr. Iniewski has published over 150+ research papers in international journals and conferences. He holds more than 25 international patents granted in the United States, Canada, France, Germany, and Japan. He wrote and edited more than 75 books for Wiley, Cambridge University Press, McGraw Hill, CRC Press, and Springer. He is a frequent invited speaker and has consulted for multiple organizations internationally.

# *Contributors*

**N. Sarzi Amadè**
IMEM - CNR
Parma, Italy

**William C. Barber**
Rapiscan Laboratories, Inc.
Northridge, California

**Muhammed E. Bedir**
University of Wisconsin-Madison
Wisconsin and Karamanoglu
Mehmetbey University, Turkey

**Manuelle Bettelli**
IMEM - CNR
Parma, Italy

**Aleksey E. Bolotnikov**
Brookhaven National Laboratory
Upton, New York

**Jeffrey J. Derby**
University of Minnesota
Minneapolis, Minnesota

**Eugen Engelmann**
Russian Academy of Sciences
Moscow, Russia

**Jan S. Iwanczyk**
Consultant
Los Angeles, California

**Ralph B. James**
Savannah River National Laboratory
Aiken, South Carolina

**Ho Kyung Kim**
Pusan National University
Busan, Republic of Korea

**Martyna Grodzicka-Kobylka**
National Centre for Nuclear Research
Otwock-Swierk, Poland

**Evgeniy Kuksin**
Rapiscan Laboratories, Inc.
Northridge, California

**Edward Morton**
Rapiscan Laboratories, Inc.
Northridge, California

**Marek Moszyński**
National Centre for Nuclear Research
Otwock-Swierk, Poland

**Elena Popova**
Russian Academy of Sciences
Moscow, Russia

**Stefan J. van der Sar**
Delft University of Technology
Delft, Netherlands

**Dennis R. Schaart**
Delft University of Technology
Delft, Netherlands

**Wolfgang Schmailzl**
Russian Academy of Sciences
Moscow, Russia

**Katherine S. Shanks**
Cornell University
Ithaca, New York

**Tomasz Szczęśniak**
National Centre for Nuclear Research
Otwock-Swierk, Poland

**Csaba Szeles**
Nious Technologies, Inc.
Wexford, Pennsylvania

**Bruce R. Thomadsen**
University of Wisconsin-Madison
Madison, Wisconsin
and
Karamanoğlu Mehmetbey University
Karaman, Turkey

**Sergey Vinogradov**
Russian Academy of Sciences
Moscow, Russia

**Jan Christopher Wessel**
Rapiscan Laboratories, Inc.
Northridge, California

# 1

# *CdZnTe and CdTe Crystals for Medical Applications*

**Csaba Szeles and Jeffrey J. Derby**

**CONTENTS**



## 1.1 CdTe and CdZnTe for Medical Applications

Cadmium Telluride (CdTe) and Cadmium Zinc Telluride (CdZnTe) semiconductor detectors are solid-state devices that provide direct conversion of the absorbed gamma-ray energy into an electronic signal. Many of the advantages of these detectors for medical applications stem from this inherent

DOI: 10.1201/9781003219446-1

energy discrimination and photon-counting capability. In addition, the high radiation stopping power and resulting high detection efficiency, low leakage current at room temperature, good charge transport of the photon-generated carriers, and the favorable chemical and mechanical properties that allow the fabrication of pixelated detectors enable the manufacture of sophisticated x-ray and gamma-ray imaging devices that are compact and can be operated at room temperature and at low voltage.

Nuclear medicine gamma cameras built for cardiac single photon emission computed tomography (SPECT) [1] and scintimammography, often referred to as molecular breast imaging (MBI) [2], take advantage of the superior energy resolution (2–5% FWHM at 140 keV) of CdZnTe detectors to improve scatter rejection and optimized pixel dimensions for high intrinsic spatial resolution that is independent of the photon energy. With the use of wide-angle collimators, these attributes lead to higher sensitivity. Combined with advanced image reconstruction techniques, this detector technology provides improved image contrast and resolution.

The high detection efficiency, energy sensitivity and good spatial resolution of pixelated CdZnTe detectors are exploited in dual energy x-ray absorptiometry (DEXA) for high performance bone mineral densitometry [3].

Photon-counting detector technology also offers benefits to digital radiography, where both the image signal to noise ratio and contrast resolution can be improved if the radiation energy information is used alongside the radiation intensity. The direct conversion detector technology enables very sharp line spread function (LSF) limited only by the pixel size. A sharp LSF together with the high absorption efficiency of CdTe and CdZnTe and low-noise readout circuitry yields high detector quantum efficiency (DQE), which ultimately determines the performance of an imaging system. CdTe detectors with fine pixelation coupled to low-noise CMOS readout chips have been developed and successfully deployed in panoramic dental imaging applications [4].

Multi-energy computed tomography (CT) is the ultimate challenge for any solid-state detector technology, including CdTe and CdZnTe detectors [5]. Just as in the previously mentioned applications, the energy discrimination capability is a key advantage of these photon-counting detectors. The fast data acquisition and high photon flux used in state-of-the-art CT imaging systems require very fast response detector technology that can operate under intense photon radiation conditions. Computed tomography applications represent a huge challenge for CdTe and CdZnTe detector technology and are the subject of intense research today.

The compact size, capability for fine pixelation, low voltage requirements and room-temperature operation enable the deployment of CdTe and CdZnTe detectors in compact gamma cameras for prostate imaging [6] and miniature ingestible imaging capsules for colorectal-cancer detection [7].

In this article we review the state of the art of CdTe and CdZnTe materials and crystal growth technologies including these technologies face for deployment in x-ray and gamma-ray detectors and the opportunities these technologies provide for both mainstream and novel medical applications.

## 1.2 CdTe and CdZnTe Materials

The binary semiconducting compound CdTe and its ternary cousin CdZnTe or CZT possess material properties that make them uniquely befitting for room-temperature, solid-state radiation detectors. The high average atomic numbers ($Z_{CT}$ = 50 and $Z_{CZT}$ = 48.2) and densities ($\rho_{CT}$ = 5.85 g/cm$^3$ and $\rho_{CZT}$ = 5.78 g/cm$^3$) provide high stopping power for x-rays and gamma-rays, thus enabling high sensitivity and high detection efficiency of the detectors.

The band gaps of CdTe and CdZnTe are $E_g$ = 1.5 eV and $E_g$ = 1.572 eV, respectively, at room temperature, making these materials ideal for room-temperature radiation detectors. In electrically compensated semi-insulating CdTe and CdZnTe crystals, the low free-carrier concentration enables devices to achieve large depletion depths, ranging from few *mm* to few *cm*, and low leakage currents, in the few *pA* to few *nA* range.

The moderately high mobility and lifetime of charge carriers in CdTe and CdZnTe allow good charge transport across detector devices depleted to several *mm* or even *cm* thickness. State-of-the-art crystal growth technology now regularly produces semi-insulating CdTe crystals with electron and hole mobility-lifetime products in the $\mu_e\tau_e = 10^{-3}$ cm$^2$/V and $\mu_h\tau_h = 10^{-4}$ cm$^2$/V range and CdZnTe crystals with electron and hole mobility-lifetime products in the $\mu_e\tau_e = 10^{-2}$ cm$^2$/V and $\mu_h\tau_h = 10^{-5}$ cm$^2$/V range.

Both CdTe and CdZnTe radiation detectors have existed for over 30 years in the industry. Acrorad, Ltd., in Japan, pioneered semi-insulating CdTe detector technology, while eV Products, Inc. (now part of Kromek Group plc), in the United States, Redlen Technologies, in Canada and Imarad Imaging Systems, Ltd., (now part of GE HealthCare), in Israel, have been commercializing CdZnTe detectors since the early nineties.

Fundamentally, there are no major differences between the two compounds. CdZnTe is an alloy of CdTe and ZnTe, with typically about 10% Zn alloyed in the ternary compound for detector applications. The alloying with Zn causes several changes to the properties of CdTe, as described here.

First, by adding Zn to CdTe, the band gap is increased. The wider band gap enables a higher maximum resistivity of the ternary compound. For CdZnTe with 10% Zn, the band gap increases from 1.5 eV to 1.572 eV and the maximum achievable resistivity increases by a factor of three, typically from $2\times10^{10}$ Ω-cm for CdTe to $5\times10^{10}$ Ω-cm for CdZnTe. It is to be noted, however, that these are only the maximum resistivity limits allowed by the band gap of the material, values that are not always achieved in practice because of incomplete electrical compensation.

Various vendors employ different doping approaches for electrical compensation to achieve high resistivity. Acrorad, Ltd., uses Cl doping of CdTe crystals to achieve *p-type* conductivity and resistivity in the $10^8$–$10^9$ Ω-cm range. This resistivity is well under the maximum possible for CdTe. The disadvantage of this approach is the relatively small depletion depth of the detectors, typically in the few *mm* range. The *p-type* conductivity of the Acrorad CdTe detectors, on the other hand, enables the use of In and Al contacts for the manufacture of high-barrier Schottky devices to achieve very low leakage-current detectors, even at very high applied bias. The high bias voltage ensures fast charge collection and fast response of the detectors, providing for significant advantages in high-flux applications.

CdZnTe vendors typically use In, Ga or Al doping to achieve *n-type* conductivity and nearly complete electrical compensation, with resistivity in the (2–3)×$10^{10}$ Ω-cm range. Because of the low free-carrier concentration in semi-insulating CdZnTe prepared with this compensation approach, the detector devices can be depleted to few *cm* in depth, enabling large active volume detectors. The detectors are typically fabricated with Pt or Au Schottky barrier contacts. However, because the Schottky barriers of Pt and Au contacts on *n-type* CdZnTe are much lower than the barrier heights of Al or In contacts on *p-type* CdTe, the leakage current of commercial CdZnTe detectors is typically an order of magnitude higher than that of commercial CdTe detectors. It is to be emphasized that the different bulk resistivity of CdTe and CdZnTe crystals and the different leakage current of the detectors stem from technology choices made by the vendors and are not the result of the different band gaps of these compounds.

The second difference between CdTe and CZT comes from the different chemical properties of CdTe and ZnTe. ZnTe has a lower iconicity and a higher binding energy than CdTe, and the bond length is shorter in ZnTe. Thus, the CdTe lattice is strengthened by the incorporation of Zn, leading to an increase of the shear modulus and solution hardening of the ternary compound, a well-known effect often used in metallurgy [8]. The solution hardening of CdZnTe reduces the propensity for plastic deformation and the formation of dislocations; however, it also reduces dislocation motion and makes the ternary compound more brittle than CdTe.

The higher binding energy and shorter bond length between Zn and Te atoms also induce a local strain into the host CdTe lattice and a relaxation of the Te atoms around the Zn atom. The local lattice distortions increase the migration barrier of interstitial atoms in the proximity of Zn atoms. As a result, diffusion and ionic migration rates are reduced in CdZnTe compared to CdTe. The effect is, however, minor, and both high-purity CdTe and CdZnTe crystals demonstrate excellent long-term stability and are inert against ionic migration when operated under high bias voltages for a prolonged period of time as radiation detectors. This critically important property of these compounds ensures the long-term, stable operation of CdTe and CdZnTe detectors under high bias, unlike many of the more ionic compounds, such as thallium bromide (TlBr), that suffer from physical polarization because of the migration of constituent atoms under applied bias.

The third difference stems from the chemical potentials of the two material systems. The addition of Zn increases the maximum deviation from stoichiometry in CdZnTe on the Te-rich side of the phase diagram. The maximum Te non-stoichiometry, or solid solubility, is about $4\times10^{18}$ $cm^{-3}$ in CdTe and about $1.2\times10^{19}$ $cm^{-3}$ in CdZnTe and is reached at about 880°C [9]. Because, in thermal equilibrium, Cd and Zn vacancies are the dominant native defects in CdTe and CdZnTe, excess Te is primarily accommodated by an increase in the number of vacancies in the lattice. The higher maximum Te solubility in CdZnTe therefore indicates that the formation energy of the Cd (and Zn) vacancy is reduced in CdZnTe relative to CdTe. Even though both CdTe and CdZnTe exhibit retrograde Te solid solubility (i.e., the maximum non-stoichiometry decreases with decreasing temperature) and much of the excess Te segregates into Te precipitates by the time the crystal is cooled to room temperature, there is a significant difference in their residual Cd and Zn vacancy concentrations. Because Cd (and Zn) vacancies introduce acceptor levels in the lower half of the band gap [10], a higher vacancy concentration causes increased hole trapping and a lower hole mobility-lifetime product in CdZnTe than in CdTe, as often observed experimentally.

The fourth difference in these compounds stems from the different solubility of Zn in solid and liquid CdTe. The ratio of the solubility, called segregation coefficient, controls the partitioning of this alloying element between the solid and liquid during crystal growth. The segregation coefficient of Zn is about $k = 1.3$ in CdTe, meaning that the Zn preferentially segregates at the solid-liquid interface into the solid rather than the liquid. This leads to an axial Zn concentration distribution in CdZnTe produced by melt-growth techniques. For a 10% Zn-doped CdZnTe, the actual Zn concentration starts at about 13–14% in the first-to-freeze section of the ingot and decreases to about 6% in the last-to-freeze section. With appropriate design of the solvent and feed material, constant Zn concentrations can be achieved with solution growth techniques, such as the traveling heater method (THM).

A changing Zn concentration causes a varying lattice constant and the development of grown-in constitutional stress and strain in the ingot [8]. This adds to the underlying thermal stresses during growth and can lead to the deformation of the crystals and a higher dislocation density in CdZnTe than in CdTe, despite the alloy-hardening of the lattice.

## 1.3 Materials Technology

Semiconductor radiation detectors require high-quality, semi-insulating single crystals for satisfactory performance. High resistivity is required in order to attain sufficient carrier depletion of the semiconductor crystal and realize active detector thickness in the few mm to few cm range and to maintain a high-electric field across the detector with low leakage current. Too low depletion limits the active depth of the device and the energy range of the detector, while excessive leakage current produces electronic noise that deteriorates the energy resolution of the detector.

Detector applications require high-quality, single crystals, because defects in the crystals cause carrier trapping, recombination and distortion of the internal electric field, all contributing to the deterioration of detector performance. Large-angle grain boundaries are typically very strong carrier traps in CdTe and CdZnTe. In order to avoid the detrimental effects of grain boundaries on detector performance, single crystals are mined from the typically polycrystalline ingots. The single-crystal yield from this mining process is usually the largest cost driver of CdTe and CdZnTe detectors. Growing as perfect single crystal ingots as possible or achieving the highest single-crystal yield from polycrystalline ingots is the primary goal and challenge of CdTe and CdZnTe crystal growth technologies. To minimize carrier trapping, the extracted single crystals themselves have to be as perfect as possible, with a low concentration of point defects and their clusters, as well as a low density of extended defects, such as dislocations, low-angle grain boundaries (subgrain boundaries), twins, and second-phase precipitates and inclusions.

In the next few paragraphs, we review the challenges and methods to achieve *high electrical resistivity and good carrier transport* in CdTe and CdZnTe crystals *simultaneously*. We start with a review of the point-defect structure of uncompensated, high-purity CdTe.

### 1.3.1 Defect Structure of High-Purity CdTe

The formation energy of native defects (vacancies, interstitial atoms and antisites) and impurities in semiconductor crystals has three components:

$$\Delta H = \left(E_{defect} - E_{host}\right) - \sum_i n_i\left(\mu_i + \mu_i^{ref}\right) + q\left(\varepsilon_{VBM} + \varepsilon_F\right). \tag{1.1}$$

In the first term, $E_{defect}$ and $E_{host}$ are the total energies of the defect-containing and the defect-free host crystal. The second term represents the energy contribution from the chemical potential of the species forming the defects. The variable $n_i$ is the difference in the number of atoms for the $i$th atomic species between the defect-containing and defect-free crystals, and $\mu_i$ is a relative chemical potential for the $i$th atomic species, referenced to $\mu_i^{ref}$. For Cd and Te, $\mu_{Cd}^{ref}$ and $\mu_{Te}^{ref}$ are the chemical potentials in bulk Cd and Te, respectively. This term captures the change in the formation energy of the defect as a function of deviation from stoichiometry and concentrations of impurities. The third term represents the change in energy due to exchange of electrons or holes with their respective carrier reservoirs. The quantity $\varepsilon_{VBM}$ is the energy of the valence band maximum (VBM) in the host system, and $\varepsilon_F$ is the Fermi energy relative to the VBM. This term captures the energy contribution from the charges residing in the defects. Once the formation energy of the defects is calculated, the defect concentration at thermal equilibrium can be evaluated using

$$N = N_{site} exp(-\Delta H / kT) \tag{1.2}$$

where $N_{site}$ is the number of available sites for the defect in the crystal, $\Delta H$ is the defect formation energy, $k$ is the Boltzmann's constant and $T$ is the absolute temperature.

With steady advancement of first-principles calculations based on density-functional theory in the past decade, computational materials science techniques today provide invaluable insights to the structure and properties of defects in semiconductor crystals. The theoretical models provide adequate estimates of the formation energies of native defects and impurities and their complexes, as well as their ionization energies in the band gap in many semiconductors, including CdTe.

Figure 1.1 shows the calculated formation energy of native defects in Te-rich CdTe as a function of the Fermi energy. Energies are plotted for

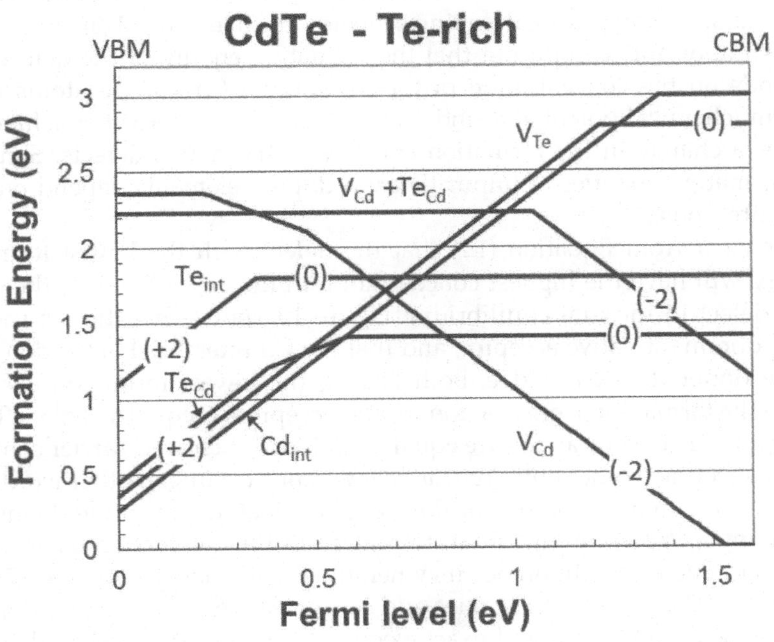


**FIGURE 1.1**
Calculated formation energy of native defects in CdTe as a function of the Fermi energy.

native donors consisting of the Cd interstitial ($Cd_{int}$), Te vacancy ($V_{Te}$), Te antisite ($Te_{Cd}$, i.e., Te sitting on a Cd site) and Te interstitial ($Te_{int}$) along with native acceptors represented by the Cd vacancy ($V_{Cd}$) and Cd vacancy – Te antisite pair ($V_{Cd} + Te_{Cd}$) [11].

It is very important to emphasize that the defect formation energy of charged defects depends on the Fermi energy. As the Fermi energy increases, the formation energy of donors linearly increases, and the formation energy of acceptors linearly decreases according to Equation (1.1). It is also important to recognize that the specific dependency of donor and acceptor formation energies on the Fermi energy gives a high degree of rigidity to the Fermi level. If we perturb the electronic system and try to move the Fermi level from its equilibrium position, dictated by the balance of the concentrations of native defect and impurities, in either direction, the Fermi level quickly returns to the equilibrium point. If we would move the Fermi level higher toward the conduction band minimum (CBM), the formation energy of Cd vacancy acceptors decreases, and their concentration rapidly rises to pull the Fermi level back to the equilibrium point. Similarly, if we would lower the Fermi level toward the VBM, the formation energy of Cd interstitial donors would decrease, and their concentration would increase rapidly to pull the Fermi level up back to the equilibrium point. Because of the exponential dependence of the defect concentrations on the formation energy in Equation (1.2), this negative feedback very strongly stabilizes the Fermi level. This rigidity of the Fermi level is a critical property of the system that enables implementation of practical doping schemes in CdTe and CdZnTe.

It is important to point out that the formation energy of the defects also depends on the concentration of the constituent Cd and Te atoms in the system (chemical potentials), and any deviation from perfect stoichiometry causes a change in the formation energies of the native defects. Similarly, the formation energies of impurities and doping elements depend on their concentrations.

It is clear from Equation (1.2) that the defect with the lowest formation energy will have the highest concentration in the crystal and be the dominant defect in thermal equilibrium. Figure 1.1 shows that the Cd vacancy is the dominant native acceptor, and that the Cd interstitial is the dominant native donor defect in CdTe, both having the lowest formation energies. When the formation energy of donors and acceptors is equal at a given Fermi energy, their concentrations are equal as well and their free-carrier contributions cancel each other out: *electrical compensation* occurs. In other words, it is energetically more favorable for the electrons liberated from the donor state to occupy an acceptor level than to remain in the conduction band as free electrons. Such a condition occurs when the Fermi energy is at $E_v + 0.7$ eV in Figure 1.1. The formation energies and concentrations of Cd interstitials and Cd vacancies are equal, and exact electrical compensation occurs between the two dominant defects. Because this point is at the middle of the band gap, the material has very low free-carrier concentration and has high resistivity.

The concentration of the various defects in the crystal and the position of the Fermi level are determined by the minimum of the total Gibbs free energy of the material and can be calculated by taking into account the energies of all defects in the crystal. However, it is not possible to tell *a priori* the position of the Fermi level in a semiconductor with a complex defect structure, where multiple donors and acceptors are present in the material.

There are nevertheless general trends that stem from the Fermi statistics governing the occupancy of defect levels in the band gap. Donor doping moves the Fermi level between the ionization level of the donor and the CBM. Acceptor doping moves the Fermi level between the ionization level of the acceptor and the VBM. When donors and acceptors compete, the Fermi level is stabilized around the middle of the band gap between the acceptor levels and donor levels, and electrical compensation occurs. Because the Fermi energy is a representation of the average electron energy in the system, its position represents the statistical average of electron energies on the various acceptor and donor energy levels, weighted by the concentrations of the defects. If only one donor and one acceptor defect are present in equal concentrations, the Fermi level settles halfway between the donor and acceptor ionization level.

### 1.3.2 Electrical Compensation

For CdTe and CdZnTe crystals to be useful for room-temperature radiation detection, spectroscopy and imaging applications, the chosen crystal-growth technology must achieve *high electrical resistivity* and *good carrier transport simultaneously.* This is not an easy task, because these two requirements are often counteracting each other. Good carrier transport requires the growth of high-purity crystals to minimize carrier trapping at impurity defects. High-purity CdTe and CdZnTe crystals are, however, typically low-resistivity *p*-type, with electrical resistivity in the $10^3$–$10^6$ Ω-cm range, because of the significant concentration of residual Cd (and Zn) vacancies that are the dominant native acceptors in these compounds. As we discussed in the previous section, there is a significant concentration of Cd interstitials in CdTe in thermal equilibrium. At high temperature, close to the solidification point of CdTe and CdZnTe, the Cd vacancies are largely compensated by Cd interstitials. However, as the crystal is cooled to room temperature, point-defect diffusion slows and eventually stops and the defects "freeze in." Because the diffusivity of Cd interstitials is much higher than the diffusivity of Cd vacancies, they freeze in at a much lower temperature than vacancies. Because the *residual* concentration of defects corresponds to their equilibrium concentration at their freeze-in temperature, the residual concentration of Cd interstitials is much lower than the residual concentration of Cd vacancies in CdTe and CdZnTe at room temperature.

To compensate the doping effect of Cd (and Zn) vacancies and reduce the free-carrier concentration to the $10^5$ cm$^3$ range, CdTe and CdZnTe crystals are doped with shallow donors, such as Al, Ga, In or Cl. The donor doping induces several effects. First, it supplies free electrons that fill the Cd (and Zn) vacancy acceptor levels and elevate the Fermi level close to the middle of the band gap. Second, the donors combine with the vacancies to form vacancy-donor pairs (such as $V_{Cd}$-$In_{Cd}$) called A-centers. This pairing reduces the concentration of Cd vacancies by converting them to A-*centers* with lower acceptor ionization energy. Because Cd (and Zn) vacancies have two ionization energy levels and contribute two holes to the valence band while the A-centers have a single ionization level, the pair formation reduces the doping effect of Cd vacancies.

Figure 1.2 shows the formation energy of the $In_{Cd}$ donor and the In A-center ($V_{Cd}$ + $In_{Cd}$), as functions of the Fermi energy, for three different In concentrations (low, medium, high). As more In is added to the crystal, the chemical potential of In increases and the formation energies decrease for both the In donor and the A-center acceptor. The circles in Figure 1.2 indicate the primary compensation points, namely the Fermi energy where the lowest-energy acceptor and donor formation energies are equal. For low In doping concentrations, the primary compensation occurs between the Cd vacancy and the Cd interstitial. At medium In concentration, the primary compensation occurs between the Cd vacancy and the In donor. At high In

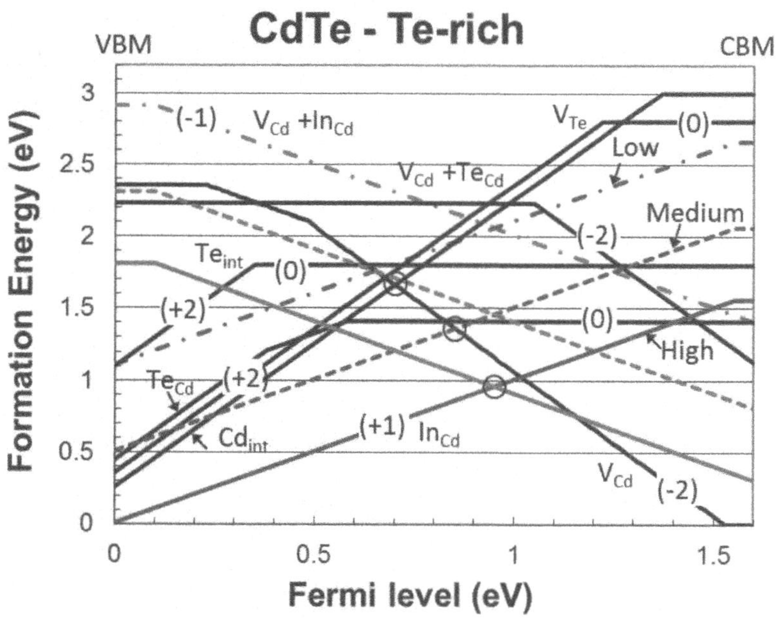


**FIGURE 1.2**
Calculated formation energy of native defects, In donor and In A-centers in CdTe as a function of the Fermi energy, after [11]. The In donor and A-center formation energies are shown for three In concentration levels: high (——), medium (– – –), and low (– · –).

concentration, the A-center formation energy becomes smaller than the Cd vacancy formation energy and the primary compensation occurs between the In donor and the A-center.

The latest theoretical analysis of electrical compensation in CdTe shows that practical electrical compensation is achievable using shallow-level donors, such as In, Al, Ga or Cl [11]. The strong dependence of the defect-formation energies on the Fermi energy provides a broad range of donor concentrations where high resistivity is achievable. Figure 1.3 shows that electrical resistivity higher than $10^9$ Ω-cm is achievable in the $2\times10^{16}$ cm$^{-3}$ to $5\times10^{17}$ cm$^{-3}$ donor concentration range both with In and Cl doping [11]. The theoretical analysis indicates that, in CdTe:In, electrical compensation primarily takes place between the $In_{Cd}$ donor and $V_{Cd}$ acceptor. In CdTe:Cl, the primary compensation is between $Cl_{Te}$ and $V_{Cd}$ in the lower Cl concentration range ($10^{16}$ cm$^{-3}$–$10^{17}$ cm$^{-3}$) and between $Cl_{Te}$ and the A-center in the high concentration range ($10^{19}$ cm$^{-3}$). The different behavior of the In-doped and Cl-doped CdTe stems from the higher binding energy of the $Cl_{Cd} + V_{Cd}$ pair than the $In_{Cd} + V_{Cd}$ pair [11].

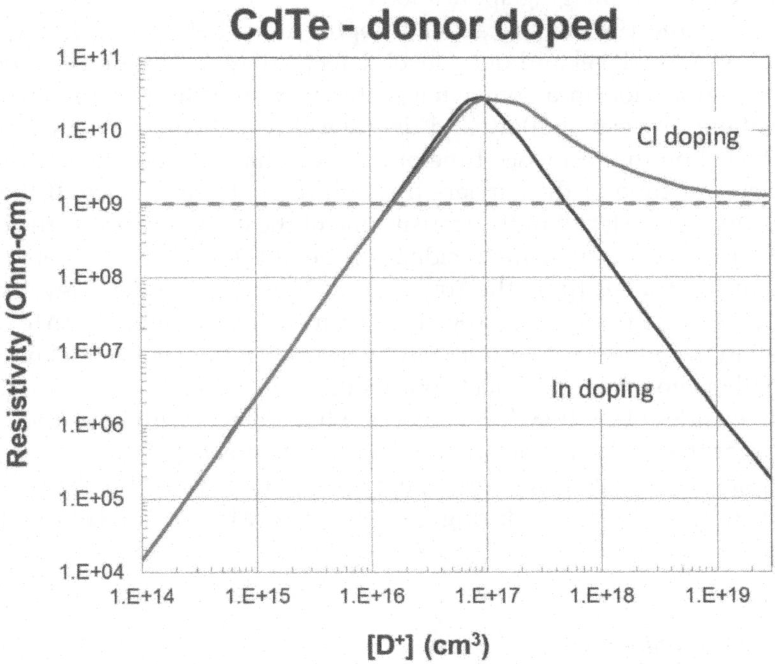


**FIGURE 1.3**
Calculated electrical resistivity of as a function of donor concentration for CdTe doped with In or Cl [11].

### 1.3.3 Carrier Transport

Gamma- and x-ray detectors require very good carrier transport though the active volume of the device to minimize carrier loss from the charge cloud generated by the photons and to preserve the proportionality of the detector signal amplitude to the energy of the photons. Crystal defects cause charge trapping and recombination that reduces the amount of collected charge and cause a low-energy tailing of the photopeaks, thereby deteriorating the energy resolution of the detector. It is therefore imperative that crystal-growth technology produces single crystals with as low defect density as possible.

Point defects, point defect clusters and extended defects all cause detrimental trapping. Point defects and small clusters (doublets, triplets) cause uniform trapping when these defects are randomly distributed in the crystal lattice. Extended defects, such as dislocations, subgrain boundaries, second-phase precipitates and inclusions, and twin boundaries, form spatially correlated structures that typically cause nonuniform trapping. Additionally, second-phase precipitates, inclusions and impurities often decorate grain boundaries, subgrain boundaries, twin planes and dislocations to further enhance nonuniform carrier trapping. Nonuniform trapping is difficult to correct by electronic or software methods and is particularly harmful for spectroscopic and imaging applications.

Although the physics of carrier trapping and de-trapping is the same for both shallow-level and deep-level defects, they induce detector performance degradation in a somewhat different way. While there is no clear delineation between shallow and deep-level defects, one can distinguish them based on the residence time of trapped charge at the defects relative to the transit time of the carriers through the detector device. If the residence time is shorter than the transit time of the charge carriers, the defect is a shallow-level trap. If the residence time of the trapped carriers is longer than the transit time, the trap can be considered a deep-level defect (Figure 1.4). The typical electron transit times in CdTe and CdZnTe detectors are in the 50 ns to 500 ns range, depending on the device thickness and applied bias voltage. The defects with ionization energies less than 0.35 eV can be considered shallow-level defects, while defects with ionization energies larger than this are considered deep-level defects [12].

Deep-level defects cause permanent charge loss from the charge cloud generated by the radiation during the collection or transit time of the charge cloud.

#### *1.3.3.1 Recombination*

If the defect level is close to the middle of the band gap, the probability of electron and hole trapping are of similar magnitude and the trap acts as a recombination center. At sufficiently high deep-level defect concentration,

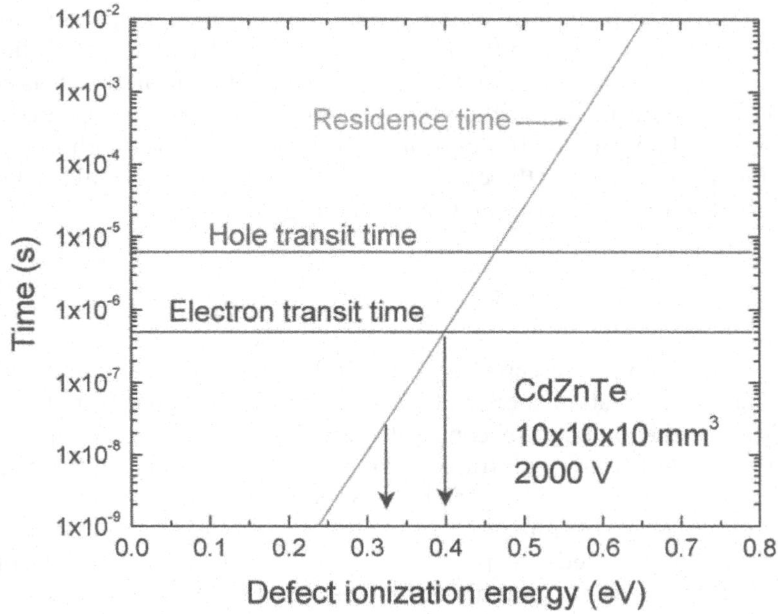


**FIGURE 1.4**
Residence time versus transit time for electrons and holes in a CdZnTe detector [12].

the recombination causes an immediate charge loss from the charge cloud generated by the radiation, even before the electron and hole clouds are separated by the electric field. Group IV elements, such as Ge, Sn, Pb and transition metals like Ti, V, Fe, Ni, are impurities known to introduce deep levels to CdTe. When the electronic level of the impurity is in the lower half of the band gap, below the Fermi level in semi-insulating CdTe or CdZnTe, it acts as a hole trap. For example, an isolated Fe impurity, with donor level 0.6 eV above the VBM, is a hole trap. If the Fe concentration is high, the trapping of holes and the subsequent recombination of the trapped hole by trapping an electron reduce the size of the charge cloud and, as a result, deteriorate the proportionality of the signal amplitude to photon energy. Despite this recombination effect and worsened detector performance, the electron mobility-lifetime product remains high (the Fe donor is not an electron trap) in Fe-doped semi-insulating CdTe and CdZnTe crystals.

#### *1.3.3.2 Uniform Trapping*

##### *1.3.3.2.1 Trapping at Deep-Level Defects*

Trapping at deep levels causes the trapped charge to be removed from the charge cloud for the duration of the transit time of the carriers. As a result, the signal amplitude is reduced, and the photo-peak suffers a low-energy tailing. Ionized defects with a net charge are stronger traps due to the Coulomb attraction between the localized charge on the defect and the free carriers of opposite sign. The capture cross section of neutral defects is about an order of magnitude lower than that of charged defects. The trapped state is metastable state, and the trapped carrier either recombines with free carriers of the opposite sign or the carrier escapes the trap by thermal excitation. The residence time of the trapped carriers $\tau_r$ is given by

$$\tau_r^{-1} = \nu \; exp\left(-\frac{E_t}{kT}\right) \tag{1.3}$$

where $\nu$ is the attempt to escape frequency of the charge, typically in the $10^{13}$ $s^{-1}$ range, $E_t$ is the ionization energy of the defect level measured from the VBM or CBM, $k$ is the Boltzmann constant and $T$ is the absolute temperature. Carriers liberated from deep traps make a small, erroneous contribution to detector signals corresponding to photons that arrived later than the photon generating the trapped charge.

Randomly distributed, deep-level defects provide a constant trapping rate at thermal equilibrium that can be described by a well-defined carrier lifetime

$$\tau^{-1} = v_{th} N_{active}^{t} \sigma \tag{1.4}$$

where $v_{th}$ is the thermal velocity of the carriers, $N_{active}^{t}$ is the density of electrically active traps and $\sigma$ is the capture cross section of the defect. It is to be pointed out that the active fraction of defects depends on the position of the Fermi level. Because the capture cross section of charged and neutral defects is very different, this is particularly important for deep-level defects close to the middle of the band gap, where relatively minor differences in the compensation condition and the resulting shift in the equilibrium Fermi level position cause a large change in the trapping probabilities and, therefore, the lifetime of the free carriers.

We can consider the case of tin (Sn) in CdTe to illustrate the behavior of deep-level defects. Tin incorporates at Cd sites to the CdTe lattice and introduces two deep-donor levels. The doubly ionized donor state $Sn^{+1/+2}$ was measured to be at $E_c - 0.85$ eV by photo-EPR at 4.2 K [13]. Considering the shift of the band edges and the reduction of the band gap to 1.5 eV, the $Sn^{+1/+2}$ donor level is expected to be between 0.85 eV and 0.8 eV from the CBM at room temperature. Theoretical calculations suggest that the singly ionized level $Sn^{0/+1}$ is a few hundredths of an eV above the doubly ionized level $Sn^{+1/+2}$ [14]. For the purpose of this analysis, we assume that the singly ionization level $Sn^{0/+1}$ is at $E_c - 0.8$ eV and the ionization level $Sn^{+1/+2}$ is at $E_c - 0.9$ eV, as shown in Figure 1.5.

It is to be emphasized that the occupation of the defect levels is governed by the Fermi distribution, and portions of the defects are in different charge

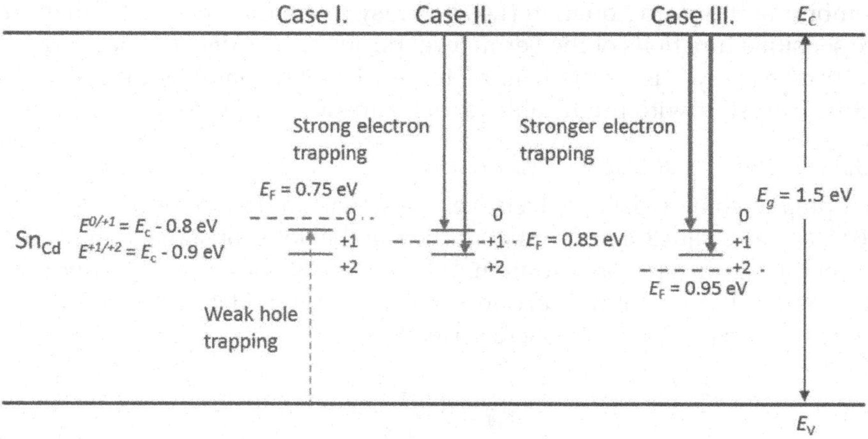


**FIGURE 1.5**
Charge state of $Sn_{Cd}$ in CdTe at room temperature for three different positions of the Fermi-level. Bold arrows indicate strong trapping while dashed arrow indicates weak trapping.

states. Figure 1.5 shows the predominant charge state of each defect level of $Sn_{Cd}$ for three different positions of the Fermi level. For perfectly compensated CdTe, the Fermi level is at midgap, around $E_c - 0.75$ eV. Both levels are occupied by electrons, and most of the Sn atoms are neutral (Case I). In this state, the defect has no effect on electron transport, because the defect is a donor and cannot trap an electron when neutral. It can, however, trap holes (i.e., release an electron to the valence band) to become positively charged. Because the defect is neutral, the trapping cross section is low, and the defect is a weak hole trap. If the Fermi level is lowered to $E_c - 0.85$ eV, the defect levels become predominantly singly ionized (+1) with a localized positive charge (Case II). The defect becomes a strong electron trap. By further decrease of the Fermi level to $E_c - 0.95$ eV, the defect becomes doubly ionized (+2), and the fraction of the defects in the ionized state significantly increases. This causes even stronger electron trapping and further decrease of the electron lifetime (Case III). The positively charged $Sn_{Cd}$ defects are repulsive to holes so do not influence hole transport.

The effect of the Fermi level position can be quite dramatic on carrier trapping at deep-level defects because of the exponential dependence of the Fermi distribution on the Fermi energy. For a donor with defect level $E_D$ below the CBM, the concentration of ionized donors $N_d^+$ is given by

$$N_D^+ = N_D \frac{1}{1 + g_D exp\left(-\frac{E_D - E_F}{kT}\right)} \tag{1.5}$$

where $N_d$ is the total concentration of donor, $g_D$ is the degeneracy factor of the donor level, $k$ is the Boltzmann constant and $T$ is the absolute temperature.

Combining this with Equation (1.4), it is easy to see that the carrier lifetimes are sensitive functions of the Fermi level position. This dependency is one of the reasons why it is so challenging to reproducibly manufacture CdTe and CdZnTe crystals with predictable carrier transport properties.

*1.3.3.2.2 Trapping at Shallow-Level Defects*

Trapping at shallow level defects, i.e., those within 0.35 eV of the CBM and VBM, can also cause significant distortion in the detector signal amplitude if present in significant concentration. When the residence time of trapped carries and the lifetime of the carriers are of the same order of magnitude, the carrier effective velocity can be described by

$$v_{eff} = v\frac{\tau}{\tau+\tau_r}, \tag{1.6}$$

where $v$ is the carrier velocity (either thermal or drift velocity), $\tau$ is the carrier lifetime, and $\tau_r$ is the residence time of the trapped carrier [15]. The carriers undergo frequent trapping and de-trapping cycles that reduces their effective speed, which causes a reduction of the signal rise time in the detector. When the signal rise time approaches or exceeds the peaking time of the amplifier, the signal amplitude is truncated, and ballistic deficit occurs.

#### 1.3.3.3 Nonuniform Trapping

The situation is significantly more complex when the carrier traps are spatially correlated rather than randomly dispersed in the lattice. This a common situation when trapping occurs at extended defects, such as dislocations, subgrain boundaries, grain boundaries, twins and second-phase precipitates and inclusions, and when the point defects and small point-defect clusters form correlated clusters or are associated with extended defects [16].

It is relatively simple to correct for uniform trapping and the resulting signal amplitude degradation in CdTe and CdZnTe detectors by electronic and software techniques. It is, however, significantly more difficult to correct for nonuniform trapping caused by spatially nonhomogeneous, three-dimensional defect distributions. Techniques are being developed to use fine detector pixelation to perform a three-dimensional signal correction and obtain a superior energy resolution from CdZnTe detectors [17]. Apart from requiring advanced custom electronics with significant software overhead, the challenge remains in that these techniques cannot correct for nonuniform charge trapping within the pixel voxel.

An elegant analysis of nonuniform trapping caused by Te inclusions in CdZnTe crystals was developed using homogenization theory that incorporates fluctuations in the induced charge, i.e., charge collection nonuniformities introduced by the random nature of the Te inclusion population [18]. The results from this general model clearly demonstrate the intricate distortions to pulse height spectra induced by nonuniform trapping, both due to nonhomogeneous spatial distribution of the defects and to the broad distribution of capture cross sections of the various defects.

#### *1.3.3.4 Carrier Transport under High Photon Flux*

State-of-the-art computed tomography uses high x-ray flux and fast-response detectors to accumulate images at very high speed to surpass physiological limits, such as the rate of a beating heart. In these applications, the semiconductor detectors must be able to operate under high-intensity radiation and must proportionally respond to changes in the photon flux nearly instantaneously.

High-flux applications represent a significant challenge for CdTe and CdZnTe detectors. The challenge stems from the charge transport properties of the crystals and, most importantly, their drift mobility. The drift mobility controls the speed of the photon-generated electrons and holes through the detector

$$v_{drift} = \mu_{drift} E \tag{1.7}$$

where $\mu_{drift}$ is the drift mobility and $E$ is the electric field in the device. If the electric field is constant across the detector, the drift velocity is also constant, and the transit time of the carriers is given by

$$t_{tr} = \frac{L}{\mu_{drift} E} = \frac{L^2}{\mu_{drift} V} \tag{1.8}$$

where $L$ is the detector thickness and $V$ is the applied voltage. If the photon flux is high and multiple photons generate charge carriers during the transit time, the charge-generation rate is higher than the removal rate and an excess charge builds up in the detector. The excess charge is proportional to the photon flux, and its magnitude changes with the radiation intensity. The detector is not in thermal equilibrium anymore but in a *dynamic equilibrium with the photon field.*

The build-up of the excess charge is not instantaneous, and the electronic system in the semiconductor undergoes a *transient* when the radiation is turned on or when the radiation intensity changes before a steady state is achieved. The temporal evolution of the transient is controlled by the carrier dynamics and the resulting changes in the internal electric-field distribution in the detector.

In the absence of trapping (ideal detector), the excess charge is in the form of free carriers and easily swept out when the radiation source is removed. In real crystals, however, there are always defects acting as carrier traps, and a significant portion of the radiation-induced charge carriers is trapped, forming a *space charge.* The excess carriers and particularly the space charge have

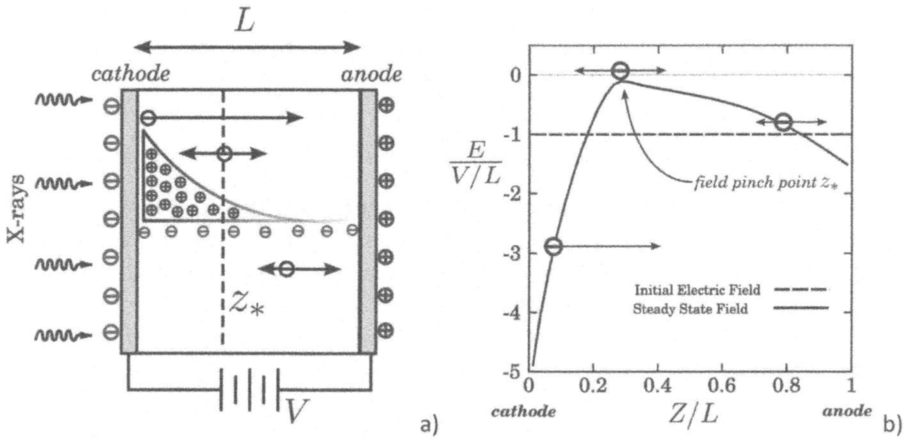


**FIGURE 1.6**
Space-charge distribution in a CdZnTe detector irradiated with high-intensity x-rays from the cathode side. (a) Steady-state electric field distribution. (b) $z_*$ indicates the location of the electric field pinch point. The circles illustrate the charge clouds generated by the x-ray photons at three different depths in the device. The arrows indicate the direction of the motion of electrons and holes [15].

significant, detrimental effect on both charge transport in the crystal and the performance of the detector under intense radiation.

First, the space charge reduces the electric field in the bulk of the detector device. Figure 1.6 shows a cartoon illustrating the charge distribution in a semiconductor detector and the calculated steady-state electric field distribution in a CdZnTe detector irradiated with high flux of 60 keV monoenergetic x-rays at the cathode side [15]. Because the x-rays have an exponential absorption profile, most of the space charge is formed by trapped holes in the proximity of the cathode. However, once a steady state is reached between the radiation and the charge transport, a low-field "pinch" point develops in the nonuniform electric field. Electrons generated in the high-field region under the cathode move fast for a short distance until the pinch point. Electrons generated in the middle of the device, beyond the pinch point, move slowly in the very weak electric field. Electrons generated deep in the detector, close to the anode, move a short distance in a region where the field strength is closest to the value of the initial, constant electric field.

The collapse of the internal electric field in the middle of the device and the much-reduced drift speed of the carriers enhances carrier trapping dramatically. As a result, the signal amplitude of the detector is dramatically reduced, and the pulse-height spectra show a striking shift toward low energies: detector *polarization* occurs [15]. This is a catastrophic loss of performance, and the detector is not a useful device in this state.

It is important to understand that electrical polarization is not a permanent change in the detector performance, and there is no physical or chemical change in the detector material and device. Because the detector is in a dynamic equilibrium with the photons, a large part of its performance recovers when the radiation source is removed. However, the recovery is not instantaneous, because it takes finite time until the excess charge carriers are swept out of the device and the electric field distribution restored. In the presence of deep-level defects, the recovery time could be many orders of magnitude longer than the transit time corresponding to the constant electric field. The thermal emission of carriers from deep-level traps generates signals much later than the moment when the radiation is removed. These delayed signals cause afterglow in imaging devices. If the deep defect levels are close to the middle of the band gap, the residence time is hundreds of milliseconds (Figure 1.4). Coupled with the longer transit times caused by the reduced electric field, the recovery of the detector may take several seconds or even minutes. The recovery may be sped up by temporarily removing the bias voltage, applying heat, or irradiating the detector with light. Numerous such techniques have been proposed and implemented in the industry to deal primarily with the catastrophic loss of performance due to polarization. It is, however, far more challenging to improve the response speed, stability and uniformity of the detectors and to minimize the detrimental effect of signal delays caused by the carrier dynamics in the semiconductor crystals.

To improve the maximum x-ray and gamma-ray flux that CdTe or CdZnTe detectors can tolerate, the removal rate of the charge carriers has to be increased, or, in other words, the transit time of the carriers has to be reduced. Because the drift mobility is a material property controlled by the carrier scattering mechanism in the crystal, it is not possible to modify. The transit time can be shortened by applying higher electric field and using thinner detectors. While higher bias voltage always comes at the price of increased leakage current and resulting detector noise, reducing the thickness of the detector is a particularly effective method of reducing the transit time because of the quadratic dependence shown in Equation (1.8). In order to preserve the detection efficiency, thin detectors are typically side irradiated, and a number of such detector configurations have been proposed and under development for CdTe and CdZnTe detectors for CT and SPECT applications [19].

Controlling the detector response transients in CdTe and CdZnTe detectors is significantly more difficult. These transients are the function of carrier dynamics occurring as the detector attains a new steady state once the photon flux is changed. When the intensity of radiation is changed, the generation rate of electron-hole pairs is changed and the electronic system has to reach a new dynamic equilibrium through the drift, trapping and de-trapping of carriers. This process is controlled by the defect structure of crystals and the position of the Fermi level set by the electrical compensation applied during crystal growth. The carrier dynamics and the temporal evolution of the concentrations of free and trapped carriers and the internal electric field is fairly complex for CdTe and CdZnTe crystals containing multiple types of native and impurity defects. Spatially nonhomogeneous defect distributions in the crystals add further complications. Not only do these defects cause pixel-to-pixel response nonuniformity in the steady state (correctable in imaging applications), but they also cause temporal pixel-to-pixel response variation that are nearly impossible to correct by electronic or software means.

Although the application of CdTe and CdZnTe detectors to high-flux applications, such as multi-energy CT, is a subject of very active research and development, it faces sizeable challenges today. Because of the complexities of carrier dynamics under high x-ray or gamma-ray fluxes, substantial improvements are needed in the perfection of the CdTe and CdZnTe crystals. It is safe to say that a few orders of magnitude reduction of the residual defect density and great improvement of the spatial uniformity of defect distribution will be needed before this detector technology can be deployed in mainstream computed tomography.

## 1.4 Crystal Growth Technology

CdTe and CdZnTe crystals are typically grown by directional solidification from their melts or from solution. In melt growth techniques, the CdTe or CdZnTe is melted at a few degrees above their melting points and then very slowly solidified in a temperature gradient. The solution growth process is very similar, except that the melt is enriched in one of the constituents to form a solution from which the material crystallizes at a lower temperature. The most often used techniques for the growth of CdTe and CdZnTe single crystals are the Bridgman, gradient freeze and electrodynamic gradient freeze (EDGF) melt-growth processes and the THM, a solution growth technique.

Crystal growth aims to achieve atomic-level perfection of the crystal lattice by controlling macroscopic parameters. This is fundamentally an impossible problem: the thermodynamics of the material system dictates the *formation of point defects* during crystallization. The unavoidable formation of defects during solidification and the ensuing *defect interactions* during cooling lead to a rich and complex defect structure in the crystals. In addition to point defects, the formation of extended structural defects is also unavoidable, because the physical system lowers its total energy under the fields and forces imposed during crystal growth. These external factors are the driving forces of extended-defect generation. Gaining an increased control over the atomic perfection of the crystals and the suppression of crystal defects are the continuing, principal challenges for CdTe and CdZnTe crystal growth technologies.

The magnitude of the crystallization challenge cannot be sufficiently emphasized: We are aiming at controlling the *multitude of microscopic interfacial phenomena* with few, far-field *macroscopic parameters* to *minimize the formation of crystal defects* at the growth interface and in the solidified crystal.

The principal crystal growth challenges of CdTe and CdZnTe can be categorized into three main areas: *parasitic nucleation, physical defect generation* and *defect interactions.*

### 1.4.1 Parasitic Nucleation

*Parasitic nucleation* is the process whereby a microscopic cluster of atoms with a different orientation than the surrounding crystal is formed at the growth interface. Depending on the relative growth rates of the parasitic grain and the surrounding matrix, the cluster either grows into a macroscopic crystallite of different orientation, or it is overcome by the surrounding crystal and retained as a microscopic defect. Solid-state recrystallization of the surrounding matrix may also occur, completely eliminating the misoriented crystallite. Parasitic nucleation is caused by a few fundamental processes.

If the imposed solidification rate (as determined by the rate of temperature gradient movement, pull rate, etc.) exceeds the *natural crystallization rate,* parasitic nucleation is guaranteed, because the newly attached atoms do not have sufficient time to diffuse along the growth interface and find their proper position in the existing crystalline structure. The resulting local, microscopic non-stoichiometry and subsequent atomic relaxation lead to the formation of parasitic crystallites.

A poor choice of the *crystal orientation relative to the direction of the temperature gradient* is another common source of parasitic nucleation. If the parasitic grains have higher natural growth rate than the surrounding crystal, they grow faster and expand into large grains. Unfortunately for CdTe and CdZnTe, most crystal orientations have similar growth rates and there is no orientation with a significantly faster growth rate that would help to stabilize the growth of a primary single crystal. In addition, the growth interface is almost never flat but typically having a varying curvature. The interface curvature is controlled by the three-dimensional spatial distribution of the temperature field; in particular, the growth interface is normal, or perpendicular, to the local thermal gradient. Because of this curved interface, the relative orientation of the crystal to the direction of the temperature gradient changes, point by point, along the growth interface. For a crystal growth process, significant interface curvature makes it unavoidable to have segments on the growth interface where the orientation of the crystal relative to the temperature gradient is unfavorable, and these sections of the interface are prone to parasitic nucleation.

High *thermo-mechanical stresses* in the solid near the growth interface are another cause of parasitic nucleation. High stress causes strain in the growing crystal near the interface. This strain is reduced if the crystal deforms and dislocations are generated. Subsequently, dislocations interact with each other to form dislocation arrays, equivalent to small-angle grain boundaries that reduce the total energy of the system. If the stress is sufficiently high and persistent for extended periods of time, the dislocation density in the vicinity of the interface may be sufficiently high for these arrays to rearrange into large-angle grain boundaries, thereby segmenting the growth interface into several parasitic crystallites.

Finally, *instability and drift of the growth parameters* are another cause of parasitic nucleation. CdTe and CdZnTe melts exhibit a significant undercooling before the initiation of nucleation. During solidification, constitutional undercooling, caused by the enrichment of the melt in Te in front of the growth interface, is a particularly common phenomenon. This enrichment is caused by the incongruent solidification of CdTe and CdZnTe, which means that the composition of the solid is different than the composition of the liquid. If the temperature or pressure control system of the crystal growth equipment is not sufficiently stable, a drop of just a few degrees in the heater temperature can initiate crystal nucleation in the undercooled melt in front of the interface. The microscopic crystal nuclei formed in the melt are obviously of random orientation; if they attach to the growth interface, parasitic grains of different orientation may grow.

Parasitic nucleation is the primary cause of the polycrystalline nature of most CdTe and CdZnTe ingots. Due to the multitude of processes that can cause parasitic nucleation, it is particularly difficult to avoid in large-diameter (50–200 mm) ingots. Figure 1.7 shows a cross section of a typical 50-mm diameter CdZnTe ingot along with the corresponding orientation map obtained by electron backscattering diffraction (EBSD). Each color represents a different orientation. Black lines show large-angle grain boundaries. Red

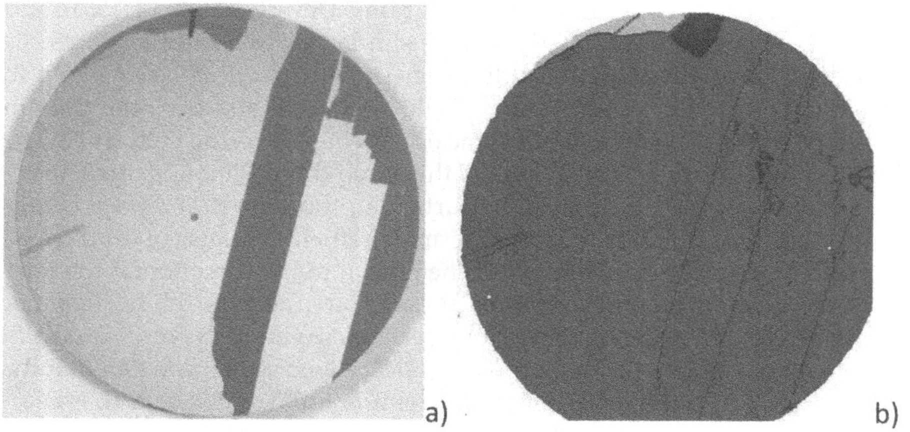


**FIGURE 1.7**
Cross section of a typical 50-mm diameter CdZnTe ingot (a) and the matching orientation map measured by EBSD (b).

lines show twin boundaries. This particular ingot consists of a large primary crystal, albeit with significant twinning, and a few parasitic grains at the periphery. There is a proliferation of densely packed grain boundaries in the top-right quadrant of the orientation map. This area probably contains a high density of low-angle grain boundaries and micro-crystallites that are not resolved with the spatial resolution of the EBSD mapping. Twins are errors in the stacking order of a crystal plane. They are typically not electrically active defects, unless they are decorated with other defects, such as impurities and second-phase particles. This ingot has a high single-crystal mining yield if the twins are not electrically active or moderate to low single-crystal yield if the twins are electrically active.

The initial and early stages of solidification pose the biggest challenge for single crystal growth. Without a seed crystal, the nucleation of the first crystallites is stochastic and may occur either heterogeneously on the wall of the growth crucible or, in case of significant *undercooling*, homogenously in the volume of molten material. The latter type of nucleation can be particularly severe in liquid CdTe and CdZnTe, since these melts are known to undergo as much as 20°C undercooling before crystallization initiates. The initial, rapid solidification of the undercooled liquid leads to the formation of many small crystallites with random orientation. Even if the undercooling of the CdTe and CdZnTe melt is minimized, numerous crystallites are often nucleated at random locations at the crucible wall with random orientation. Crystallites that are oriented favorably with respect to thermal gradients may attain higher growth rates and evolve into dominant, larger grains as crystallization proceeds.

*Seeding* is a time-tested technique to achieve controlled crystallization from the beginning of solidification and is used for the growth of many semiconductor crystals. Seeding helps to minimize the detrimental effects from undercooling and provides a single, perfect crystal over the entire growth interface. With a distinct, favorable fast-growth orientation of the seed, stable crystallization is realized from the very beginning of the solidification process. The one caveat is that any defects present in the seed crystal are replicated in the grown crystal. Seed crystal selection, characterization and preparation are critically important for successful seeded growth. The presence of defects and strain and poor choice of seed orientation and preparation can easily diminish all the benefits of seeding and may result in ingots with high defect concentration and even with polycrystalline structure.

### 1.4.2 Physical Defect Generation

It is impossible to produce crystals free from defects in practical bulk crystal growth processes that employ directional solidification from a melt or solution at high temperature. Equilibrium point defects are always present in substantial concentration at the growth temperature, as dictated by the thermodynamics of the material. Some fraction of these defects is frozen-in to the crystals during cooling, because at lower temperatures the defects have insufficient mobility to redistribute and attain thermal equilibrium. In addition, extended structural defects are nearly always generated because of the physical and thermal conditions imposed during crystal growth.

Instabilities in the thermodynamic parameters such as temperature, pressure and constituent concentrations, such as occur during interfacial breakdown during growth, may cause the formation second-phase particles, such as Te inclusions.

Thermo-mechanical and constitutional stresses induced by the imposed temperature gradient and by the compositional variation in the ternary compound CdZnTe due to inhomogeneous Zn distribution induce considerable strain to the growing crystal. If these exceed the yield stress of the crystal, dislocations are generated to reduce the strain energy. The barrier to the nucleation of dislocations is low in CdTe and CdZnTe, due to the low stacking fault energy and low yield stress of the materials. Zinc alloying increases the yield stress of the material, but this is not sufficient to completely suppress dislocation generation and it comes at the price of the additional constitutional stress. The density of the dislocations depends on various formation, multiplication and annihilation mechanisms, and their interactions lead to dislocation networks and small-angle, subgrain boundaries. As the density of dislocations in subgrain boundaries grows, they may rearrange to large-angle grain boundaries, thereby releasing some strain energy.

The presence of insoluble contaminant particles (such as particles from crucible and ampoule materials: quartz, graphite, alumina, etc.) in the CdTe and CdZnTe liquid may lead to complex dislocation structures if the particles are attached to the growth interface and grown into the crystal. Cd vacancy supersaturation may lead to the precipitation of vacancies and the formation of stacking faults and dislocation loops upon cooling.

Numerous growth parameters have conflicting effects on crystal growth – beneficial for some aspects of the growth process but detrimental for others. The temperature gradient is a pertinent example. A high-temperature gradient is usually beneficial to the stability of the solidification interface and helps to suppress parasitic nucleation and the formation of Te inclusions. On the other hand, high thermal gradients induce large thermo-mechanical stresses that drive the generation of dislocations and subgrain boundaries.

### 1.4.3 Defect Interactions

Defects formed at the crystallization interface or within the body of the crystal during cooling migrate under the influence of thermodynamic factors and imposed outside fields and forces. As the defects move across the crystal, they intersect each other's paths and interact with each. Often, the defect interactions lower the energy of the system by forming a new combined defect. The stress fields of grain boundaries, subgrain boundaries and

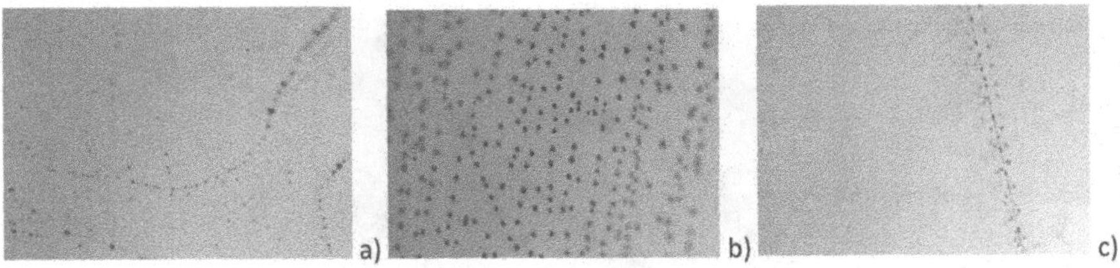


**FIGURE 1.8**
Correlated clusters of Te inclusions in CdZnTe trapped along dislocations (a); grain boundaries (b) and twin planes (c). The size of the Te inclusions in these infrared microscopy images is in the 10–70 μm range.

dislocations attract excess alloy constituents, dopants and impurities to form boundary decorations.

For example, Te inclusions, extensively studied extended defects in CdTe and CdZnTe, are typically pinned to other defects, such as grain boundaries, subgrain boundaries and twin planes, forming correlated clusters of inclusions and causing nonuniform trapping and distortion of charge transport. Figure 1.8 shows correlated clusters of Te inclusions in CdZnTe trapped at dislocations (a), grain boundaries (b) and twin planes (c). The inclusions seen in the infrared microscopy images are dark triangular or hexahedral shape features and their size is in the 10–70 μm range.

Figure 1.9 shows a transmission electron microscopy (TEM) image of a dislocation network surrounding a Te inclusion in CdZnTe (the Te inclusion is outside of the visible area in the direction of the top-left corner of the image). The dense network of dislocations is probably formed when the Te inclusion froze out during cooling of the crystal. Because the melting temperature of Te is low (450°C) the Te inclusions are liquid droplets at higher temperatures. Because the thermal expansion coefficient of the Te is larger than the host CdZnTe matrix the large stress induced to the crystal during Te solidification gave way to dislocation generation. Notice, that many of the dislocations are terminated by dark spots. These are Te precipitates of about 50 nm size. Unlike to their much larger cousins Te inclusions that are generated at the crystallization interface, Te precipitates are formed during cooling. The excess Te atoms dissolved in the matrix at high temperature are rejected by the crystal as the temperature lowered: CdTe and CdZnTe have a retrograde solubility of Te. The excess Te atoms form the small precipitates. If there are other defects in the crystals the Te precipitates preferentially nucleate along those defects. As illustrated in Figure 1.9, the end result is a correlated cluster of dislocations and Te precipitates.

There is a high-equilibrium concentration of native point defects at the growth temperature. As the CdTe and CdZnTe ingots are cooled from the solidification temperature the crystals get supersaturated by the native point defects. The diffusion rate of the defects also rapidly decreases with temperature. At certain temperature, the defects cannot diffuse sufficient

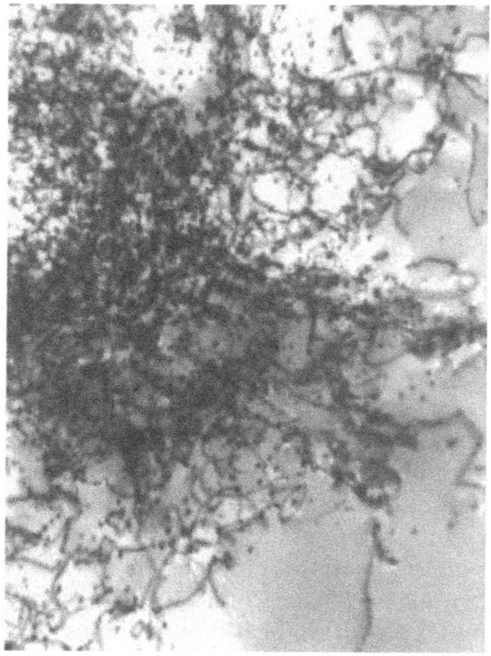

**FIGURE 1.9**
Dislocation network surrounding a Te inclusion in CdZnTe shown by TEM. The dislocations are decorated with Te precipitates seen as dark spots. The area of the image is about 3 μm × 5 μm.

distances anymore to reach the ingot surface and maintain the equilibrium defect concentration. Because the average defect to defect distances are much shorter the point defects form pairs, triplets and precipitate to small clusters to maintain their equilibrium concentration in the host matrix. Similarly, they are attracted to extended structural defects where they get trapped. By the time the crystals are cooled to room temperature there is an excess of point-defect pairs and small clusters of native defects above their equilibrium concentrations.

Impurities undergo a similar process, and one typically finds that a significant fraction of the total impurities are trapped at extended defects, such as inclusions, precipitates, grain boundaries and dislocations.

Cooling the crystals from the growth temperature to room temperature is an important element of the crystal growth process. It is easy to see that cooling too rapidly freezes in a higher concentration of defects than slow cooling. Typically, slow cooling or annealing at low temperature (100–300°C) is required to provide sufficient time for defects to reach the crystal surface or extended defect structures, so that they may be removed from the pool of electrically active defects. This defect relaxation is a critical phase of the crystal growth process for achieving good charge transport in the crystals.

### 1.4.4 Annealing

Annealing of crystals following growth is often proposed to reduce the concentration of defects and improve the charge transport properties of CdTe and CdZnTe crystals. Annealing can be implemented as part of the cooling procedure (in-situ ingot annealing) or performed in a separate process after the ingots are sliced to wafers (wafer-level annealing). While annealing in controlled vapor pressures of the constituents is an effective way to dissolve and eliminate Te inclusions and precipitates and adjust the residual concentration of native defects, it is ineffective in eliminating most extended defects, such as dislocation networks, subgrain boundaries, twins and grain boundaries. It must also be pointed out that annealing may cause deterioration of the charge transport properties of the crystals, because impurities trapped at Te-inclusions, precipitates and other extended defects are liberated and released to the host crystal, which may perturb electrical compensation and enhance carrier trapping.

In summary, CdTe and CdZnTe crystals typically have a rich and complex defect structure stemming from the numerous defect generation mechanisms and defect interactions operating during growth and cooling of the crystals. Many of these defects are electrically active and trap electrons and holes, thus hampering the transport of carriers. It is also more typical than not that these defects form correlated clusters that result in nonuniformities in trapping, electric field and carrier transport, factors that ultimately lead to nonuniform detector response.

### 1.4.5 Computational Materials Science

Computational materials science has made tremendous progress in the past three decades and has led to exciting breakthroughs in a large variety of scientific and technological fields. In particular, crystal growth poses notoriously difficult challenges for computational materials science, because the relevant physical mechanisms and process parameters span a multitude of scales, from atomic to macroscopic. In addition, many of the physical properties needed for computation are extremely difficult to measure experimentally or to calculate theoretically with sufficient accuracy.

Nevertheless, the modeling of CdTe and CdZnTe crystal growth has progressed impressively over the past two decades. The results from these computational material science efforts have either already been implemented into CdTe and CdZnTe crystal growth technology or provide the platform for further experimental technology development efforts.

Crystal growth theoretical modeling focusing on macroscopically accessible growth parameters, such as those affecting heat and mass transport, has provided the most direct help to advance CdTe and CdZnTe crystal growth technology. For example, significant progress has been achieved by improved understanding of convection patterns in the liquid in front of the

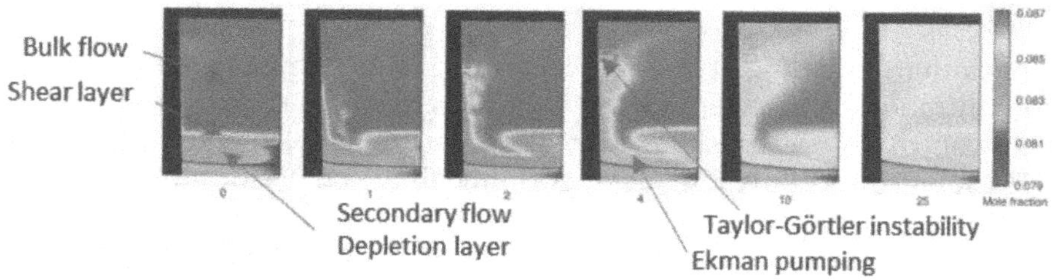


**FIGURE 1.10**
Active mixing of CdZnTe melt by the accelerated crucible rotation technique (ACRT). Zn concentration in the crucible is shown after given number of ACRT cycles.

crystal growth interface and its effects on the compositional uniformity of the melt. Heat transport has been analyzed to understand the effect of furnace profiles on the shape of the growth interface. Advancement in understanding of three-dimensional heat transport in the melt and in the solidified ingot is another significant accomplishment that is directly transferable to CdTe and CdZnTe crystal growth technology. Without doing proper justice to the vast body of computational material science work relevant to CdTe and CdZnTe crystal growth, we highlight two examples to illustrate the progress in the field.

#### *1.4.5.1 Melt Mixing*

Insufficient mixing of the liquid during compound semiconductor growth can lead to segregation of the components and nonuniform liquid composition in front of the crystal growth interface. In the case of CdZnTe growth by the vertical Bridgman (VB) technique, modeling results demonstrated that a weak, secondary convection cell is formed adjacent to the growth interface and separated from a much stronger flow in the bulk of the melt by a shear layer, as shown in the leftmost image of Figure 1.10 [20]. The figure shows the calculated spatial distribution of Zn concentration in the solidified ingot (bottom) and liquid in the crucible (top). The black line shows the solid-liquid interface. The Zn concentrations in the top and bottom vortices are clearly very different and, while the upper cell is well mixed, there is a strong radial concentration variation in the liquid along the growth interface caused by incomplete mixing in the lower cell. By applying the accelerated crucible rotation technique (ACRT), a set of complex flow transitions are activated that lead to nearly complete mixing of the entire melt volume after several cycles of the ACRT agitation.

#### *1.4.5.2 Interface Shape Control*

The shape of the crystal-liquid interface is indicative of the heat flow distribution during crystallization and should be nearly flat or slightly convex

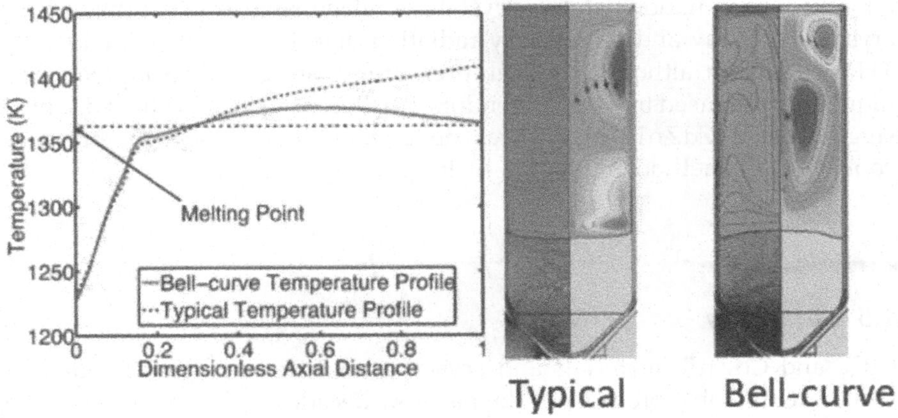


**FIGURE 1.11**
A convex growth interface shape is achieved during CdZnTe crystal growth by employing a bell-curve shaped furnace temperature profile.

(toward the liquid). A concave interface indicates too high heat loss through the crystal in the radial direction and crystallization from the periphery of the ingot. This growth mode stimulates parasitic grain nucleation along the crucible wall and competing grain growth from the ingot periphery inward at the expense of the main crystal grains.

It was shown theoretically and verified experimentally that during the growth of CdZnTe by the EDGF technique it is possible to convert the unfavorable concave interface to a convex interface by application of a bell-curve shaped axial furnace temperature profile as shown in Figure 1.11 [21, 22]. The change in the interface shape is the result of a significant change in the distribution of heat flows into and out of the charge, which is accompanied by an extensive rearrangement of the flow structure in the liquid. This result paves the way for a significant improvement in the single crystal yield of CdZnTe grown by the EDGF method.

### 1.4.6 Status of Crystal Growth Technology

CdTe and CdZnTe crystal growth has steadily progressed over the past three decades. The growth of 100-mm diameter CdTe ingots that are nearly entirely single-crystal have been demonstrated by the THM technique [22]. The growth of large CdZnTe single crystals proved to be considerably more difficult due to the complexities induced by Zn. The highest perfection CdZnTe single crystals are grown by the vertical gradient freeze (VGF) technique. CdZnTe ingots as large as 125-mm diameter with high single crystal yield and low defect density have been demonstrated with the VGF technique, mostly for infrared detector substrate applications [23]. Interestingly, despite numerous development efforts, the VGF technique has not yet gained a foothold in radiation detector crystal manufacturing. Much of the CdZnTe crystals for x-ray and gamma-ray radiation detectors are produced by the THM technique, although the high-pressure and horizontal Bridgman techniques are also used by various vendors [24]. Recently the growth of oriented, single crystal CdZnTe ingots was demonstrated by a seeded, pressure-controlled VB method employing ACRT [25, 26].

## 1.5 Summary

CdTe and CdZnTe materials and crystal growth technology have demonstrated remarkable progress in the past few decades and enabled the development and deployment of room-temperature, photon-counting x-ray and gamma-ray detectors and imaging arrays in several medical applications. Utilization in low and moderate flux applications, such as cardiac SPECT, MBI, and DEXA, is expected to grow as the performance and availability of these detectors improve and their cost decreases. The performance requirements in numerous medical imaging applications, particularly in applications where high photon fluxes are used and high-speed detectors are needed, are still challenging for this detector technology.

As with other semiconductor detector technologies, the performance of CdTe and CdZnTe detectors is controlled by electrically active point defects and extended structural defects in the crystals. Our understanding of the nature of these defects, their formation, diffusion and electrical properties, as well as their effects on the charge transport properties of the crystals, has significantly expanded in the past decade. Similarly, impressive progress is occurring in understanding crystallization phenomena and defect formation processes and interactions during solidification and cooling of crystals to room temperature. The maturing scientific understanding of crystallization and defect phenomena provided by computational materials science and modeling tools fuel further advancement of materials and crystal growth technologies and facilitate the development of exciting new CdTe and CdZnTe detectors with enhanced performance.

## References

1. K. Erlandsson, K. Kacperski, D. van Gramberg and B. F. Hutton, "Performance evaluation of D-SPECT: A novel SPECT system for nuclear cardiology," *Physics in Medicine and Biology*, vol. 54, pp. 2635–2649, 2009.
2. C. B. Hruska and M. K. O'Connor, "CZT detectors: How important is energy resolution for nuclear breast imaging?," *Medical Physics*, vol. 21 (Supplement 1), pp. 72–75, 2006.

3. J. Wear, M. Buchholz, R. K. Payne, D. Gorsuch, J. Bisek, D. L. Ergun, J. Grosholz and R. Falk, "CZT detector for dual-energy x-ray absorptiometry (DEXA)," in *Proc. SPIE 4142, Penetrating Radiation Systems and Applications II*, 2000.
4. K. Spartiotis, J. Havulinna, A. Leppanen, T. Pantsar, K. Puhakka, J. Pyyhtia and T. Schulman, "A CdTe real time imaging sensor and system," *Nuclear Instruments and Methods A*, vol. 527, pp. 478–486, 2004.
5. J. P. Schlomka, E. Roessl, R. Dorscheid, S. Dill, G. Martens, T. Istel, C. Baumer, C. Herrmann, R. Steadman, G. Zeitler, A. Livne and R. Proksa, "Experimental feasibility of multi-energy photon-counting K-edge imaging in pre-clinical computed tomography," *Physics in Medicine and Biology*, vol. 53, pp. 4031–4047, 2008.
6. Y. G. Cui, T. Lall, B. Tsui, J. H. Yu, G. Mahler, A. Bolotnikov, P. Vaska, G. De Geronimo, P. O'Connor, G. Meinken, J. Joyal, J. Barrett, G. Camarda, A. Hossain, K. H. Kim, G. Yang, M. Pomper, S. Cho, K. Weissman, Y. H. Seo, J. Babich, N. LaFrance and R. B. James, "Compact CdZnTe-based gamma camera for prostate cancer imaging," in *Proc. SPIE 8192, International Symposium on Photoelectronic Detection and Imaging: Laser Sensing and Imaging; and Biological and Medical Applications of Photonics Sensing and Imaging*, 2011.
7. Check Cap Ltd., Mount Carmel, Israel, [Online]. Available: http://www.check-cap.com.
8. R. Triboulet, "Fundamentals of the CdTe and CdZnTe bulk growth," *Physica Status Solidi (c)*, vol. 2, no. 5, pp. 1556–1565, 2005.
9. J. H. Greenberg, V. N. Guskov, M. Fiederle and K.-W. Benz, "Experimental study of non-stoichiometry in Cd1-xZnxTe1-δ," *Journal of Electronic Materials*, vol. 33, no. 6, pp. 719–723, 2004.
10. S. H. Wei and S. B. Wang, "Chemical trends of defect formation and doping limit in II-VI semiconductors: The case of CdTe," *Physical Review B*, vol. 66, pp. 155211–155221, 2002.
11. K. Biswas and M. H. Du, "What causes high resistivity in CdTe," *New Journal of Physics*, vol. 14, p. 063020, 2012.
12. C. Szeles, "Advances in the crystal growth and device fabrication technology of CdZnTe room temperature radiation detectors," *IEEE Transactions on Nuclear Science*, vol. 51, no. 3, pp. 1242–1249, 2004.
13. W. Jantsch and G. Hendorfer, "Characterizaion of deep levels in CdTe by photo-EPR and related techniques," *Journal of Crystal Growth*, vol. 101, pp. 404–413, 1990.
14. J. E. Jaffe, "Computational study of Ge and Sn doping of CdTe," *Journal of Applied Physics*, vol. 99, pp. 33704–33708, 2006.
15. D. S. Bale and C. Szeles, "Nature of polarization in wide-bandgap semiconductor detectors under high-flux irradiation: Application to semi-insulating Cd1–xZnxTe," *Physical Review*, vol. B 77, pp. 035205–0352021, 2008.
16. A. E. Bolotnikov, S. O. Babalola, G. S. Camarda, H. Chen, S. Awadalla, Y. Cui, S. U. Egarievwe, P. M. Fochuk, R. Hawrami, A. Hossain, J. R. James, I. J. Nakonechnyj, J. Mackenzie, G. Yang, C. Xu and R. B. James, "Extended defects in CdZnTe radiation detectors," *IEEE Transactions on Nuclear Science*, vol. 56, no. 4, pp. 1775–1783, 2009.
17. F. Zhang, Z. He and C. E. Seifert, "A prototype three-dimensional position sensitive CdZnTe detector array," *IEEE Transactions on Nuclear Science*, vol. 54, no. 4, pp. 843–848, 2007.

18. D. S. Bale, "Fluctuations in induced charge introduced by Te inclusions within CdZnTe radiation detectors," *Journal of Applied Physics*, vol. 108, pp. 24504–24512, 2010.
19. I. Takahashi, T. Ishitsu, H. Kawauchi, J. Yu, T. Seino, I. Fukasaku, Y. Sunaga, S. Inoue and N. Yamada, "Development of edge-on type CdTe detector module for gamma camera," *IEEE Nuclear Science Symposium Conference Record (NSS/MIC)*, pp. 2000–2003, 2010.
20. A. Yeckel and J. Derby, "Effect of accelerated crucible rotation on melt composition in high-pressure vertical Bridgman growth of cadmium zinc telluride," *Journal of Crystal Growth*, vol. 209, pp. 734–750, 2000.
21. J. J. Derby, N. Zhang and A. Yeckel, "Improving the growth of CZT crystals for radiation detectors: A modeling perspective," *SPIE Proceedings Volume 8507, Hard X-Ray, Gamma-Ray, and Neutron Detector Physics XIV*, vol. 8507, p. 850707, 2012.
22. J. J. Derby and A. Yeckel, "Heat Transfer Analysis and Design for Bulk Crystal Growth: Perspectives on the Bridgman method," in *J.J. Derby and A. Yeckel, "Heat Transfer Analysis and Design for Bulk Crystal Growth: PersHandbook of Crystal Growth, Vol. II, edited by T. Nishinaga and P. Rudolph*, Amsterdam, Elsevier, 2015, pp. 793–843.
23. H. Shiraki, M. Funaki, Y. Ando, A. Tachibana, S. Kominami and R. Ohno, "THM growth and characterization of 100 mm diameter CdTe single crystals," *IEEE Transactions on Nuclear Science*, vol. 54, no. 4, pp. 1717–1723, 2009.
24. R. Hirano and H. Kurita, "Bulk Growth of CdZnTe/CdTe Single Crystals," in *Bulk Crystal Growth of Electronic, Optical and Optoelectronic Materials*, John Wiley & Sons, 2005, pp. 241–268.
25. H. Chen, S. A. Awadalla, J. Mackenzie, R. Redden, G. Bindley, A. E. Bolotnikov, G. S. Camarda, G. Carini and R. B. James, "Characterization of traveling heater method (THM) grown Cd0.9Zn0.1Te crystals," *IEEE Transactions on Nuclear Science*, vol. 54, no. 4, pp. 811–816, 2007.
26. F. Yang, W. Jie, M. Wang, X. Sun, N. Jia, L. Yin, B. Zhou and T. Wang, "Growth of single-crystal Cd0.9Zn0.1Te ingots using pressure controlled bridgman method," *Crystals*, vol. 10, p. 261, 2020.

# 2

# *A First Principles Method to Simulate the Spectral Response of CdZnTe-Based X- and Gamma-Ray Detectors*

**M. Bettelli and N. Sarzi Amadè**

**CONTENTS**



## 2.1 Introduction and Historical Background

Nowadays, the design phase and engineering of CdZnTe (CZT) based X- and γ-ray room detectors is becoming increasingly essential. On the one hand, the research in this field has reached a point where the serial production is possible (high reproducibility of crystal and contact quality, low

DOI: 10.1201/9781003219446-2

failure rate during detectors manufacturing, continuous reliable operation of the devices). On the other hand, the fabrication of CZT is still a delicate procedure owing to the problems that may arise during the cutting and polishing of the crystals and the deposition and patterning of metal contacts, which makes certain types of detectors arduous to fabricate. Therefore, finely tuning all the aspects which may affect, directly or indirectly, the performances of the device without the actual construction is even more necessary with great savings in time and costs. Furthermore, CZT-based devices still have some limitations in terms of energy and spatial resolution and charge collection efficiency. Depending on the final application, it may be desirable to prioritize a certain feature by reducing or suppressing distortions like charge sharing and hole tailing: definitely, a proper design helps to achieve this goal. Ultimately and most importantly, the capability of predicting the spectral response of the device is essential to reconstruct the real energy distribution of the incoming radiation (spectral unfolding or deconvolution). If, given the detector characteristics and operating conditions, the simulated spectra accurately match the experimental ones, implementing an algorithm able to reverse the direction and come back to the ground truth by removing and correcting the distortions affecting the measurements is to some extent possible. In this framework, a simulation tool able to reproduce the spectral response of these devices by first principles methods is extremely powerful.

The simulation of the spectral response of semiconductor detectors dates back to the pre-CZT era, where only lithium-drifted silicon or germanium devices were present on the scene [1–4]. At the time, the procedure was limited in simulating the interaction of the incoming high-energy photons with the crystal and did not consider the degradation effects inherent in the signal generation and processing. Since the beginning, Monte Carlo calculations were preferred due to the natural stochastic processes of the main physical interaction mechanisms in the energy range of interest of room temperature semiconductor detectors (RTSD). When reliable CdTe-based detectors started appearing, the simulation of the radiation-matter interaction alone was not sufficient to describe the response of the device. The worse transport properties of this compound (short lifetime and low mobility of carriers) in comparison to Si or high-purity Ge inevitably led to incomplete collection of the photogenerated charge. Therefore, several works introduced the modeling of carrier transport and the processing of charge pulses to reproduce the charge collection efficiency as a function of the interaction position within the detector [5–8]. During the years, the development of sophisticated photolithographic techniques allowed to abandon the simple planar-planar geometry and to realize complex contact patterns (e.g., matrix of small pixels surrounded by a guard ring electrode, coplanar grid) with the purpose of neglecting the contribution of slow carriers (single polarity charge sensing). Consequently, charge induction on the read-out contact was necessarily included in the simulation of the response function [9, 10]. Despite the computational power available nowadays, reproducing the exact behavior of this class of detectors presents limitations and is still an object of research.

In this chapter, the simulation toolkit described in [11] is presented. Strengths, limitations and possible improvements of this approach are discussed. Finally, the experimental validation is reported through the comparison between simulated and real current transient and pulse height spectra in a wide range of radiation energies and for detectors with different geometries and dimensions. This simulator has been conceived for the specific case of CZT-based devices, but its usage can be extended to other semiconductors compounds.

## 2.2 Spectral Response

The detection process of X- and γ-rays from a CZT-based detector can be described by Fredholm integral function of the first kind [12]:

$$S_{meas}(E) = \int_0^\infty R(E,E')S_{inc}(E')dE' \tag{2.1}$$

where $S_{inc}(E')$ and $S_{meas}(E)$ represent the incident and measured spectra, respectively, and $R(E,E')$ is the spectral response function of the detector: it represents the probability density function that a photon of energy $E'$ is measured with an energy $E$. In a real measurement, the detector measures the number of events in a finite energy range with a fixed number of channels; hence Equation (2.1) is discretized as follows:

$$S_{meas} = R \cdot S_{inc} \tag{2.2}$$

where each column of $R$ corresponds to the ideal measured spectrum for a monoenergetic incident radiation (that is a vector of zeros except one at the position of the corresponding energy). The dimension of $R$ depends on the energy range of interest and the desired resolution of the final spectra. The validity of Equations (2.1) and (2.2) is ensured in absence of non-linear effects such as pulse pile-up which is present at high count rates. In principle, given an arbitrary radiation source, the knowledge of $R$ allows to predict the shape of the measured spectra. Each step of the signal generation process must be included to obtain $R$:

1. interaction of the photon with the semiconductor and generation of carriers,
2. transport and collection of carriers and charge induction on the read-out electrodes, and
3. amplification and processing of the charge pulse.

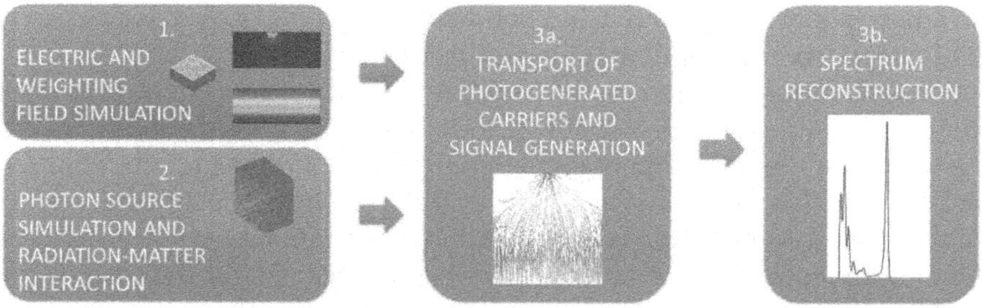


**FIGURE 2.1**
Simulator concept scheme.

Since these processes can be treated independently, the simulator is composed of multiple blocks, each having different tasks. The simulator workflow is shown in Figure 2.1, detailed explanation of each block is reported here.

## 2.3 Radiation-Matter Interaction

The first block is devoted to the simulation of the radiation-matter interaction and is based on Geant4 [13], which has become during the year the most common reference tool for Monte Carlo calculations in the field of X- and γ-ray spectroscopy [10, 14–17]; Penelope is another widely used Monte Carlo-based system [18–20]. Geant4 is a toolkit for simulating the transport of particles in arbitrary materials and allows to select among several packages the ones that describe the desired physical processes (electromagnetic, hadronic or optical) and which include the corresponding cross sections (calculated via formulas, parametrizations or interpolation of databases). The geometrical configuration of the whole system can be defined (photon source, possible collimators or filters, sensing material) and particles are efficiently tracked during the passage through materials according to optimized steps. The information relative to each collision (interaction position, energy released, secondary particles) is recorded according to a branching scheme to relate each event to its primary particle.

The interaction of X- and γ-radiation with CZT in the energy range from few keV up to tens of MeV occurs through four main mechanisms:

- Electron/positron pair production
- Compton scattering
- Rayleigh scattering
- Photoelectric effect

These processes may lead to the production of fast electrons or positrons within the crystal which release energy in the medium via

- Ionization
- Bremsstrahlung
- Multiple scattering
- Annihilation (only positrons)

G4EmStandardPhysics and G4EmLivermorePhysics are Geant4 libraries specialized in electromagnetic interactions and consider all the aforementioned mechanisms. Usually, the photon source is defined as planar whose shape and dimensions correspond to the ones of the irradiated surface of the detector. The only exception is when the simulated radiation source consists of synchrotron light which is generally collimated to a small spot. Photons are emitted orthogonally in respect to the planar source that is located at a distance of 50 cm. The flux is considered as uniform in the whole source area. Collimators, filters or passive layers can be placed between the source and the detector. Dimensions and chemical composition of each volume present in the simulation can be defined in this step. The CZT crystal is considered the sensitive volume and only the interactions occurring within it are monitored. The coordinates of each collision are saved in a CSV file as well as the energy released at that point. Particles are tracked until their mean free path (i.e., their energy) falls below a defined threshold. In this case, the distance cut-off limit for tracking secondary particles in "multiple scattering" process type has been set to 500 µm. The number of photons generated in each simulation is typically in the order of $10^6$ to obtain sufficient statistics. Different types of radiation sources such as radioactive isotopes ($^{241}$Am, $^{57}$Co, $^{137}$Cs), X-ray tubes or synchrotron light have been simulated with this tool.

## 2.4 Electric and Weighting Fields

The task of the second block is calculating the electric and weighting fields within the detector volume by finite elements method. Given the geometry of the system, the properties of the material (spatial charge density, electric permittivity) and the boundary conditions (applied bias voltage at each electrode), the electric potential can be obtained by solving the Poisson's equation:

$$\nabla^2\varphi = -\frac{\rho}{\varepsilon} \tag{2.3}$$

This approach is valid only if the electric field is not significantly perturbed by the photogenerated charges and, therefore, the Poisson's equation can be decoupled from the continuity equations for electrons and holes (quasi steady-state condition). In the case of CZT, these assumptions are valid since it is not affected by bias induced polarization effects which would, otherwise, create a progressive accumulation of positive or negative charge near one of the contacts, hence resulting in a dead layer region. Furthermore, high-flux grade CZT can hold high-radiation fluxes and it is not subjected to radiation induced polarization [21]. These assumptions have been proven experimentally with different techniques [22–25].

Analogously, the weighting potential, which describes the charge induction on a given electrode by charges moving within the semiconductor, can be calculated by solving the Laplace's equation with Dirichlet boundary conditions in which all electrodes are grounded except the considered one which is set at unitary potential (Shockley-Ramo theorem [26, 27]). The electric and weighting fields can be successively calculated as the gradient of their respective potentials.

In practice, electric and weighting fields are calculated by using the "electrostatics" package of COMSOL MultiPhysics, a versatile simulation software for finite element analysis. Depending on the geometry of the detector and contacts, COMSOL automatically creates an adaptive mesh in which the element size spans from a minimum of 50 nm and a maximum of 100 μm, dividing the crystal in up to ~$10^6$ volume elements, according to the required accuracy. The grid is denser in the high-gradient regions where the field lines quickly converge toward the electrodes (usually the anode) and coarser elsewhere. Additionally, COMSOL permits to specify the presence of a spatial charge within the crystal which, in case of detector with full-area contacts, leads to a linearly increasing/decreasing electric field along the detector thickness. After the calculation, electric and weighting potentials are extrapolated and exported on a uniform 3D square mesh with about 5 μm steps, depending on the detector geometry and dimensions. Thanks to the flexibility of the software, the tri-dimensional model of detectors with any desired electrode patterning can be simulated. The choice of detector geometry is strongly dependent on the transport properties of the selected semiconductor as well as the final application. During the years, different contact geometries have been developed to limit the extent of distortions caused by severe hole trapping in CdTe and CZT (single polarity charge sensing) [28]. In the following sections, the most common geometries are described. A qualitative behavior of electric and weighting potentials for each geometry is shown in Figure 2.2 to provide an overview.

### 2.4.1 Planar-Planar Electrodes

In a planar-planar configuration, two full-area metal contacts are deposited on the opposite surfaces of the semiconductor as illustrated in the first row of Figure 2.2. Generally, the signal is read from the grounded contact (typically the anode), whereas the bias voltage is applied to the other electrode

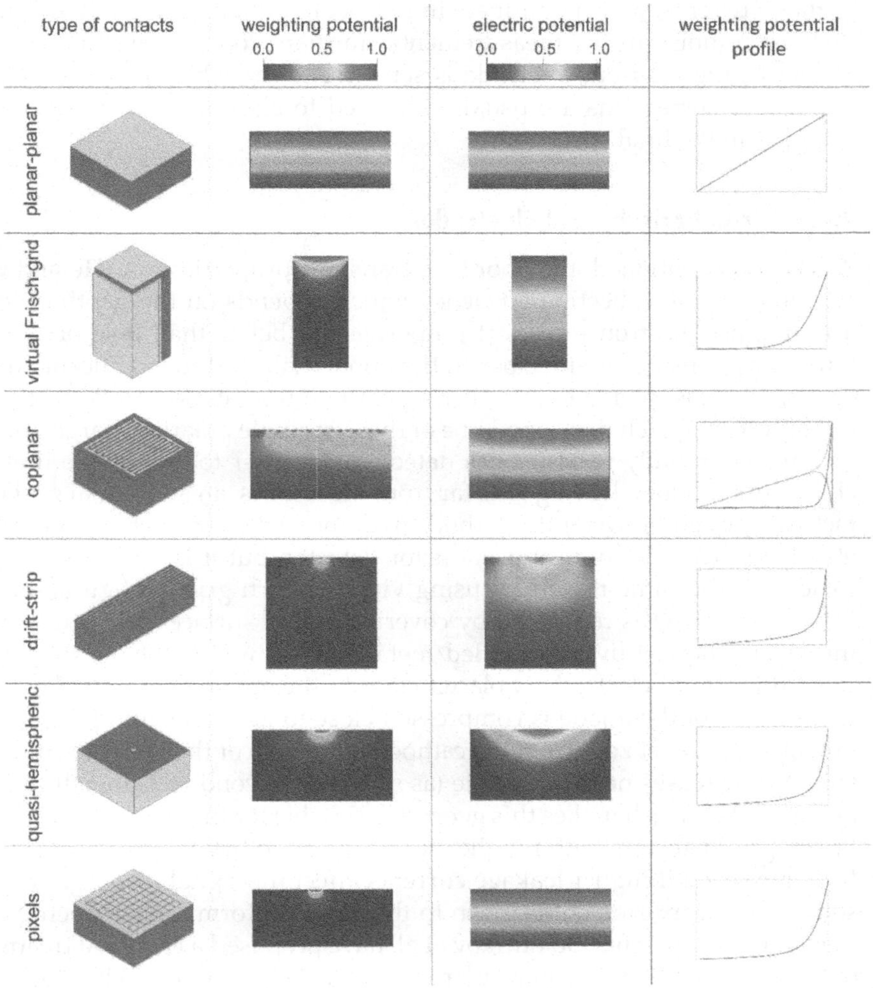


**FIGURE 2.2**
Qualitative results of the simulation on different kinds of example detectors: contacts type and detectors shape are represented in the first column; weighting potential maps of a cross-section of each detector are in the second column; normalized electric potential maps of a cross-section of each detector are in the third column; weighting potential profile in the fourth column (further details in the text) [11].

(cathode). In this configuration, the weighting potential is a linear function of depth (as shown in the fourth column of Figure 2.2) and, consequently, the resulting weighting field is uniform along the entire detector thickness ($L$). The profiles of weighting and electric fields are identical, and the value of the latter is simply equal to $V/L$, where $V$ is the applied bias voltage. This is the most simple geometry and also the worst one for two main reasons:

surface currents, which dominate in respect to bulk currents in CZT, introduce large noise in the measurements since are collected by the read out electrode; the read out electrode is sensitive to the whole detector volume and signal corrections are usually required to eliminate the effect of hole trapping in the final spectrum.

### 2.4.2 Virtual Frisch-Grid Electrodes

As already mentioned, the poor hole transport properties in CdTe and CZT lead to a charge collection efficiency which depends on the depth of interaction. Since electron transport properties are better than hole ones, photon absorptions occurring close to the anode will lead to significant charge losses, whereas interactions occurring close to the cathode are less affected by this effect. Frisch developed one of the first single polarity charge sensing techniques initially used for gas detectors. In order to reduce the induced charge due to slow drifting ions far from the anodes, an additional grid contact was placed between the cathode and anode. Unfortunately, this geometry is not feasible in a semiconductor detector, but it is however possible to achieve the same results by using virtual Frisch grid configuration [29]. Frisch grid effect is replicated by covering lateral surfaces with an insulating layer followed by a grounded metallic sheet which acts as shield: the more this lateral electrode is placed close to the anode, the more the sensitive volume of the anode is compressed close to it. The resulting weighting potential is almost zero from the cathode to the end of the metallic layer and strongly increases near the anode (as shown in second row, fourth column of Figure 2.2) which makes this geometry suitable for thick detectors. Virtual Frisch-grid detectors with rougher surfaces lead to better spectroscopic performance even if higher leakage currents affect the signal-reading. The reason for this improvement is given to the more uniform sensor electric field due to higher currents. Bolotnikov et al. have proposed a chemical treatment of crystal edges that can make more reproducible this effect [30]. A constant voltage gradient is imposed on lateral surfaces of simulated detectors to replicate better the detector response.

### 2.4.3 Coplanar Grid Electrodes

Single-polarity charge sensing can be achieved also with coplanar grids electrodes, which were firstly implemented by Luke [31, 32]. In this electrode design, the anode consists of two sets of interdigitated grids labeled as grids "A" and "B" whereas the cathode is full area (an example is shown in the third row of Figure 2.2). A voltage difference is applied between grid "A" and "B", each of which is connected to a different readout electronics. The weighting potential of electrode A ($V_w^A$), considered here as the collecting electrode, is calculated by setting it to 1 V whereas both the cathode and grid "B" are set to 0 V [33]. Viceversa, the weighting potential of electrode "B" ($V_w^B$) can

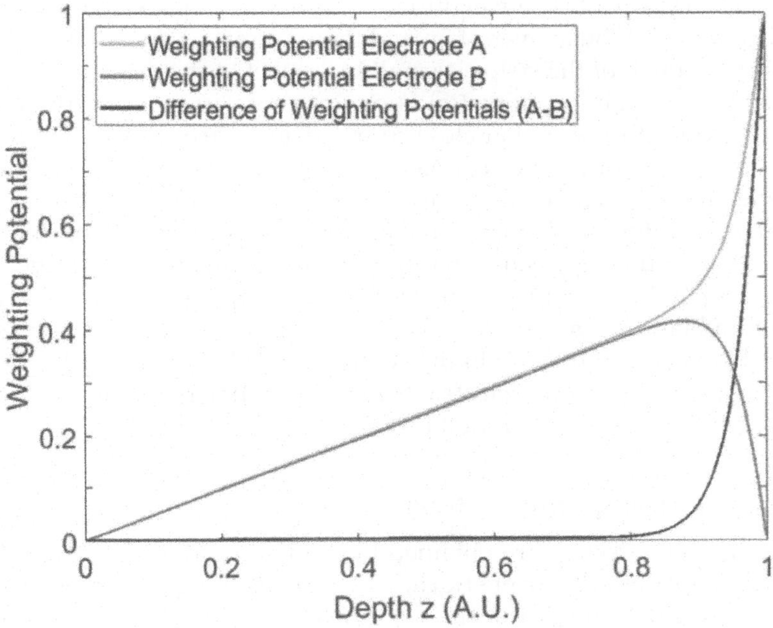


**FIGURE 2.3**
Weighting potential calculated along the dotted line showed in Figure 2.2 for electrode A and B. The black line represents the difference between the two [11].

be calculated by setting the potential of electrode "B" to 1 V and the other ones to 0 V. The profile of $V_w^A$ and $V_w^B$ along the cross-section perpendicular to the anode surface intersecting electrode "A" (white dotted line in the third row of Figure 2.2) is reported in Figure 2.3. The behavior as a function of the thickness is almost identical for both electrodes in the range $0<z<L-P$ where $z$ is the linear coordinate along the line, $P$ is the pitch of the coplanar grids and $L$ is the sample thickness. Instead, in the proximity of the anode, $V_w^A$ quickly rises to 1 whereas $V_w^B$ falls to 0. The trend of the difference between $V_w^A$ and $V_w^B$ (shown as a black curve in Figure 2.3) is similar to virtual Frisch-grid configuration. Therefore, single polarity charge sensing is obtained by reading the difference of the signals induced on both grid electrodes, since only carriers moving near the anode produce a substantial signal.

### 2.4.4 Drift-Strip Electrodes

Semiconductor detectors using strip electrodes were developed for two-dimensional position sensing in which the spatial coordinates are given by reading the signal produced by two sets of orthogonal strips on the opposite surface of the detector [34, 35]. This geometry is also suitable for single-polarity charge sensing by reading out signals from individual strip electrodes [33, 34, 36]. The weighting potential of a specific strip is calculated applying 1 V on a given strip and then setting all the other electrodes to 0 V. If the pitch of the strips is small in respect to the detector thickness, the resulting weighting potential is close to zero in most of the detector volume, except in the region close to the strip (fourth row in Figure 2.2). Qualitative shape of weighting potential profile calculated along a cathode-perpendicular line that pass through the center of collecting anode is shown in the fourth column of Figure 2.2. A drift detector for high-resolution spectroscopy can be realized by modifying the bias at the anode electrodes [35, 37]: both the cathode and the outermost strips are held at the maximum bias voltage, whereas the other strips are set to a progressively reduced voltage down to 0 V in the central strip which is the collecting one (Figure 2.2). The resulting electric field drives carriers toward the central strip regardless of the generation position.

### 2.4.5 Quasi-Hemispheric Electrodes

Hemispherical detectors are obtained by using a small collecting anode at the focus point of the hemispheric cathode. The resulting weighting potential is almost null within most of the detector volume but rapidly rises to 1 in the proximity of the anode. Analogously to the "drift strip" configuration, the electric field lines within the detector converge toward the small anode. Consequently, the photogenerated charges flow toward it regardless of the interaction position. The use of this geometry allows the creation of detectors with large active volumes for high-energy X-ray and $\gamma$-ray spectroscopy and, at the same time, the limitations of degradation effects related to holes trapping. However, the fabrication of CdTe- or CZT-based hemispheric detectors is not feasible owing to the mechanical properties of these materials and the technological limits in processing such a particular crystal shape. The "quasi-hemispheric" detector geometry was thus proposed to overcome this limitation [38]. The idea is to realize the hemispheric detector on a parallelepiped shaped CZT block, as shown in the fifth row of Figure 2.2, by maintaining the same anode shape and by extending a continuous metallization on the lateral surfaces. The profile of electric and weighting potential are similar to that of the hemispheric cathode configuration. Qualitative shape of weighting potential profile calculated along the cathode-perpendicular line that passes through the center of collecting anode is shown in the fourth column of Figure 2.2.

### 2.4.6 Pixel Electrodes

Similarly, single-polarity charge sensing is possible when the pixel dimensions are small in respect to the device thickness (from which the so-called "small-pixel effect") [39]. Here, the induced charge on a considered pixel is negligible when the charge cloud moves far from the pixel and rapidly increases as it moves close to the electrode (sixth row of Figure 2.2). Qualitative shape of weighting potential profile calculated along a cathode-perpendicular line that passes through the center of the considered pixel is shown in the fourth column of Figure 2.2. Typically, pixel detectors also include a guard-ring electrode surrounding the whole array. The role of this additional electrode is to minimize the contribution of surface currents to the signal induced on each pixel. Because of the presence of defects on the lateral surfaces of the crystal, the resistivity is lower in respect to the bulk material and the corresponding current may be up to two orders of magnitude greater. This represents a noise source which would hide the signal produced by photons and completely degrade the performance of the device. Instead, thanks to the presence of a guard ring, which must be biased at either the same or near the same voltage as that of the array of pixels, the surface leakage current is collected by the guard ring, thus eliminating its contribute to the front-end electronics and greatly improving the signal-to-noise ratio on each pixel. The presence of a matrix of electrodes allows us to realize position-sensitive detectors for imaging applications [40, 41].

## 2.5 Charge Transport and Signal Induction

By combining the information provided by the two previous blocks, the charge transport and signal induction can be calculated. Electrons and holes are separated by the electric field and move toward the electrodes according to the electric field calculated with COMSOL and the initial location of the charge cloud corresponds to the interaction position provided by the Geant4 block. From the moment carriers start moving, a signal is induced on the read-out electrode. The induction depends also on the spatial charge distribution of the cloud which evolves because of charge diffusion and Coulomb repulsion. However, if self-shielding effects are neglected and a homogeneous electric field within the volume of the cloud is assumed, the equation which describes the trajectory of the barycenter can be decoupled from the broadening of the cloud and can be written as:

$$r_i(t) = r_{i0} + \int_{t_0}^{t} v_i(t')dt' = r_{i0} + \mu_i \int_{t_0}^{t} E(r_i(t'))dt' \tag{2.4}$$

where $i = e, h$, $r_{i0}$ is the initial position, $r_i(t)$ and $v_i(t')$ are the instantaneous positions of the barycenter, $\mu_i$ is the electron and hole mobility and $E(r_i(t'))$ is the electric field experienced by the carriers at the time $t'$ at the position $r_i(t')$. $\mu_i$ is assumed to be independent from the electric field since its saturation occurs at value greater than 15000 V/cm [34], whereas typical field strengths in CZT-detectors are of the order of 5000 V/cm. The software calculates trajectories of both carriers by means of an ODE (ordinary differential equation) function. Assuming a Gaussian distribution of the charge density, the broadening of the carrier cloud as a function of the time can be calculated with the following equation:

$$\sigma^2(t) = \sigma_0^2 + 2Dt + 2\alpha \int_{t_0}^{t} \frac{dt'}{\sqrt{\sigma^2(t')}} \tag{2.5}$$

$$\text{with } D = \frac{\mu k_B T}{q} \text{ and } \alpha = \frac{3\mu N q}{4\pi\epsilon} \frac{\sqrt{5}}{15}$$

where $\sigma(t)$ is the variance at time $t$, $D$ is the diffusion coefficient, $\mu$ is the mobility, $k_B$ is the Boltzmann's constant, $T$ is the temperature, $N$ is the number of charges, $q$ is the absolute electron charge and $\epsilon$ is the electric permittivity of CZT. The initial cloud dimension $\sigma_0$ has been obtained via Geant4 simulations in which secondary interactions of a photoelectron of a given energy were tracked. The variance has been obtained by fitting the energy weighted interaction positions with a Gaussian model. The time evolution of $\sigma$, obtained with Equation (2.5), is shown in Figure 2.4 at various energies.

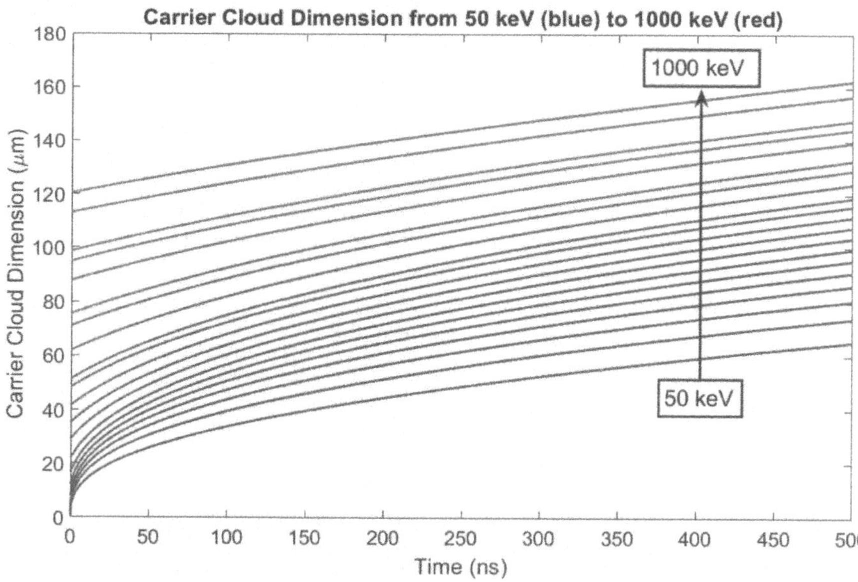


**FIGURE 2.4**
Time evolution of carrier cloud size generated by photoelectrons in the range 50–1000 keV due to Coulomb repulsion and charge diffusion [11].

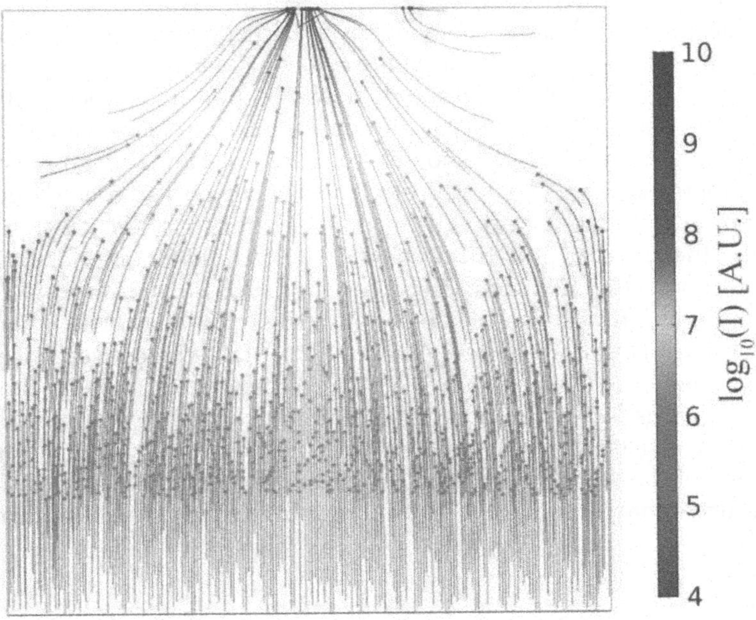


**FIGURE 2.5**
Representation of carrier transport and signal induction in a randomly chosen moment of carrier travel simulation. The actual position of carriers and their route are respectively represented with dots and routes. The color and intensity of the line represents the electric current induced on the anode. Simulation concerns only the electron transport in a quasi-hemispherical detector [11].

Once motion equations are solved and trajectories are available, the software calculates the induced current for both electrons and holes via the Shockley-Ramo theorem. An example of carrier positions, trail paths and induced signals is reported in Figure 2.5. In this plot, the actual position of the charge clouds and their routes are respectively shown (the color indicates the intensity of the induced current).

By integrating the induced current on a given electrode as a function of time, it is possible to determine the net induced charge. The induced current on a given electrode at time $t$ is given by:

$$I_i^{ind}(t) = Q_0\, exp\,exp\left(-\frac{t}{\tau_i}\right)\mu_i E(r_i(t)) \cdot W(r_i(t)) \tag{2.6}$$

where $\tau_i$ is the electron lifetime, $Q_0$ is the initial charge and $W(r_i(t))$ is the weighting field at the position $r_i(t)$. The $\tau_i$ values used are obtained experimentally (see Table 2.2) and, thus, they consider both the effect of crystal defects and secondary phase like Te inclusions, a well-known recombination center. Instead of integrating the charge distribution over the cloud volume, the value of $W(r_i(t))$ can be obtained by weighted averaging the weighting field on the whole charge cloud volume to consider the effects of its finite dimension. This approach, although slower, is powerful since it allows to include the effect of the cloud size on the current transient which becomes important at high energy as shown in Figure 2.4. However, in most cases, the geometric dimensions of the electrodes in respect to the charge cloud and typical times of flight in CZT (about few tens of ns at most) allow to consider the cloud as punctiform below a certain threshold, depending on the detector: in the case of pixelated detectors this value is around 200 keV for pitch down to 0.2 mm, whereas for drift-strips/hemispheric detectors the threshold is around 1 MeV. As a matter of fact, the convergence of the electric field lines toward the collecting electrode counteracts the cloud spreading, thus facilitating a better signal induction. The main limit of this approach consists in not considering possible cloud splitting since only the trajectory of the barycenter of the charge distribution is calculated. Finally, the induced charge is obtained by integrating Equation (2.6) with respect to time:

$$Q_i^{ind} = \int_{t_0}^{TOF_i} I_i^{ind}(t)dt = \int_{t_0}^{TOF_i} Q_0\, exp\, exp\left(-\frac{t}{\tau_i}\right)\mu_i E(r_i(t')) \cdot W(r_i(t'))dt \qquad (2.7)$$

where $TOF_i$ is the carrier time of flight, that is the time required to complete the drift from the initial position to the collecting electrode. The operations performed by the electronic read-out chain can be added in this step (pulse shaping). Indeed, more accurate results can be achieved by exploiting the knowledge of the transfer function of the read-out system since possible ballistic deficit can also be considered. The contribution of events generated by the same photon are added together because they occur on a timescale of a few picoseconds from each other and can be treated as a single event. For example, in the case of the generation of Cd and Te fluorescence photons, the charge induced by the interaction of these photons is added to the one produced by the primary interaction. In a pixelated detector, the charge induced by fluorescence photons interacting far from the volume below the pixel is almost null, hence producing the so-called escape peaks since a fraction of the energy is lost. The reconstruction of the energy spectrum (right side of Figure 2.6). can be performed in two ways. The first method consists in adding the electronic noise by convolving the simulated spectrum with a Gaussian kernel which is characteristic of each read-out system:

$$\frac{1}{\sqrt{2\pi}\sigma(E_0)} exp\left(-\frac{(E-E_0)^2}{2\sigma(E_0)^2}\right) \qquad (2.8)$$

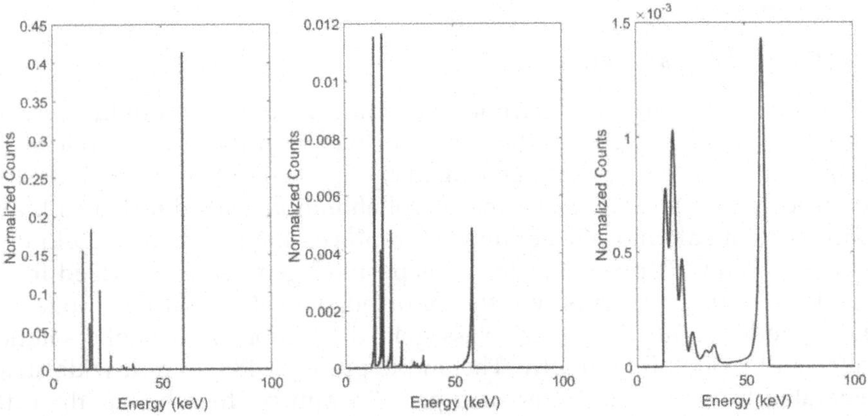


**FIGURE 2.6**
Left: Spectrum obtained by adding together the deposited energy of multiple interactions of each single photon (only Geant4 simulation). Center: Spectrum after the electrode charge collection. Right: Spectrum after the Gaussian broadening [11].

where the variance is usually energy dependent. Instead, if the noise spectral density of the system is known, the noise can be added to each single pulse which is later processed. The former method is usually preferred due to its simplicity since the convolution can be performed with an arbitrary kernel (the simulated spectra reported in this chapter are obtained in this way). This block has been implemented in MATLAB.

Figure 2.6 Shows intermediate results of simulation. The left spectrum is related to the deposited energy of photons (obtained by using results of Geant4). The central spectrum is achieved by simulating the induced charge of photogenerated carriers on the collecting electrode. The right spectrum is the final result of the simulator and is obtained by adding the electronic noise.

## 2.6 Experimental Validation

As already mentioned, these simulations can provide information both on the variety of possible current transients produced by the detector and on the shape of the final pulse height spectra. In this section, the accuracy of the simulation toolkit is presented through the comparison between synthetic and measured spectra in case of different types of detectors and radiation sources to prove the reliability of the system. Furthermore, it will be shown how this tool can be used to understand the physical origin of certain features observed in real measurements.

## 2.7 Current Transients

Detectors with complex electrode patterning require an equally complex signal processing. Recently, the simulator has been used to reproduce the shapes of current transients generated by a 3Ddrift-strip detector system realized at IMEM-CNR of Parma in collaboration with due2lab s.r.l. and University of Palermo [42], obtained by reprocessing a 19.4 mm × 19.4 mm × 6 mm Redlen CZT pixel detector. This peculiar geometry, described in the previous section, presents two sets of orthogonal strips on the anode and the cathode obtained by electroless gold deposition in alcoholic solution followed by photolithography. The anode presents 48 0.15 mm wide strips separated by a 0.25 mm inter-strip gap (0.4 mm pitch), whereas the cathode is characterized by ten 1.9 mm wide strips with a 0.1 mm inter-strip gap (2 mm pitch). Anode strips are organized in 12 drift cells (1.6 mm × 19.4 mm × 6 mm) in which the grounded collecting strip is surrounded by two negatively biased strips on each side: –100 V and –200 V for the immediately adjacent and the outermost ones, respectively, where the latter are shared among two cells; the cathodes are biased at –350 V. The signal is read, amplified and digitized in temporal coincidence from the 12 collecting strips, the 10 cathodes and from the short-circuited left and right adjacent drift strip with a total of 24 read-out channels (schematic diagram of frontend readout system is shown in the bottom part of Figure 2.7).

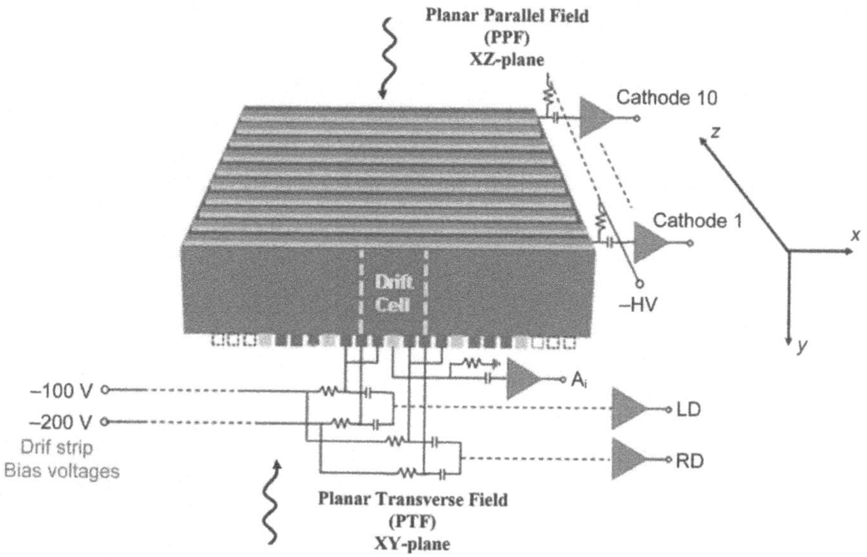


**FIGURE 2.7**
Layout of a single drift cell of the 3D drift-strip CZT detectors. Each drift cell (1.6 mm × 19.4 mm × 6 mm) contains a collecting anode strip and four drift strips, two RDs and two LDs of the collecting strip. The bias voltage values are optimized to collect electron charges on the collecting anode. The PTF and PPF irradiation geometries are also shown [42].

Depending on the interaction position and the trajectories of carriers, a wide variety of collected and induced-charge pulses can emerge owing to the complex electrodes structure and the read-out scheme. While the fine segmentation of anode strips allows to achieve the "small pixel" effect, the coarser pitch of the cathode strips makes them sensitive to the whole detector volume. As reported in Figure 2.4, the pulse shapes reflect the behavior of the weighting field. If a photon interacts in the volume below the collecting strip near the cathode (top of Figure 2.8), the fraction of trapped holes is negligible resulting in a full charge collection. On the other hand, if the primary interaction occurs near the anode, the amount of charge collected by the cathode is small due to severe hole trapping. On the other hand, full charge collection on the anode is ensured by the focused weighting potential. An important consequence is the negative saturation levels of non-collecting electrodes which result in an effective charge collection (bottom of Figure 2.8). Even if CZT presents good transport properties for electrons, also the reverse situation may occur in which holes are fully collected and a small fraction of electrons is trapped (Figure 2.9). In this case, the saturation level of lateral strips is positive. Interestingly, this leads also to a positive induced charge and negative saturation level on the nearby cathode (weighting potential cross talk). Both spatial and energy resolution can be improved by correctly interpreting these pulses [42].

## 2.8 Energy Spectra

In the following sections, the characteristics of each experimental setup, the devices as well as their corresponding readout systems and the radiation sources are described (details are summarized in Table 2.1).

Experimental validation of the described simulation toolkit through the comparison of the simulated spectra with the measured ones. Both spectra are normalized by area and are divided in 8192 channels. Depending on the electrode patterning, crystal dimension and the energy of the incident photons, different distortions may arise in the final spectrum. Therefore, the comparative criteria differ in each examined case and are listed here.

- Normalized Standard Deviation (NSD):

$$\sigma = \sqrt{\sum_{i=1}^{N} \frac{(E_i - S_i)^2}{N}}$$

where $E_i$ and $S_i$ are the number of counts in the experimental and simulated spectra in the $i$-th bin, respectively, and $N$ is the total number of channels (8192).

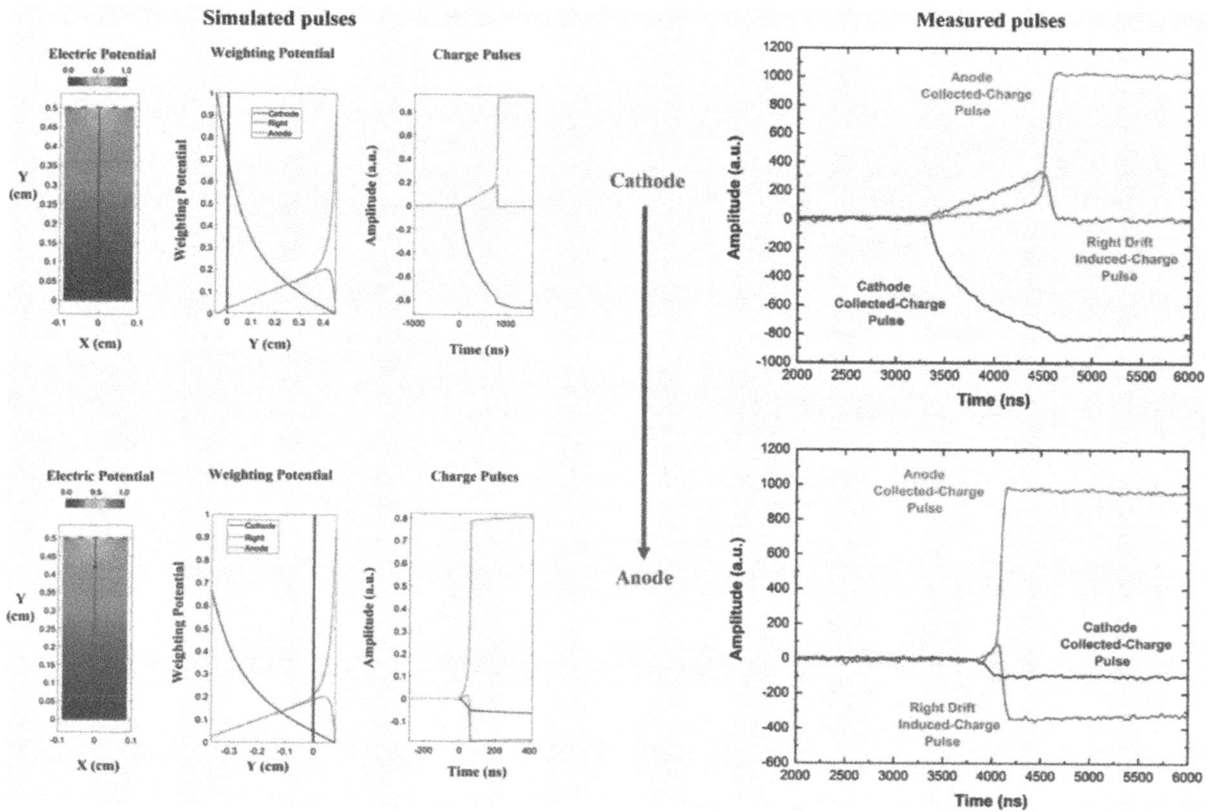


**FIGURE 2.8**
Measured and simulated pulses from the anode, cathode and drift strips. Calculated electric and weighting potential profiles are also reported. Collected and induced-charge pulses are clearly visible, related to photon interaction near the cathode (top) and near the anode (bottom) [42].

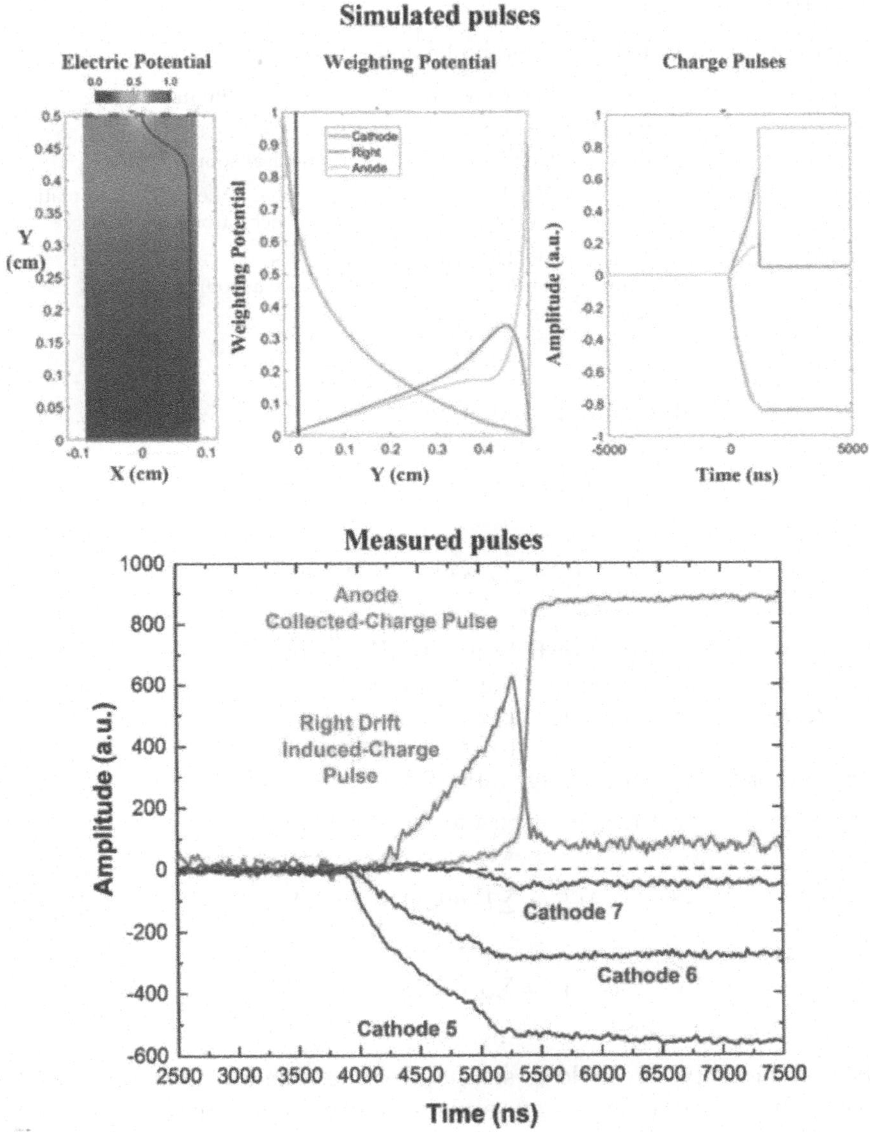


**FIGURE 2.9**
Simulated and measured pulses from the anode, cathode and drift strips. Induced-charge pulses with positive saturation levels can be attributed to the electron trapping. Mixed induced-collected charge pulses (cathode 7) are also observed, due to both charge sharing and weighting potential cross talk [42].

**TABLE 2.1**

Summarization of Tested Detectors and Photon Source [11]

| Geometry | Dimension | Photon Source |
|---|---|---|
| Large pixel detector | Detector $4 \times 4 \times 2.8$ mm$^3$<br>Pixel $2 \times 2$ mm$^2$ | $^{241}$Am<br>nuclear source |
| Small pixel detector | Detector $5.5 \times 19.2 \times 2.5$ mm$^3$<br>Pixels $0.2 \times 0.2$ mm$^2$ | 25 and 50 keV monochromatic synchrotron radiation |
| Drift-strip detector | Detector $20 \times 4.5 \times 6$ mm$^3$<br>Strips 0.25 mm<br>Gap 0.55 mm | $^{137}$Cs<br>nuclear source |

- Area under the Compton edge (AC):

$$AC_E = \Sigma E_i \text{ with } E_i < 500 \text{ keV}$$

$$AC_S = \Sigma S_i \text{ with } S_i < 500 \text{ keV}$$

where the energy threshold is selected for the case of $^{137}$Cs spectra (662 keV); in this energy range Compton scattering predominates in respect to photoelectric effect.

- Photopeak/Compton edge ratio (P/C)
- Area under the fluorescence and escape peaks (AFE):

$$AFE_E = \Sigma E_i \text{ with } E_i < 40 \text{ keV}$$

$$AFE_S = \Sigma S_i \text{ with } S_i < 40 \text{ keV}$$

where the energy threshold is selected for the case of $^{241}$Am spectra (59.5 keV); in this energy range, escape and fluorescence peaks represent the main distortions.

### 2.8.1 Large-Pixel Detector

Large single-pixel detector represents one of the simplest cases since both electric and weighting fields are nearly uniform, similarly to the planar-planar configuration, and it is well suited as a first validation example. The sample was realized with Redlen standard spectroscopic grade CZT material. Gold contacts were fabricated on the CZT crystal using an

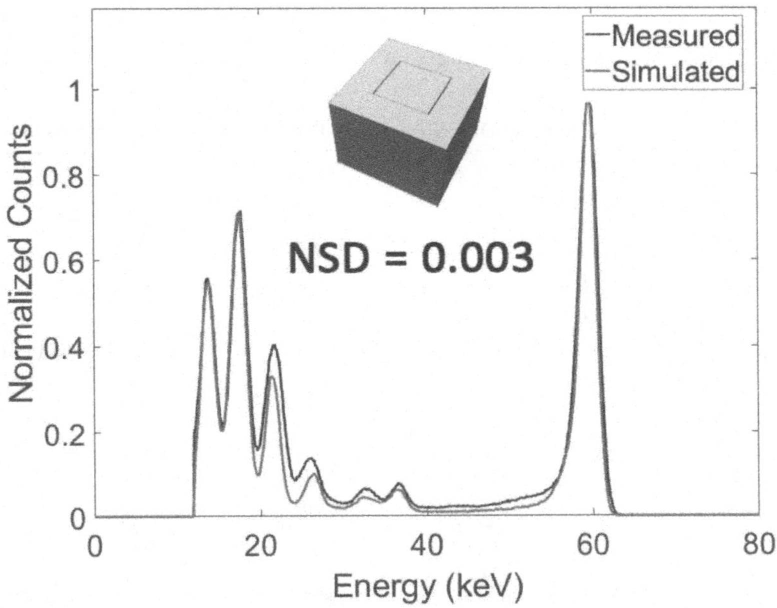


**FIGURE 2.10**
Comparison between experimental and simulated energy spectra of $^{241}$Am nuclear source.

electroless deposition process [43, 44]. The dimension of the cathode electrode is 4 × 4 mm² and the anode consists of a central pixel of 2 × 2 mm² surrounded by a guard ring. The CZT crystal thickness is 2.8 mm and the pixel-guard ring gap is 50 μm. The measurement was carried out at 25°C biasing the cathode at –800 V and grounding both the guard-ring and the anode and irradiating the detector with a $^{241}$Am source from the cathode side. The signal is read from the anode pixel and digitized after the charge sensitive preamplifier stage; the overall electronic noise was 2.1 keV. Digital pulse processing was used to perform pulse shaping [45, 46]. The comparison between the experimental and the simulated energy spectrum is shown in Figure 2.10. The simulation was carried out for spectroscopic grade CZT material using the values for the charge carrier transport properties listed in Table 2.2. Both the simulated and the measured spectra demonstrate a high degree of agreement with an NSD value of 0.003 calculated across the energy range 0–80 keV. The experimental and simulated AFE values are 0.596 and 0.598, respectively, with a difference of 0.3%.

### 2.8.2 Small-Pixel Detector

The second validation of the simulation framework was carried out for a 2D pixelated detector. This detector was fabricated from 2.5 mm thick standard Redlen spectroscopic grade CZT material. The cathode area was 5.5 × 19.2 mm²

**TABLE 2.2**

Overall Table Containing Simulation Parameters for All Experiments (Carriers Parameters Measured by Thomas et al. for Standard Redlen CZT [11, 47])

| Parameter | Large-Pixel Detector | Small-Pixel Detector | Stripes Detector |
|---|---|---|---|
| Electron mobility | 1000 $cm^2/Vs$ | 1000 $cm^2/Vs$ | 1000 $cm^2/Vs$ |
| Electron lifetime | 11 µs | 11 µs | 11 µs |
| Hole mobility | 88 $cm^2/Vs$ | 88 $cm^2/Vs$ | 88 $cm^2/Vs$ |
| Hole lifetime | 0.2 µs | 0.2 µs | 0.2 µs |
| Cathode bias voltage | –800 V | –750 V | –450 V |
| Drift-strips bias voltages | Not present | Not present | –450 V; –300 V; –150 V |
| Anode bias voltage | 0 V | 0 V | 0 V |
| Photon source | $^{241}Am$ | 25 and 50 keV Monoenergetic beam, 10×10 $\mu m^2$ large and aligned with pixel center | $^{137}Cs$ |
| Source position | 5 cm far along the main detector axis | Not relevant, collimated beam | 10 cm far along the main detector axis |
| Gaussian electronic noise | 2.1 keV | 0.7keV | 6 keV |

and the anode consisted of an array of 20 × 75 pixels having a pitch of 250 µm with a pad side width of 200 µm and an inter-pixel gap of 50 µm. The detector was flip-chip bonded directed to the spectroscopic X-ray imaging Hexitec ASIC. The measurement was carried out at 28°C; further details about the electronics setup are reported by Jones et al. [48]. The detector was tested at the Diamond Light Source synchrotron (UK) using a monochromatic beam collimated to 10 × 10 µm using two sets of tungsten slits and focused on the pixel center. The energy of the beam was set to 25 and 50 keV (below and above the K-edge of Cd and Te) to study the effect with and without fluorescence and escape effects. The FWHM of the photopeak was 2.7% and 1.5% at 25 and 50 keV, respectively, corresponding to an electronic noise of 0.7 keV. The comparison between the spectra measured using the HEXITEC ASIC and the simulation outputs are shown in Figure 2.11. Simulation parameters are listed in Table 2.2. In both cases a remarkable agreement has been achieved with NSD values of 0.012 and 0.018. In the case of the 50 keV measured spectrum a significant tail is present in the low energy side of the peak which can tentatively be ascribed to charge sharing effects due to fluorescence photons absorbed below the gaps among pixels. The experimental and simulated AFE values are 0.198 and 0.202, respectively, with a difference of 0.2%.

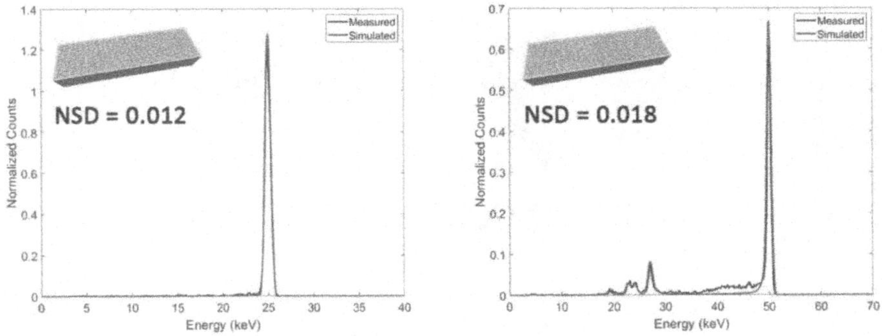


**FIGURE 2.11**
Comparison between simulated and measured spectra of 25 (left) and 50 keV (right) monochromatic synchrotron radiation.

### 2.8.3 Strip Detector

As last validation we report a comparison with a detector that was fabricated as part of a flagship project and was mounted on an Unmanned Aerial Vehicle (UAV). The aim of this project was to develop a technology that could aid human operators during the examination of radioactively contaminated areas [37, 49, 50]. Contrary to previous cases, here the energy range of interest spans up to ~1 MeV to detect various radioactive sources. The detector is a modular system composed of four 20.0 mm × 4.5 mm × 6.0 mm CZT detectors with a drift-strip electrode geometry. The detectors were fabricated from standard Redlen spectroscopic grade CZT and consisted of a planar cathode and a segmented anode of seven 250 μm strips with an inter-gap of 550 μm (3D model of detector is shown in the top left corner of Figure 2.12). Each module was connected to an analog readout system developed by due2Lab s.r.l. (Parma, Italy). Each readout channel includes CSP (Cremat CR110, 140 μs of decay time), shaper amplifier (2 μs of shaping time) and peak detector. The measurement was performed at room temperature with a collimated $^{137}$Cs radioactive source (see Table 2.2). The NSD value is 0.0003 (Figure 2.12). The experimental and simulated AC values are 0.815 and 0.824, respectively, with a difference of 1%. The experimental and simulated P/C ratio values are 2.44 and 2.3, respectively, with a difference of 5%. In this case the electronic noise seems to be underestimated in the low energy region which leads to an excessive broadening. This is suggested by the discrepancy of the peak-to-valley ratio. The backscattering peak at ~200 keV in the measured spectrum is due to the presence of objects in the physical surroundings of the detector (e.g., case, electronics). Since a bare crystal was assumed in the simulations, this feature does not appear in the simulated spectrum although it could be easily reproduced by adding suitable scattering volumes in the Geant4 block.

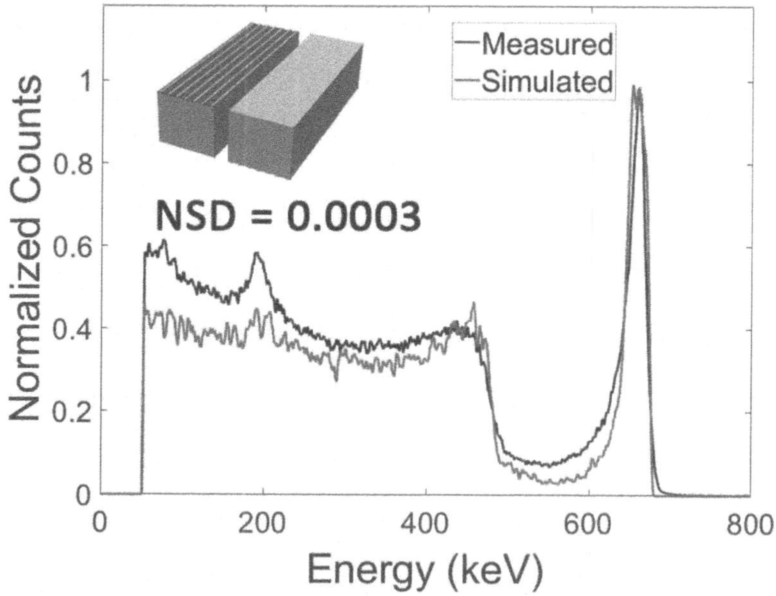


**FIGURE 2.12**
Comparison between simulated and measured spectra achieved with stripe detector and $^{137}Cs$ nuclear source.

## 2.9 Conclusion

This chapter describes a framework based on first principles calculations for simulating the spectral response of CZT-based detectors in different operative conditions. The system has been validated by comparing current transients and spectra obtained with actual detectors having different geometries in a wide range of energies with the corresponding simulated ones. The effects related to the radiation–matter interaction mechanisms (photoelectric absorption, Compton scattering, CZT fluorescence edges and escape peaks) and charge transports (electron and hole trapping) and signal induction are correctly reproduced. The discrepancies can be attributed to secondary effects such as scattering from other materials not considered in the simulations and charge sharing. In principle, spectral distortions due to backscattering and external fluorescences could be easily reproduced by properly modeling the physical surroundings of the detector. On the contrary, modeling charge sharing is complex because the transport of carriers in the inter-gap between electrodes in CZT is still not clear. As a matter of fact, simply summing the collected charges of each electrode in the case of a shared event does not return the total charge [51]. Depending on the distance from the electrodes, a fraction of the deposited energy is not collected which, however, does not depend uniquely on the geometry but also from the surface processing during the detector fabrication. Consequently, simulating this effect by first principles methods is not possible, although it is through experiments [51]. If the inter-gap area is small in respect to the electrodes area, a possible simplification could be neglecting the effect of surface trapping: in this case it is sufficient to project the 3D Gaussian cloud on the contact plane and to integrate the charge density within each collecting electrode [17]. Alternatively, if computation power and running time permit it, the charge transport should consider the whole charge distribution which automatically includes the possibility of splitting among multiple electrodes.

In conclusion, this simulation framework can produce excellent results without the need of experimental measurements, hence allowing the designing and optimization prior the fabrication, granting large benefits in terms of development costs. Furthermore, the information provided by the tool (e.g., electric and weighting fields) can be extremely useful for the simulation-led design of readout electronics and correction algorithms both on current transients (digital pulse processing) and pulse height spectra (spectral unfolding).

## References

1. H. I. Israel, D. W. Lier and E. Storm, "Comparison of detectors used in measurement of 10 to 300 keV X-ray spectra," *Nuclear Instruments and Methods*, vol. 91, pp. 141–157, 1971.
2. B. Grosswendt and E. Waibel, "Determination of detector efficiencies for gamma ray energies up to 12 MeV II. Monte Carlo calculation," *Nuclear Instruments and Methods*, vol. 131, pp. 143–156, 1975.
3. W. W. Seelentag and W. Panzer, "Stripping of X-ray bremsstrahlung spectra up to 300 kVp on a desk type computer," *Physics in Medicine and Biology*, vol. 24, no. 4, pp. 767–780, 1979.
4. C. S. Chen, K. Doi, C. Vyborny, H.-P. Chan and G. Holje, "Monte Carlo simulation studies of detectors used in the measurement of diagnostic X-ray spectra," *Medical Physics*, vol. 7, pp. 627–635, 1980.
5. E. Di Castro, R. Pani, R. Pellegrini and C. Bacci, "The use of cadmium telluride detectors for the qualitative analysis of diagnostic X-ray spectra," *Physics in Medicine and Biology*, vol. 29, pp. 1117–1131, 1984.
6. F. Olschner, J. C. Lund and I. Stern, "Monte Carlo simulation of gamma ray spectra from semiconductor detectors," *IEEE Transactions on Nuclear Science*, vol. 36, no. 1, pp. 1176–1179, 1989.
7. M. G. Scannavini, P. Chirco, G. Baldazzi, G. Guidi, E. Querzola, P. Partemi, M. Rossi, M. Zanarini, F. Casali and E. Caroli, "Computer simulation of charge trapping and ballistic deficit effects on gamma-ray spectra from CdTe semiconductor detectors," *Nuclear Instruments and Methods in Physics Research A*, vol. 353, pp. 80–84, 1994.

8. M. Jung, J. Morel, P. Fougères, M. Hage-Ali and P. Siffert, "A new method for evaluation of transport properties in CdTe and CZT detectors," *Nuclear Instruments and Methods in Physics Research A*, vol. 428, pp. 45–57, 1999.
9. T. H. Prettyman, "Method for mapping charge pulses in semiconductor radiation detectors," *Nuclear Instruments and Methods in Physics Research A*, vol. 422, pp. 232–237, 1999.
10. B. P. F. Dirks, O. Limousin, P. R. Ferrando and R. Chipaux, *3D modeling of Cd(Zn) Te detectors for the Simbol-X space mission*, 2004.
11. M. Bettelli, N. S. Amadè, D. Calestani, B. Garavelli, P. Pozzi, D. Macera, L. Zanotti, C. A. Gonano, M. C. Veale and A. Zappettini, "A first principle method to simulate the spectral response of CdZnTe-based X- and gamma-ray detectors," *Nuclear Instruments and Methods in Physics Research A*, vol. 960, p. 163663, 2020.
12. L. Bouchet, "A comparative study of deconvolution methods for gamma-ray spectra," *Aaps*, vol. 113, p. 167, 1995.
13. S. Agostinelli and others, "GEANT4–a simulation toolkit," *Nuclear Instruments Methods A*, vol. 506, pp. 250–303, 2003.
14. M. Moralles, D. A. B. Bonifácio, M. Bottaro and M. A. G. Pereira, "Monte Carlo and least-squares methods applied in unfolding of X-ray spectra measured with cadmium telluride detectors," *Nuclear Instruments and Methods in Physics Research A*, vol. 580, pp. 270–273, 2007.
15. M. Gerlach, M. Krumrey, L. Cibik, P. Müller and G. Ulm, "Comparison of scattering experiments using synchrotron radiation with Monte Carlo simulations using Geant4," *Nuclear Instruments and Methods in Physics Research Section A*, vol. 608, pp. 339–343, 2009.
16. J. Barylak, P. Podgórski, T. Mrozek, A. Barylak, M. Stęślicki, J. Sylwester and D. Ścisłowski, *Geant4 simulations of detector response matrix for Caliste-SO*, 2014.
17. D. Wu, X. Xu, L. Zhang and S. Wang, "A hybrid Monte Carlo model for the energy response functions of X-ray photon counting detectors," *IAEA*, vol. 830, pp. 397–406, 2016.
18. O. Alirol, F. Glasser, E. G. d'Aillon and J. Tabary, *Simulation and measurements of the internal electric field of a CZT (or CdTe) detector under high X-ray flux for medical imaging*, 2009.
19. A. Tomal, J. C. Santos, P. R. Costa, A. H. L. Gonzales and M. E. Poletti, "Monte Carlo simulation of the response functions of CdTe detectors to be applied in X-ray spectroscopy," *Alejandro Heyner Lopez Gonzales*, vol. 100, pp. 32–37, 2015.
20. A. Makeev, M. Clajus, S. Snyder, X. Wang and S. J. Glick, "Evaluation of position-estimation methods applied to CZT-based photon-counting detectors for dedicated breast CT," *J Med Imaging (Bellingham)*, vol. 2, p. 023501, 2015.
21. M. C. Veale, C. Angelsen, P. Booker, J. Coughlan, M. J. French, A. Hardie, M. Hart, J. Lipp, T. Nicholls, A. Schneider, P. Seller, D. Sole, B. Thomas, M. Wilson, G. A. Carini, P. Hart, K. Nakahara, T. Sato, C. Hansson, K. Iniewski, P. Marthandam and G. Prekas, "Cadmium zinc telluride pixel detectors for high-intensity X-ray imaging at free electron lasers," *Journal of Physics D: Applied Physics*, vol. 52, p. 085106, 2019.
22. Š. Uxa, E. Belas, R. Grill, P. Praus and R. B. James, "Determination of electric-field profile in CdTe and CdZnTe detectors using transient-current technique," *IEEE Transactions on Nuclear Science*, vol. 59, pp. 2402–2408, 2012.

23. J. Fink, H. Krüger, P. Lodomez and N. Wermes, "Characterization of charge collection in CdTe and CZT using the transient current technique," *IAEA*, vol. 560, pp. 435–443, 2006.
24. A. Zumbiehl, M. Hage-Ali, P. Fougeres, J. M. Koebel, R. Regal and P. Siffert, "Electric field distribution in CdTe and Cd1–xZnxTe nuclear detectors," *Journal of Crystal Growth*, vol. 197, pp. 650–654, 1999.
25. M. Pavesi, A. Santi, M. Bettelli, A. Zappettini and M. Zanichelli, "Electric field reconstruction and transport parameter evaluation in CZT X-ray detectors," *IEEE Xplore*, vol. 64, pp. 2706–2712, 2017.
26. S. Ramo, "Currents induced by electron motion," *IEEE Xplore*, vol. 27, pp. 584–585, 1939.
27. W. Shockley, "Currents to conductors induced by a moving point charge," *Journal of Applied Physics*, vol. 9, pp. 635–636, 1938.
28. Z. He, "Review of the Shockley–Ramo theorem and its application in semiconductor gamma-ray detectors," *Nuclear Instruments and Methods in Physics Research Section A*, vol. 463, pp. 250–267, 2001.
29. A. E. Bolotnikov, N. M. Abdul-Jabbar, S. Babalola, G. S. Camarda, Y. Cui, A. Hossain, E. Jackson, H. Jackson, J. R. James, A. L. Luryi and R. B. James, *Optimization of virtual Frisch-grid CdZnTe detector designs for imaging and spectroscopy of gamma rays*, 2007.
30. A. E. Bolotnikov, G. S. Camarda, G. A. Carini, G. W. Wright, D. S. McGregor, W. McNeil and R. B. James, *New results from performance studies of Frisch-grid CdZnTe detectors*, 2004.
31. P. N. Luke, "Single-polarity charge sensing in ionization detectors using coplanar electrodes," *Applied Physics Letter*, vol. 65, pp. 2884–2886, 1994.
32. P. N. Luke, "Unipolar charge sensing with coplanar electrodes-application to semiconductor detectors," *IEEE Transactions on Nuclear Science*, vol. 42, no. 4, pp. 207–213, 1995.
33. Z. He, "Potential distribution within semiconductor detectors using coplanar electrodes," *Nuclear Instruments and Methods in Physics Research A*, vol. 365, pp. 572–575, 1995.
34. C. M. Stahle, A. M. Parsons, L. M. Bartlett, P. Kurczynski, J. F. Krizmanic, L. M. Barbier, S. D. Barthelmy, F. B. Birsa, N. A. Gehrels, J. L. Odom, D. M. Palmer, C. Sappington, P. K. Shu, B. J. Teegarden and J. Tueller, *CdZnTe strip detector for arcsecond imaging and spectroscopy*, 1996.
35. S. Fatemi, C. Gong, S. Bortolussi, C. Magni, I. Postuma, X. Tang, S. Altieri and N. Protti, *Preliminary Monte Carlo simulations of a SPECT system based on CdZnTe detectors for real time BNCT dose monitoring*, 2018.
36. J. M. Ryan, J. R. Macri, M. L. McConnell, B. K. Dann, M. L. Cherry, T. G. Guzik, F. P. Doty, B. A. Apotovsky and J. F. Butler, *Large-area submillimeter resolution CdZnTe strip detector for astronomy*, 1995.
37. J. Aleotti, G. Micconi, S. Caselli, G. Benassi, N. Zambelli, M. Bettelli, D. Calestani and A. Zappettini, *Haptic Teleoperation of UAV Equipped with Gamma-Ray Spectrometer for Detection and Identification of Radio-Active Materials in Industrial Plants*, 2019, pp. 197–214.
38. T. H. Prettyman, "Theoretical framework for mapping pulse shapes in semiconductor radiation detectors," Nuclear Instruments and Methods in Physics Research A, vol. 428, pp. 72–80, 1999.

39. H. H. Barrett, J. D. Eskin and H. B. Barber, "Charge transport in arrays of semiconductor gamma-ray detectors," *Physical Review Letters*, vol. 75, no. 1, pp. 156–159, 1995.
40. F. P. Doty, H. B. Barber, F. L. Augustine, J. F. Butler, B. A. Apotovsky, E. T. Young and W. Hamilton, "Pixellated CdZnTe detector arrays," *Nuclear Instruments and Methods in Physics Research A*, vol. 353, pp. 356–360, 1994.
41. H. B. Barber, D. G. Marks, B. A. Apotovsky, F. L. Augustine, H. H. Barrett, J. F. Butler, E. L. Dereniak, F. P. Doty, J. D. Eskin, W. J. Hamilton, K. J. Matherson, J. E. Venzon, J. M. Woolfenden and E. T. Young, "Progress in developing focal-plane-multiplexer readout for large CdZnTe arrays for nuclear medicine applications," *Nuclear Instruments and Methods in Physics Research A*, vol. 380, pp. 262–265, 1996.
42. L. Abbene, G. Gerardi, F. Principato, A. Buttacavoli, S. Altieri, N. Protti, E. Tomarchio, S. D. Sordo, N. Auricchio, M. Bettelli, N. S. Amadè, S. Zanettini, A. Zappettini and E. Caroli, "Recent advances in the development of high-resolution 3D cadmium–zinc–telluride drift strip detectors", *Journal of Instrumentation*, vol. 27, 2020.
43. M. Bettelli, G. Benassi, L. Nasi, N. Zambelli, A. Zappettini, E. Gombia, L. Abbene, F. Principato and D. Calestani, *Mechanically stable metal layers for ohmic and blocking contacts on CdZnTe detectors by electroless deposition*, 2015.
44. G. Benassi, L. Nasi, M. Bettelli, N. Zambelli, D. Calestani and A. Zappettini, "Strong mechanical adhesion of gold electroless contacts on CdZnTe deposited by alcoholic solutions," *Journal of Instrumentation*, vol. 12, pp. P02018–P02018.
45. L. Abbene, N. Zambelli, G. Gerardi, G. Raso, G. Benassi, M. Bettelli, F. Principato and A. Zappettini, *High bias voltage CZT detectors for high-flux measurements*, 2016.
46. L. Abbene, G. Gerardi, G. Raso, F. Principato, N. Zambelli, G. Benassi, M. Bettelli and A. Zappettini, "Development of new CdZnTe detectors for room-temperature high-flux radiation measurements," *Journal of Synchrotron Radiation*, vol. 24, no. Pt 2, pp. 429–438, 2017.
47. B. Thomas, M. C. Veale, M. D. Wilson, P. Seller, A. Schneider and K. Iniewski, "Characterisation of Redlen high-flux CdZnTe," *Journal of Instrumentation*, vol. 12, pp. C12045–C12045.
48. L. Jones, P. Seller, M. Wilson and A. Hardie, "HEXITEC ASIC—a pixellated readout chip for CZT detectors," *Nuclear Instruments and Methods in Physics Research A*, vol. 604, pp. 34–37, 2009.
49. J. Aleotti, G. Micconi, S. Caselli, G. Benassi, N. Zambelli, M. Bettelli and A. Zappettini, "Detection of nuclear sources by UAV teleoperation using a Visuo-Haptic Augmented Reality Interface," *Sensors*, vol. 17, p. 2234, 2017.
50. J. Aleotti, G. Micconi, S. Caselli, G. Benassi, N. Zambelli, D. Calestani, M. Zanichelli, M. Bettelli and A. Zappettini, *Unmanned aerial vehicle equipped with spectroscopic CdZnTe detector for detection and identification of radiological and nuclear material*, 2015.
51. L. Abbene, F. Principato, G. Gerardi, G. Benassi, N. Zambelli, A. Zappettini, M. Bettelli, P. Seller, B. Thomas and M. C. Veale, *Microscale X-ray mapping of CZT arrays: spatial dependence of amplitude, shape and multiplicity of detector pulses*, 2017.

# 3

# *Application Specific Geometric Optimization of CdTe and CdZnTe Detector Arrays*

**W. C. Barber, E. Kuksin, J. C. Wessel, J. S. Iwanczyk, and E. Morton**

**CONTENTS**



## 3.1 Introduction

This chapter discusses the development of innovative room-temperature x-ray semiconductor detector technologies based on cadmium telluride (CdTe) and cadmium zinc telluride (CdZnTe) pixellated arrays used as direct conversion sensors and electrically connected to application specific integrated circuit (ASIC) readouts for energy-dispersive photon counting and spectroscopic x-ray imaging applications. This chapter begins with a description of two distinct readout methods, used to perform either energy-dispersive photon-counting pulse height discrimination (PHD) at high flux, or spectroscopic pulse height analysis (PHA), for transmission or emission and scattering imaging respectively. Then devices and results from linear and area arrays of both types of x-ray imaging detectors, with the transmission detectors providing highly efficient high flux photon-counting capabilities

DOI: 10.1201/9781003219446-3

with sufficiently good energy-dispersive characteristics (energy resolution) for multi-energy attenuation imaging, and the emission and scattering detectors providing high spatial resolution spectroscopic capabilities with extremely good energy resolution for radionuclide and/or diffraction imaging, are presented. Finally, a discussion of the geometries involved in the optimization of the semiconductors, ASICs, and readout schemes for each application is provided along with future efforts to integrate more readout functions closer to the pixels.

## 3.2 Compound Semiconductor Charge Collection and Readout Schemes

The array detectors presented here use CdTe and/or CdZnTe, and operate at room temperature with a large negative bias voltage across a continuous metal cathode and a pixellated metal anode pattern as direct conversion x-ray sensors. X-ray energy is directly converted to charge which can be detected by connecting each of the pixellated anodes to amplifiers. The high bulk resistivity of CdTe and CdZnTe can produce low noise performance in single photon counting charge sensitive amplifiers at room temperature due to the relative low dark current. The high effective Z of CdTe and CdZnTe provides efficient stopping power for x-rays in the range of energies used in many medical, security, and industrial applications within a few millimeters of material providing a compact sensor design. Direct conversion producing charge that drifts with the electric field in CdTe and CdZnTe provides high spatial resolution in relatively thick sensors, whereas scintillators can suffer from degrading spatial resolution as a function of increasing thickness, and therefore stopping power, due to light spread in the scintillator. We discuss here the development of two very distinct direct-conversion solid-state x-ray imaging array technologies using CdTe and/or CdZnTe for the very different modalities of x-ray transmission imaging requiring very high output count rates (OCRs), and x-ray diffraction imaging requiring very good energy resolution. These distinct technologies use either, as defined in this chapter, 'photon-counting' or 'spectroscopic' readout schemas optimized for transmission or diffraction imaging respectively. The 'photon-counting' technology uses PHD where multiple discriminators create multiple energy bands, whereas the 'spectroscopic' technology uses PHA where a multi-channel analyzer creates a histogram of event energies in list mode with hundreds or thousands of energy bins. Both these technologies use CdTe and/or CdZnTe as direct-conversion sensors, and they both perform single photon counting but with different charge collection schemes optimized for the application, so they should not be conflated because they use the same sensor material. Performing 'photon-counting' using PHD or performing 'spectroscopic' imaging using PHA requires very different pixel structures, analogue electronics, and digital readout schemas to optimize the performance of these sensor materials for these different imaging modalities.

It should be noted that a small number of parallel single channel analyzers (SCAs) per pixel, as used in the PHD scheme, provide insufficient sampling in energy space to measure the energy response and the detailed spectra of sources; however, a series of many measurements (one for each energy bin in the desired output spectrum) can be made with the energy settings of the SCAs adjusted to record an integrated spectrum. This is due to the fact that there are counters connected to each SCA that iterate up one for every pulse height above the energy setting for that SCA. Digitally integrating the result generates the spectrum and this is the method used to create the spectra presented here for the PHD detectors. Iteration of the energy settings of the SCAs is impractical due to dose and scan time limitations and such schemes are intended for fixed SCA level scanning.

## 3.3 Array Devices and Geometric Optimization

Various x-ray imaging applications require different performance metrics from the detector system including; the detective quantum efficiency (DQE) across a certain dynamic range, the spatial and energy resolutions, the OCR across a certain input count rate (ICR), and the bandwidth of the readout to the image formation and/or reconstruction system. The sensors must cover a certain field of view (FoV), which requires the tiling of modules when using CdTe and CdZnTe due to the limited size of a few centimeters for high quality single crystals. Additionally each pixel must be intimately connected electrically to amplifiers which must be provided using complementary metal-oxide-semiconductor (CMOS) ASICs due to the large number of small pixels within the active area that creates space and power constraints. All of these considerations taken together present a primarily geometric optimization problem, of the sensor (for charge collection and tiling considerations), of the ASIC (for noise performance and tiling considerations), and of the readout (for bandwidth and tiling considerations).

### 3.3.1 Photon Counting with Pulse Height Discrimination

For x-ray transmission imaging in medical and luggage security imaging applications the performance metrics include; a high DQE for mean beam energies up to 160 keV, a good intrinsic spatial resolution of ~1 mm, moderate energy resolution of ~10 keV full width at half maximum (FWHM) to take advantage of the dose and/or contrast improvements possible with energy-dispersive photon counting in these applications, OCRs in excess of a million counts per second per square millimeter (Mcps/mm$^2$) for most applications with half a million Mcps/mm$^2$ for some low dose screening procedures, and a readout with sufficient bandwidth to handle a larger data stream as compared to energy integrated x-ray detectors. To achieve high OCR above 1 Mcps/mm$^2$ the sensor's pixel geometry can make use of the small pixel effect wherein an induced signal is apparent on the preamplifier's input once electrons generated in the semiconductor travel only a small distance in the vicinity of the anode rather than the entire thickness of the semiconductor from the incident origin to the anode. This allows for the use of fast peaking times in the analog preamplifiers on the order of ten nanoseconds that can provide the required output counting rates when connected to an analog shaping stage and analog discriminators that iterate digital counters. Each pixel is connected to a single preamplifier and shaper and each shaper is connected to a parallel set of discriminators that can be independently set to different energies and the simplest description of the readout scheme would be that each pixel contains a parallel set of SCAs. This allows for a number of contiguous energy bins to be read out equal to the number of thresholds with the highest energy bin unbounded at high energy.

#### 3.3.1.1 Photon Counting Line Scan Detectors

Line scanning in medical and luggage security imaging applications by x-ray transmission, for example dual-energy x-ray absorptiometry (DEXA) for bone mineral densitometry (BMD) and/or single-view or dual-view conveyor belt driven scanning for baggage inspection, can be performed with semiconductor detectors that achieve two-side butt-ability so that modules can be tiled in one dimension (1D) without significant gaps to achieve sufficient length to cover the object being scanned.

DEXA systems are primarily used to scan the hip at an x-ray tube voltage of 90 kVp, with a linear array detector with an intrinsic spatial resolution of between 1 mm and 2 mm, and with a system magnification of about two. ICRs of 0.5 Mcps/mm$^2$ are typical and a linear OCR range is required, and very good energy resolution is desired for accurate quantitative bone area mineral content determination. Most DEXA systems use energy integrating detectors and perform a high and low energy scan with different tube filtering and kVp settings; however, higher performance systems use photon counting energy-dispersive detectors and perform a single scan to obtain a high and low energy image for calcium material decomposition. The metrics required from the detector, to perform the scanning, presents the geometric optimization problem for the use of direct conversion semiconductor detectors in this application. The relatively low dynamic range allows a relatively thin sensor in the incident direction while maintaining sufficient DQE, and a thin sensor along with the relatively low ICR allows a pulse shaping time in the amplifiers sufficiently short enough to produce a linear OCR across the ICR range and at the same time near complete charge collection for excellent energy resolution. At the same time the pulse shaping, at the required OCR range in this application, can tolerate more stray capacitance on the amplifiers' inputs compared to higher ICR applications, which allow for more length in the electrical connections between the pixels and the amplifiers. This along with the 1D nature of the array allows for the positioning of the ASICs alongside the sensors rather than being vertically integrated within the active area of the pixels as will be seen in other applications requiring two dimensional (2D) detector arrays. For low dose screening applications uniform statistical sampling across the patient's body and in the image is desired for dose consideration and this requires that the active area of the sensors extends to the edge of the crystals and that they can be tiled where the gaps between modules match gaps between the pixels within each crystal. Line scanner detectors can have multiple rows and the frames can be summed in a 'shift and add' to increase statistics and/or decrease the scan time. With these considerations we have developed a line scan DEXA detector using a two-side butt-able tile of ~1 mm thick CdTe sensors with four rows of ~1 $mm^2$ pixels each. Four sensors are tiled onto a single printed circuit board (PCB) providing an 8-cm FoV perpendicular to the scanning direction. The substrate PCB also contains many mixed-signal ASICs performing a PHD readout with four SCAs per pixel. These ASICs have a linear channel configuration that comports with the linear geometry of the sensors. This is allowed due to the lack of required vertical integration thus allowing an intimate (short length and therefor low stray capacitance) connection between the pixel and the amplifier inputs by placing the ASICs alongside the sensors. This placement of the ASICs will in practice limit the number of rows and the linear tile of sensors can have (in this case four rows) to maintain a short enough pixel connection for the required OCR in this application.

The top of Figure 3.1 has a picture of the DEXA detector from the incident direction showing four pixellated CdTe sensors tiled in a row on a substrate PCB. There are eight ASICs tiled in two rows on the back side of the PCB and to either side of the sensors. The bottom of Figure 3.1 has a picture of the backside of the detector showing a back-plane readout PCB containing a field programmable gate array (FPGA) that the substrate PCB is plugged into using board to board (B2B) connectors. The back-plane readout provides power and input output (I/O) function. Figure 3.2 shows the OCR of a typical pixel of the 256 pixel DEXA detector. The OCR is linear to 0.6 Mcps/$mm^2$ and saturates above 1 Mcps/$mm^2$ which is sufficient for the application. Figure 3.3 shows the energy resolution of a typical pixel of the 256 pixel DEXA detector for various radioactive sources with photopeaks within the dynamic range of the detector that is between 10 keV and 90 keV, which is also sufficient for the application. The energy resolution is 2.3 keV

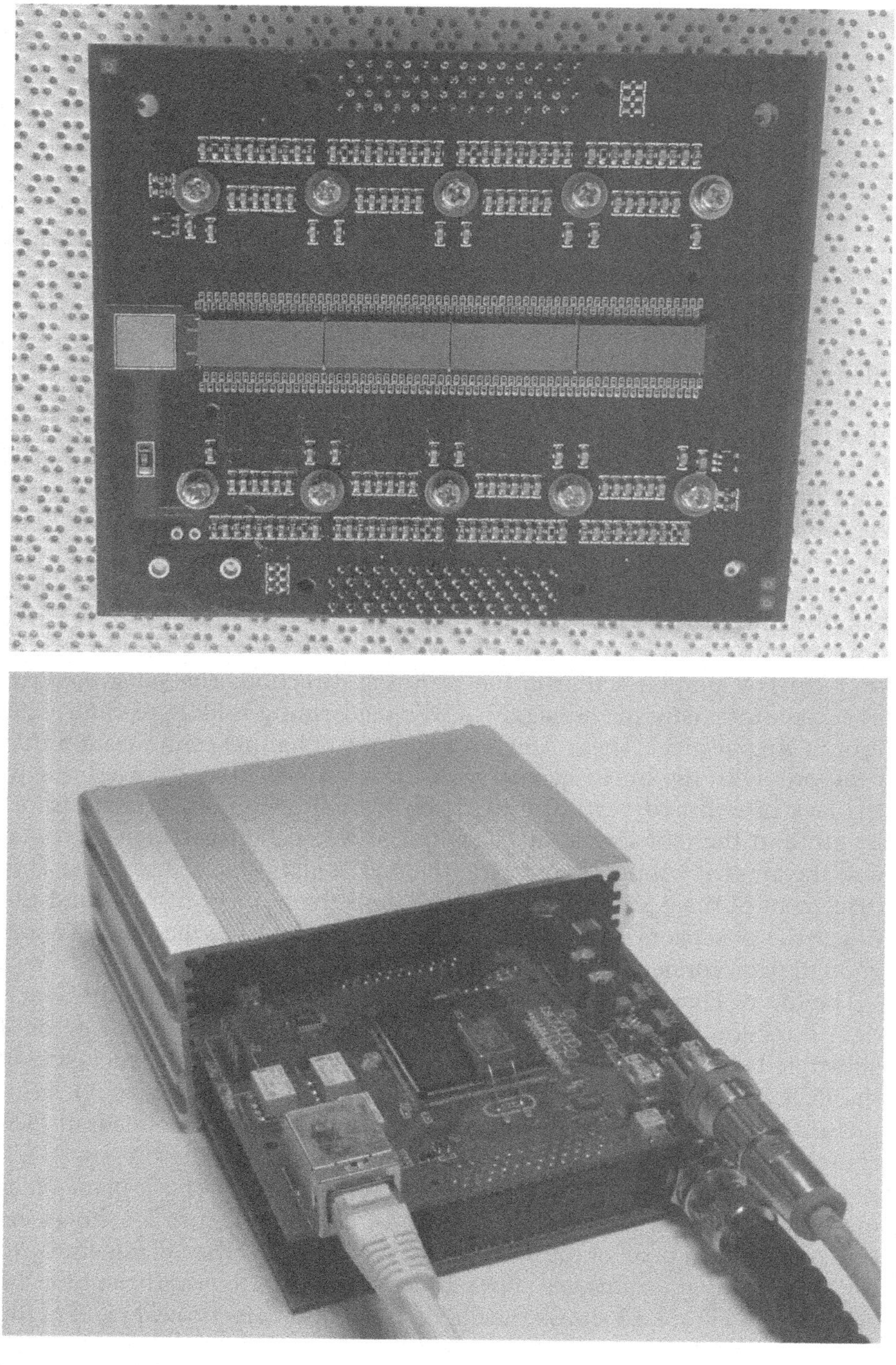

**FIGURE 3.1**
Images of the DEXA detector.

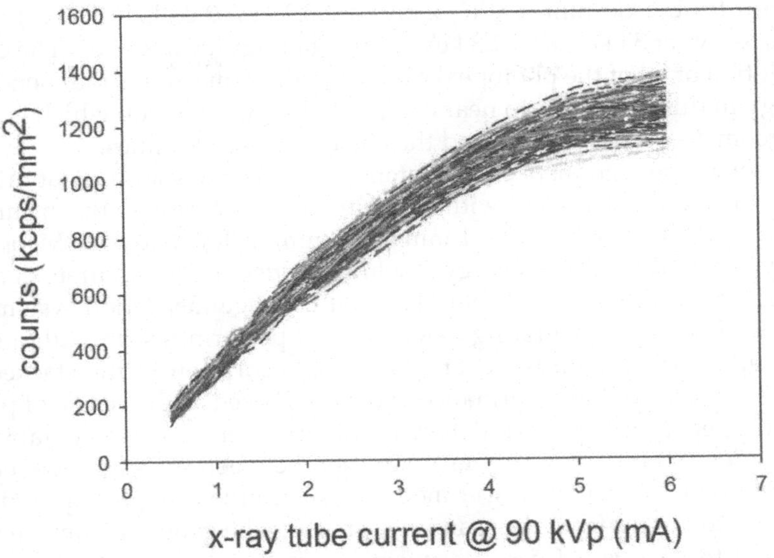


**FIGURE 3.2**
OCR for the photon counting PHD DEXA detector.

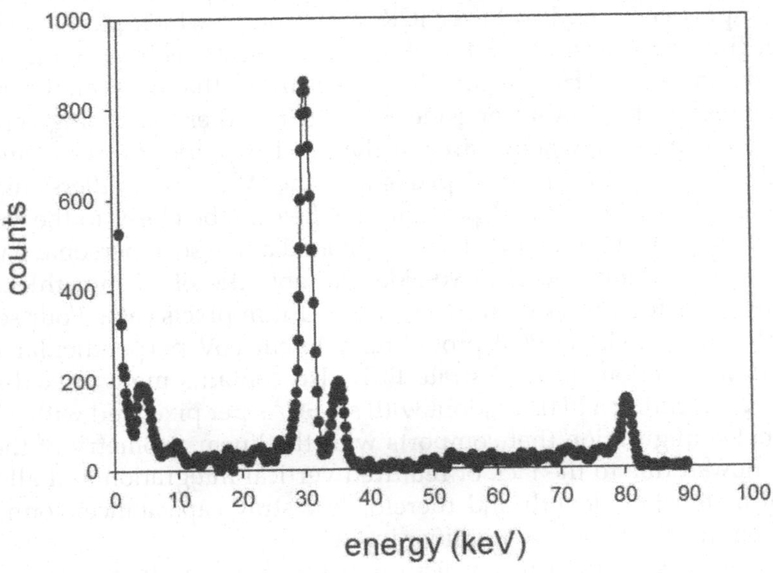


**FIGURE 3.3**
Spectra from $^{133}$Ba with the photon counting PHD DEXA detector.

FWHM for the Cesium (Cs) $K_\alpha$ x-rays at 30.9 keV, 2.28 keV FWHM for the Cs $K_\beta$ x-rays at 35 keV, and 2.3 keV for the gamma-ray at 81 keV. The energy resolution of all of the photopeaks is nearly the same and is independent of energy in this range due to near complete charge collection which is due to the geometry of the sensors and the choice of shaping times.

Personal baggage inspection systems are used to scan bags at an x-ray tube voltage of ~160 kVp, with a linear array detector with an intrinsic spatial resolution of between 1 mm and 2 mm. ICRs of above 1 Mcps/mm$^2$ are typical and moderate energy resolution is desired for accurate quantitative $Z_{eff}$ determination of materials. Most baggage inspection systems use dual-layer energy integrating detectors and perform a scan with a single kVp setting relying on the energy specific absorption in the stacked layers. However, higher performance may be achieved with the use of photon counting energy-dispersive detectors that perform a scan to obtain more energy levels and with a better separation between high and low energy images compared to dual-layer methods. Again the metrics required from the detector, to perform the scanning, presents the geometric optimization problem for the use of direct conversion semiconductor detectors in this application. The higher dynamic range requires a relatively thick sensor in the incident direction to maintain sufficient DQE, and a thick sensor along with the higher ICR allows the use of the small pixel effect to perform pulse shaping time in the amplifiers to achieve a higher OCR and at the same time produce moderately good energy resolution in a nearly electron-only charge collection scheme. Here the pulse shaping, at the required OCR range in this application, can tolerate less stray capacitance on the amplifiers' inputs compared to lower ICR applications, which places a smaller upper limit on the length of the electrical connections between the pixels and the amplifiers. Here again the 1D nature of the array allows for the positioning of the ASICs alongside the sensors rather than being vertically integrated within the active area of the pixels but the shorter connection lengths between the sensors' pixels and the ASICs' amplifiers' inputs a smaller number of channels per ASIC so they can be closer to the sensors. With these considerations we have developed a line scan personal baggage inspection detector using a two-side butt-able tile of ~3 mm thick CdTe and/or CdZnTe sensors with two rows of ~1 mm pixels each. Four sensors are tiled onto a single PCB providing a 10-cm FoV perpendicular to the scanning direction. The substrate PCB also contains many mixed-signal ASICs performing a PHD readout with six SCAs per pixel and with a linear channel configuration that comports with the linear geometry of the sensors allowed due to the lack of required vertical integration and allowing an intimate (short length and therefor low stray capacitance) connection between the pixel and the amplifier inputs.

The top of Figure 3.4 has a picture of the dual-row line scan detector. There are four pixellated CdZnTe sensors tiled in a row on a substrate PCB. There are fourteen ASICs tiled in one row on the PCB and next to the

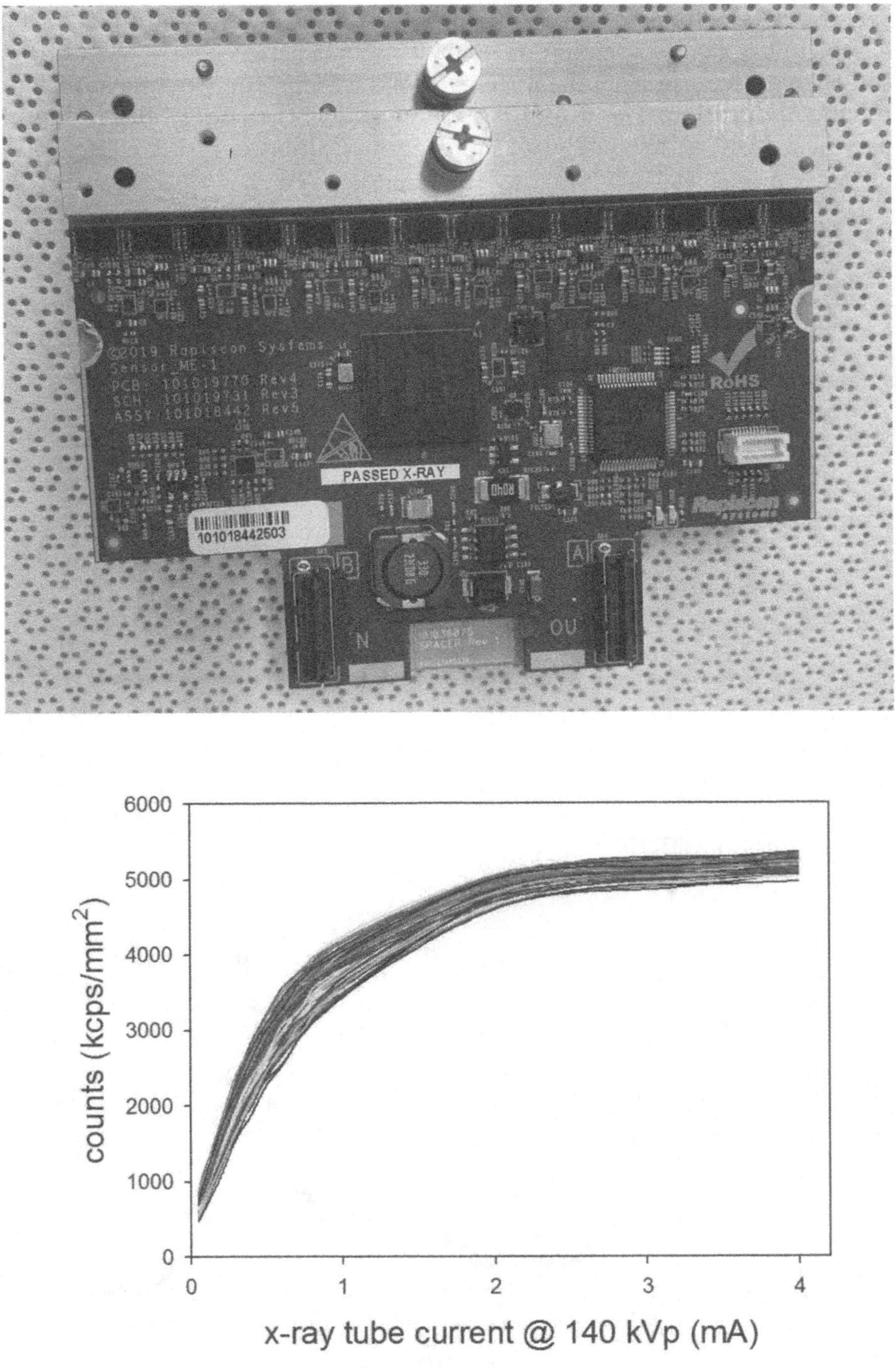


**FIGURE 3.4**
Images of the line scan detector and OCR photon counting PHD baggage inspection detector.

sensors. The bottom of Figure 3.4 shows the OCR of a typical pixel of the 224 pixel baggage inspection detector using CdZnTe sensors. The OCR is linear to 2 Mcps/mm$^2$ and saturates above 5 Mcps/mm$^2$ that is sufficient for the application. The top of Figure 3.5 shows the energy resolution of a typical pixel of the 224 pixel baggage inspection detector using CdTe

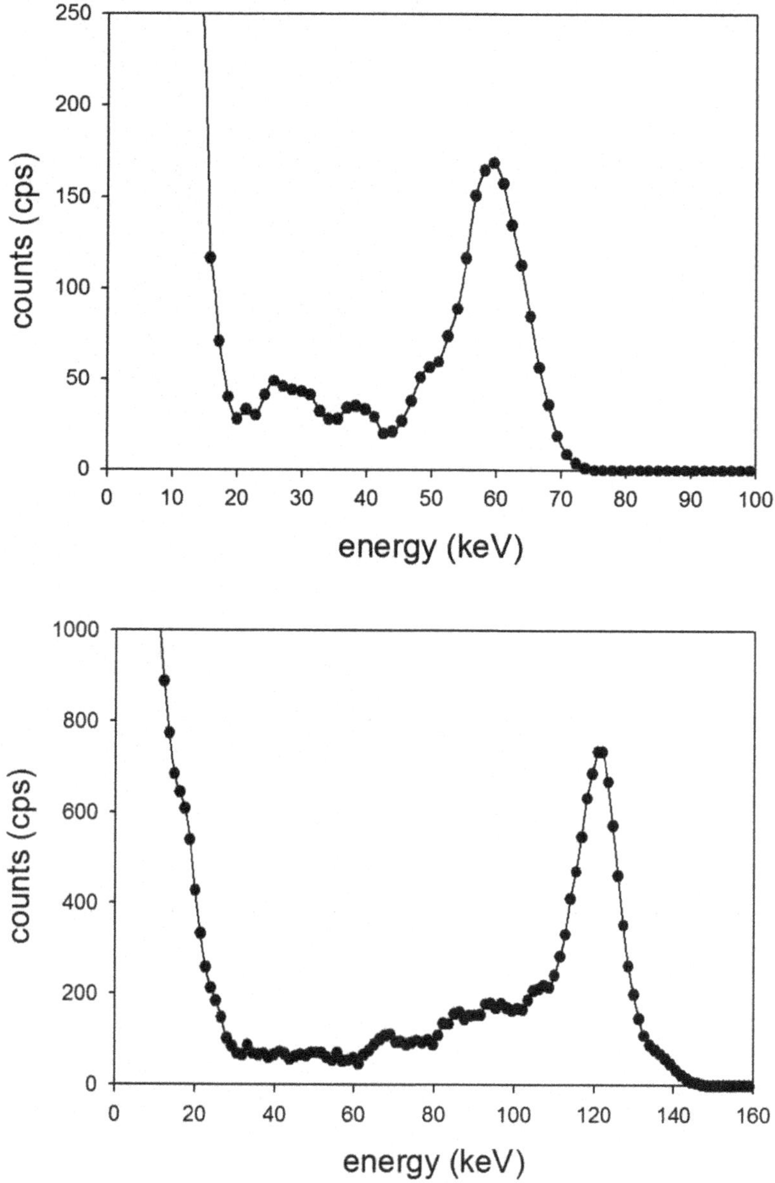


**FIGURE 3.5**
Spectra from $^{241}$Am and $^{57}$Co with the photon counting PHD baggage inspection detector.

sensors for a $^{241}Am$ radioactive source. The energy resolution is 11.7 keV FWHM for the photopeak. The bottom of Figure 3.5 shows the energy resolution of a typical pixel of the 224 pixel baggage inspection detector using CdZnTe sensors for a $^{57}Co$ radioactive source. The energy resolution is 13.8 keV FWHM for the photopeak.

#### 3.3.1.2 Photon Counting Computed Tomography Detectors

Computed tomography (CT) scanning in medical and luggage security imaging applications by x-ray transmission can also be performed with semiconductor detectors that achieve two-side butt-ability so that modules can be tiled in 1D without significant gaps around the object being scanned; however, three-side or ideally four-side butt-ability is desired to produce a sufficient number of slices for rapid scanning.

Medical CT systems are used to scan patients at an x-ray tube voltage between 110 kVp and 140 kVp, with an area array detector with a certain number of slices (pixels in the scanning direction) with an intrinsic spatial resolution of between 0.5 mm and 1 mm. ICRs of above 100 $Mcps/mm^2$ are typical for the unattenuated beam and up to 20 $Mcps/mm^2$ for the maximum attenuated beam in the most demanding procedures such as cardiac imaging. Moderate energy resolution is desired for optimal energy weighting to reduce dose and/or improve contrast and to take advantage of iodine identification and subtraction contrast enhanced images. Some dual-energy CT systems use dual-layer energy integrating detectors and perform a scan with a single kVp setting relying on the energy specific absorption in the stacked layers or use rapid kVp switching to obtain high and low energy images. However, higher performance may be achieved with the use of photon counting energy-dispersive detectors to obtain more than two energy levels and with a better separation between high and low energy images compared to dual-layer or dual-kVp methods. And here again the ranges and resolutions required from the detector, to perform the scanning, presents the geometric optimization problem for the use of direct conversion semiconductor detectors in this application. As with the higher dynamic range required for baggage line scanning, a relatively thick sensor in the incident direction to maintain sufficient DQE and the higher ICR allows the use of the small pixel effect to perform pulse shaping with moderately good energy resolution. In this case the faster pulse shaping, at the required higher OCR range in this application, can tolerate even less stray capacitance on the amplifiers' inputs compared to both the lower ICR applications listed previously, which places an even smaller upper limit on the length of the electrical connections between the pixels and the amplifiers. But in this case the need for many slices with the shorter connections leads to the use of 2D arrays with a vertical integration of the ASIC(s) within the active area of the pixels. Geometric optimization in this case leads to a 2D cell-like structure for the amplifiers producing shorter connection lengths between the sensors' pixels and the ASICs' amplifiers' inputs.

Personal baggage inspection CT generally uses higher x-ray tube voltage settings of 160 kVp or slightly more with similar ICRs on the detector. Whereas in medical CT x-rays less than 30 keV are filtered to reduce patient dose, baggage inspection generally uses lower x-ray tube currents to reduce the shielding required to limit dose to people in the security inspection lane without filtering low energy x-rays from the beam. The results of these different dose considerations are differently shaped beams with respect to energy but with similar ICRs on the detectors.

With these considerations we have developed a 2D area detector for both medical CT and personal baggage inspection CT using a three-side buttable tile of ~2.5 mm thick CdTe and/or CdZnTe sensors with 16 rows (in the slice direction) and 32 columns of ~0.4 $mm^2$ pixels. A single sensor is tiled onto a single PCB providing a 1 cm by 2 cm FoV. The substrate PCB also contains mixed-signal ASICs performing a PHD readout with four SCAs per pixel, and with a cellular channel configuration that comports with the area geometry of the sensors which in this case requires vertical integration to produce the shortest possible connection between the pixel and the amplifier inputs and allow 2D tiling.

The top of Figure 3.6 has a picture of the CT detector. There is one pixellated CdZnTe sensor on a substrate PCB. There are two ASICs on the PCB and behind the sensors. The bottom of Figure 3.6 shows the OCR of a typical pixel of the 512 pixel CT detector using CdZnTe sensors. The OCR is linear to 15 Mcps/$mm^2$ and saturates above 20 Mcps/$mm^2$ which is sufficient for the application and the pixel uniformity is good. The top of Figure 3.7 shows the energy resolution of a typical pixel of the 512 pixel CT detector using CdZnTe sensors for a $^{241}$Am radioactive source. The energy resolution is 8.2 keV FWHM for the photopeak. The bottom of Figure 3.7 shows the energy resolution of a typical pixel of the 512 pixel CT detector using CdZnTe sensors for a $^{57}$Co radioactive source. The energy resolution is 8.3 keV FWHM for the photopeak. The energy response (gain and energy resolution) is preserved up to high flux in the linear portion of the OCR range since the same pulse shaping is used when obtaining integrated spectra by iteration of the discriminators in the SCAs or obtaining energy weighted images with fixed SCAs. To confirm this we have obtained simultaneous x-ray and $^{57}$Co spectra under increasing ICR from an x-ray tube operated with an 80 kVp setting and a fixed $^{57}$Co source with unchanging decay rate during the experiments. The top of Figure 3.8 shows the 80 kVp bremsstrahlung x-rays and the 122 keV gamma-rays from the $^{57}$Co source. The bottom of Figure 3.8 shows integrated spectra from the x-ray tube operated with a tube current of 0.25 mA, 0.50 mA, and 1.00 mA, and a voltage setting of 80 kVp generating OCRs of 1.8 Mcps/pixel, 3.6 Mcps/pixel, and 7 Mcps/pixel, in the presence of the $^{57}$Co source. Note that the integrated spectra above 80 keV is due to the gamma-rays from the $^{57}$Co source and the response is essentially the same regardless of the ICR from the x-ray tube.

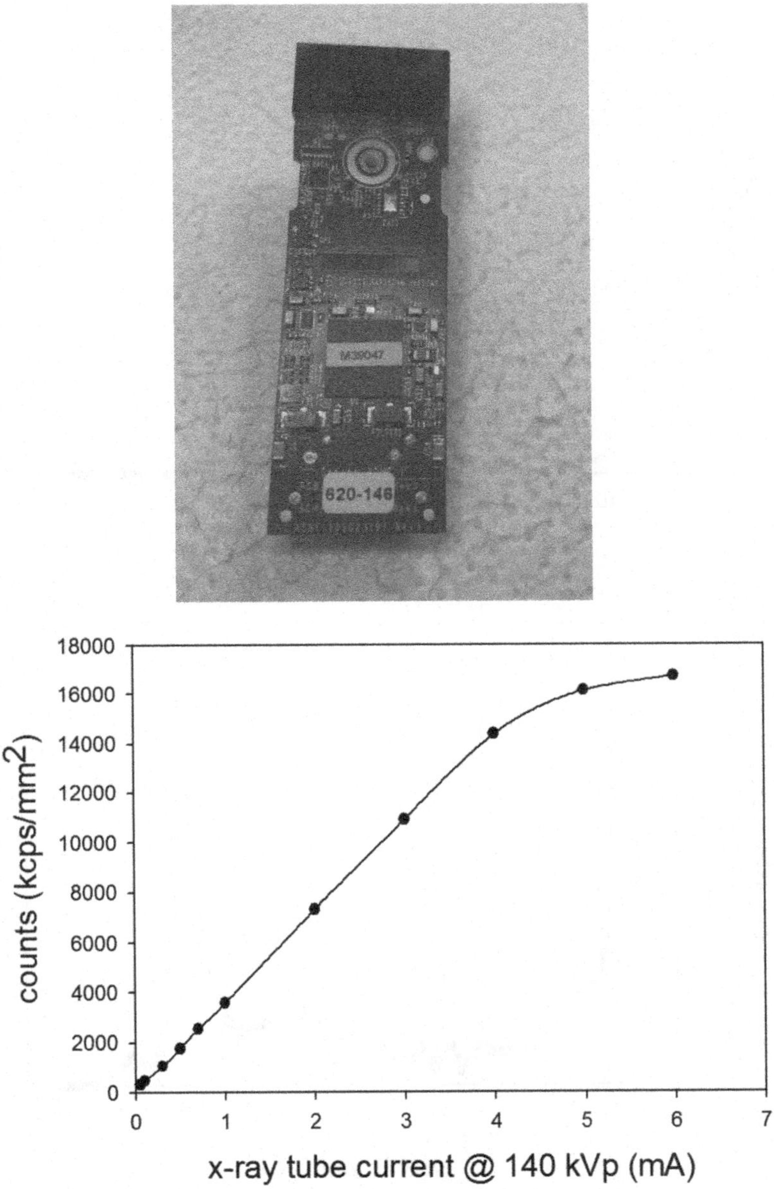


**FIGURE 3.6**
Image of the CT detector and OCR for the photon counting PHD CT detector.

### 3.3.2 Spectroscopy with Pulse Height Analysis

For x-ray diffraction imaging in security and material science imaging applications the performance metrics include; a high DQE for mean beam energies up to 160 keV, a good intrinsic spatial resolution of ~1 mm, excellent

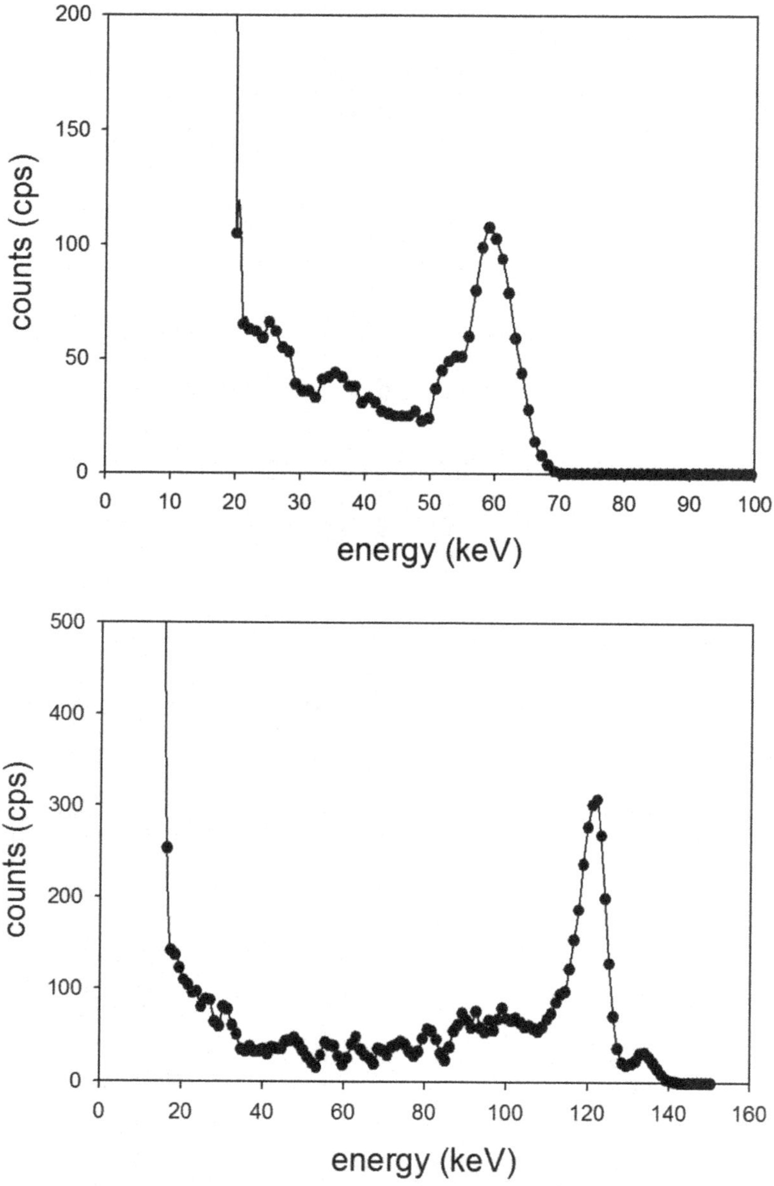


**FIGURE 3.7**
Spectra from $^{241}$Am and $^{57}$Co with the photon counting PHD CT detector.

energy resolution of ~1 keV to 2 keV full FWHM, relatively low OCRs below 0.1 Mcps/mm$^2$, and a multi-channel analyzer (MCA) readout with sufficient bandwidth to handle a larger number of energy bins as compared to photon counting PHD detectors that do not perform full PHA. The need

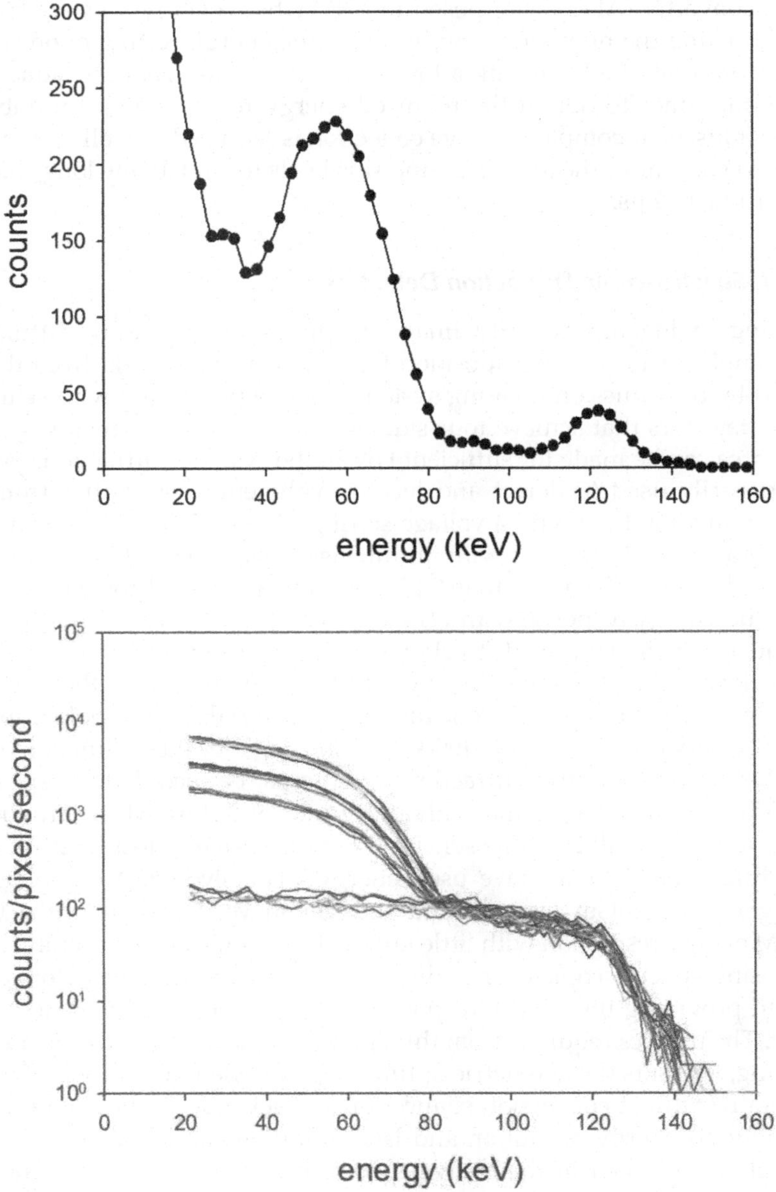


**FIGURE 3.8**
Simultaneous x-ray and $^{57}$Co spectra with the photon counting PHD CT detector.

for excellent energy resolution and for hundreds of energy bins within the dynamic range are such that each individual x-ray photon's energy must be recorded and readout in list more. This requires PHA with an MCA channel for each pixel, and where each pulse is analyzed with an analog to digital converter (ADC) as opposed to PHD where energy bins are created by subtracting the number of counts above the energy setting of one of the SCAs from that of a SCA with a lower energy setting from the same pixel and frame time. To obtain the required energy resolution and number of narrow bins, near complete charge collection is required as well as a sample and hold stage after the amplifier, and this leads to a relatively long shaping time of 1 μs to 2 μs.

#### *3.3.2.1 Spectroscopic Diffraction Detectors*

Scanning in luggage security imaging applications by x-ray diffraction, for example chemical decomposition for false alarm rejection from threats flagged by transmission imaging systems, can be performed with semiconductor detectors that achieve four-side butt-ability so that a relatively large active area can be made for sufficient FoV in the system. Diffraction systems are primarily used to detect and localize coherent x-ray scatter from the beam of an x-ray tube with a voltage setting of ~160 kVp or slightly higher, with an area array detector with an intrinsic spatial resolution of between 1 mm and 2 mm. ICRs of less than 0.1 Mcps/mm$^2$ are typical and a linear OCR range and energy response with very good energy resolution is required for accurate localization of the diffraction source in 1D and collimators are used for localization in the other direction both perpendicular to the scanning direction. Diffraction peaks from the threat materials being looked for are generally between 35 keV and 100 keV at the angle of the collimators used. The photopeaks from the diffraction signal must be recorded at very good FWHM energy resolution and without significant tailing which produces a background that will limit the sensitivity of the system. Most luggage scanning diffraction systems have used energy-dispersive spectroscopic cryogenic germanium (Ge) detectors and perform an MCA readout with ~1 keV FWHM energy resolution with little tailing; however, room temperature systems using spectroscopic energy-dispersive semiconductor detectors could be used providing the energy response is of sufficient quality as described earlier. The metrics required from this spectroscopic detector, to perform the scanning, presents the geometric optimization problem for the use of room-temperature direct conversion semiconductor detectors in this application. The required energy resolution and lack of tailing can be achieved with a relatively thin sensor in the charge collection direction while maintaining sufficient DQE by edge illumination with the incident coherent x-rays. An edge illuminated thin sensor along with the relatively low ICR allows a pulse shaping time in the amplifiers sufficiently long enough to produce near complete charge collection for excellent energy resolution and a sufficient sample and hold time for the ADC, while at the same time producing a linear energy response across the dynamic range at high DQE. The charge collection efficiency is now decoupled from the energy of the detected x-ray producing low photopeak tailing across the dynamic range.

With these considerations we have developed an area diffraction detector using a four-side butt-able tile of ~1 mm thick Schottky barrier CdTe sensors with one row of ~1 mm pixels each with 4 mm of sensor material in the incident direction. Three sensors on single substrate PCBs are tiled onto a 1D array providing a 6-cm 1D FoV and 60 1D arrays are stacked to provide a 6 cm by 6 cm 2D FoV. The substrate PCB also contains an analog ASIC performing a PHA readout with a 1024 channel MCA per pixel. The ASICs and MCA readouts are contained within the active area of the pixels as viewed from the incident direction (vertical integration). This is true for individual segments and also for tiles and stacks of segments producing 2D area detectors. Inherently, due to the edge illumination and required connections to the pixel anodes, the detector's incident area will have necessary gaps between the stacked 1D panels created by tiling the two-side butt-able individual segments. The gaps can be aligned with the parallel slat collimators commonly used in diffraction scanning systems. Another result of the edge illumination and vertically integrated readout is an increase in the back-plane footprint of the area detector.

The left side of Figure 3.9 shows two individual edge illuminated modules of the diffraction detector. The edge illuminated geometry provides for the use of a thin sensor between the cathode and anode for efficient charge collection, and at the same time produce a thick sensor in the incident direction. There are analog ASICs and ADCs providing the front-plane portion of the readout for the pixels. The front-plane readout is passed from module to module in each stack. The right side of Figure 3.9 shows a picture of a 6 cm by 6 cm FoV diffraction or fluorescence camera for hard x-rays created by tiling and stacking modules as shown on the left side of Figure 3.9. The stacks of modules are supported by a motherboard PCB that contains an internal high voltage supply and an FPGA-based back-plane readout providing a real-time list-mode readout. Figure 3.10 shows the energy resolution of a good pixel from the 2700 pixel 6 cm by 6 cm diffraction camera

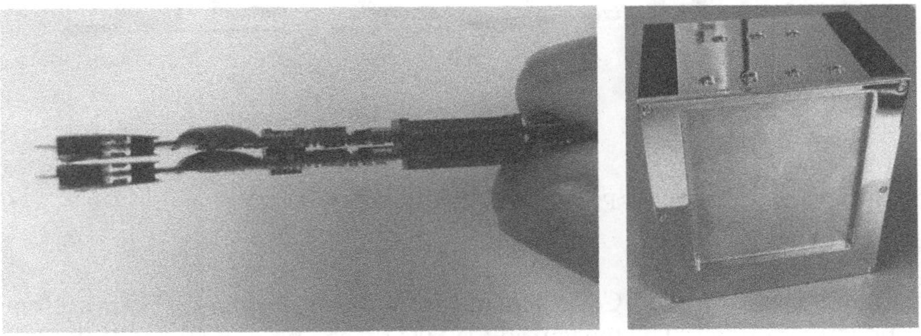

**FIGURE 3.9**
Images of the diffraction detector.

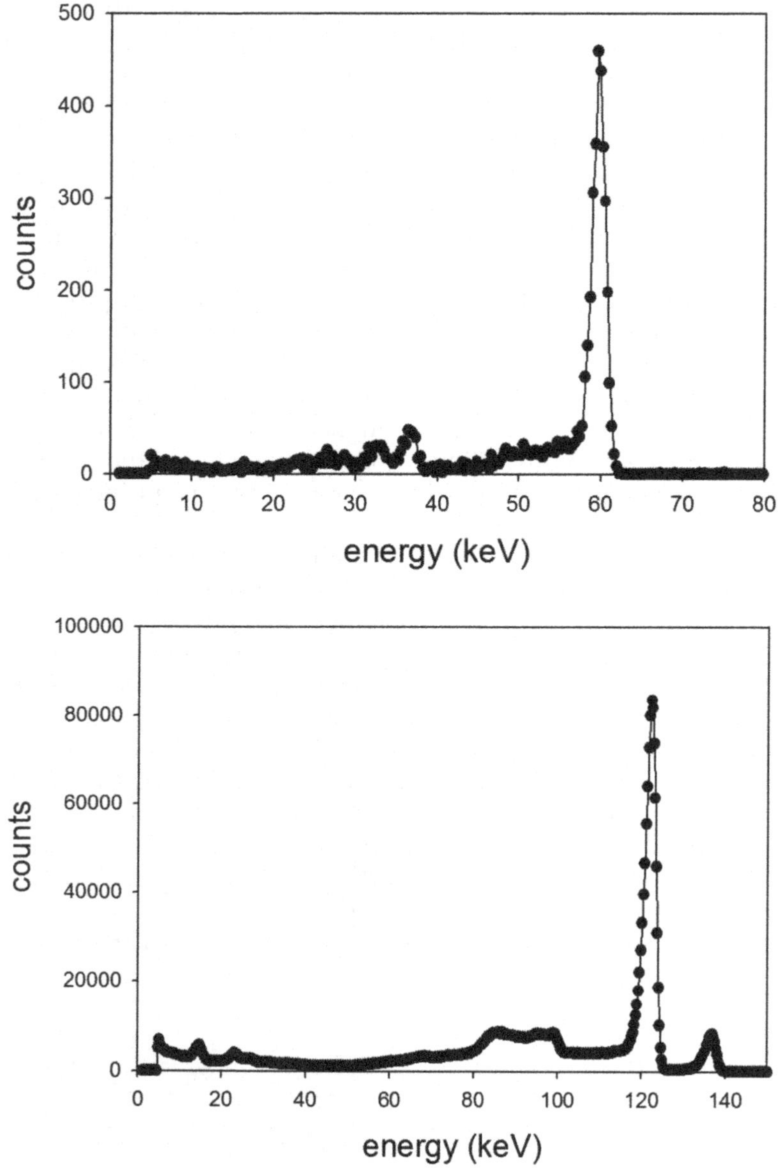


**FIGURE 3.10**
Spectra from sources with the spectroscopic PHA diffraction detector.

using edge illuminated CdTe sensors from $^{241}Am$ (top) and $^{57}Co$ (bottom) radioactive sources. The measured FWHM for the photopeaks is 2.1 keV for the $^{241}Am$ source and 2.3 keV for the $^{57}Co$ source. Figure 3.11 shows absorption and emission spectra of thin films of gold (Au) and lead (Pb) excited

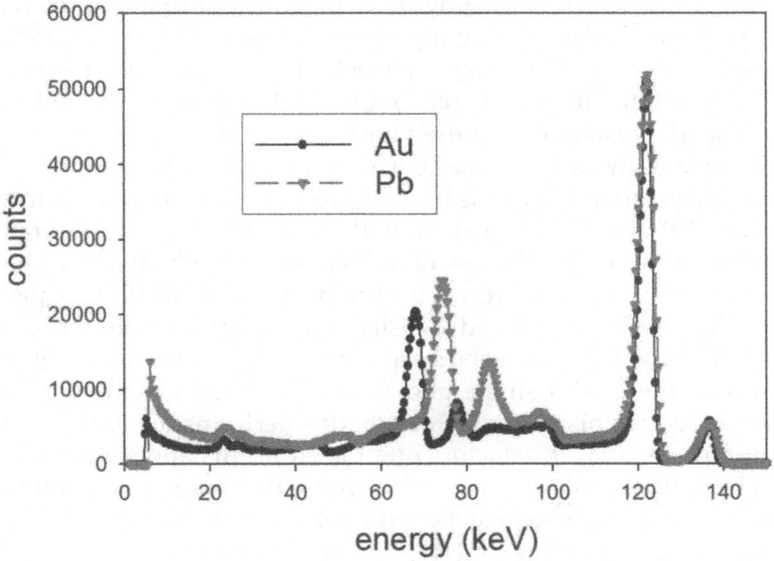


**FIGURE 3.11**
Fluorescence with the spectroscopic PHA diffraction detector.

with a $^{57}$Co source. The low tailing as a result of near complete charge collection along with the length of CdTe in the incident direction can provide good performance for large FoV x-ray diffraction and x-ray fluorescence applications.

## 3.4 More on Geometry for CdTe and CdZnTe Pixel Detectors

As exemplified in the previous sections, each application's specific requirements lead to a different optimization of the detector's structure, in large part due to geometric consideration. With respect to the use of CdTe and CdZnTe, as pixellated direct conversion x-ray sensors to form relatively large length or area arrays, for systems with sufficient FoV to scan people, luggage, and packages, we consider four performance requirements to be key. These can be described as 'ranges' and 'resolutions' and are the; OCR range, the dynamic range, the spatial resolution, and the energy resolution. Additionally we consider whether or not vertical integration of the ASIC containing the analog front end electronics and in the case of PHD a portion of the digital readout is needed. These are to one degree or another all related.

For example in the case of the DEXA detector, a relatively low dynamic range up to 90 keV with high DQE at the mean beam energy and the relatively low OCR range for x-ray imaging up to 0.6 Mcps/mm$^2$ allow the use of a thin sensor and a pulse shaping time suitable to achieve the OCR range and at the same near complete charge collection. This gives rise to very good energy resolution with a PHD readout with ~1 mm$^2$ pixels suitable for the intrinsic spatial resolution required for the application.

In the case of the line scanner detector for luggage screening, a higher dynamic range up to 160 keV is needed precluding the use of a thin sensor in the incident direction, and along with the relatively high OCR range up the 3 Mcps/mm$^2$ allows the use of a thick sensor with small pixels and a faster shaping time. In this case, for the purpose of calculating the effective atomic number ($Z_{eff}$) and density ($\rho_{eff}$), energy resolution of about 10 keV FWHM will produce good separation between low and high energy images and along with single photon counting and six energy bins (six parallel SCAs per pixel) could offer better performance as compared to conventional dual-layer detectors often used in this application. Here the small pixel effect produces a near electron only charge collection schema allowed due to the less stringent energy resolution requirement. And since this is a line scanner the segments only need to be two-side butt-able and along with the higher but still moderate OCR range needed we can place the ASICs along the side of the sensors and close to the pixels. In this case we choose a linear array of channels in the ASIC with inputs close to the sensors as with the DEXA detector, which is used as a linear scanning system for BMD scoring.

In the case of the CT detector we also need a relatively high dynamic range up to 160 keV, and along with a high OCR range requirement above 10 Mcps/mm$^2$ we again use thick sensor and the small pixel effect for a near electron only charge collection, but in this case a shorter shaping time is needed and this requires less stray capacitance on the inputs. This along with the desire for 2D arrays leads to a vertical integration of the sensor and ASIC which minimizes the distance between the pixels and ASIC inputs. The energy resolution requirement is deemed moderate (~10 keV FWHM) and the faster peaking time needs to be paired with a smaller pixel to produce sufficient signal. In this case we choose a 2D cell-like structure for the channels in the ASIC for making the interconnections between the pixels and the ASIC inputs as small as possible.

Higher OCRs can be achieved with smaller pixels and/or faster pulse shaping. In the case of smaller pixels, charge sharing due to characteristic x-rays from Cd and Te being absorbed in nearest neighbor pixels increases dramatically for pixel sizes below 0.5 mm and this increases tailing that increases the cross-talk between high and low energy weighted images reducing the benefit of the energy information. In the case of faster pulse shaping, since the characteristic x-rays are emitted randomly and isotropically, reabsorption vertically in the same pixel will produce two charge clouds separated in space and this places a lower bound on the peaking time in the pulse shaping amplifiers in the ASICs to trigger all the SCAs set below the full energy of the x-ray. For these reasons we place a lower bound on the pixel size of 0.5 mm and a lower bound on the peaking time of 10 ns in our PHD devices.

In the case of the spectroscopic diffraction detector we need both a high dynamic range up to 160 keV and very good energy resolution. In this case we also need the energy of each x-ray output that requires PHA and a list mode readout, but the OCR range is relatively low at < 20,000 cps/mm$^2$ expected from coherent scatter given the power of the x-rays tubes used in these applications. Additionally a relatively large 2D area is often needed to cover the objects scanned. Here we use an edge illumination to get both a high dynamic range and complete charge collection at the same time.

## 3.5 More on Readouts for CdTe and CdZnTe Pixel Detectors

For the purpose of discussion we define three stages of reading out signal generated in the sensors in the energy-dispersive single photon counting PHD and PHA schemes, starting with the analog pulse shaping, continuing to the digitization by either PHD with parallel SCAs or PHA with MCAs, and finally finishing with the parsing of the data with FPGAs for preparation to send to the data acquisition system (DAS). We define the second stage as the detector's 'front-plane readout' and the third and last stage as the detector's 'back-plane readout', not to be confused with the general term 'readout' or 'back-plane readout' which most often refers to the DAS, whereas we are focused here on the detector. In the case of our PHD scheme pulse shaping (first stage) and the front-plane readout (second stage) are contained within the mixed signal PHD ASICs, and FPGAs are used to readout the number of counts above each SCA acquired in each frame-time completing the detector's back-plane readout (third stage) with data prepared and sent to the DAS. In the case of our PHA scheme the pulse shaping is contained in an analog ASIC and the front-plane readout is contained within multi-channel ADCs which data is prepared in list mode in FPGAs completing the detector's back-plane MCA readout.

Following the evolution of the integration of the ASICs in close proximity to the pixel, and in the case of area detectors for both the PHD-based CT detector and the PHA-based diffraction camera, the vertical integration of the pulse shaping and front-plane readout within the active area of the pixels in tile-able modules, we intend next to vertically integrate the detector's back-plane readout. We are currently working to move the FPGAs involved in the back-plane readout closer to the pixel with the goal of vertically integrating such function into the active area of the pixels. This has advantages in reducing the additional band-width load on the DAS from new multi-energy CT detectors as shown here. For example the CT detector presented here transmits four times the data of an energy integrating detector with the same pixel density and FoV, and if optimal energy weighting for a particular imaging task can be performed in the detector then both the DAS bandwidth could be reduced as well as the burden to the image reconstruction system for example. The PHD ASICs' channels are built with the same counter depth for all SCAs and therefore an FPGA can do on chip subtraction to send only energy binned values to the DAS. Algorithms can be implemented in the FGPAs to reduce the amount of data from multi-energy detectors without losing the benefit of the additional energy information by using higher number of SCAs to correct and optimize two energy bins to send to the DAS. This will reduce the data rate by factor of two from a four-SCA per pixel PHD ASIC for example. These features, possible now due to new ultra-small package, ultra-low power FPGAs, could allow the design the ASICs to have a much higher number of SCAs per channel than the DAS can handle, because the FPGA can perform pixel-based analysis on a larger amount of information that can be handled by the DAS, and send only a few optimized data numbers that DAS can easily handle. These low power and high computation FPGAs can now be found in packages less than 4 mm by 4 mm and are low cost making it not worth inserting these digital algorithms into the ASICs.

In addition to data reduction, pixel-based corrections and some ASIC management can be performed in these FPGAs. Corrections, as long as they are pixel based and deterministic, can be applied including pulse pile-up corrections based on a model of channel response for a known number of counts. This could be done by several methods including look-up tables and/or calculation. Many ASIC tasks could be off-loaded to the FPGA such as fast channel to channel calibrations and bias optimization.

## 3.6 Summary

We describe the geometric optimization of solid state direct conversion compound semiconductor x-ray and gamma-ray detectors using CdZnTe and/or CdTe sensors for imaging in the hard x-ray and soft gamma-ray range. We note that different imaging applications often have unique dynamic and OCR ranges, energy and spatial resolutions, and intrinsic FoVs. This gives rise to a geometric optimization of the sensors, ASICs, and PCBs in a unique way for each application, including the charge collection scheme. A holistic approach, keeping all the minimum performance requirements in mind, is needed to develop a geometric solution for the optimization. Due to the limited size of high-quality single crystal compound semiconductor sensors, detector modules containing and supporting electrically and mechanically the sensors, ASICs, and on-module FPGAs and power must be designed to tile to large area to cover the FoVs required by the imaging system. We demonstrate this holistic approach in the descriptions of the various detectors presented. Included are examples of two and three side butt-able modules with cathode side illumination and using either an electron only charge collection scheme with the small pixel effect in relatively thick sensors or a complete charge scheme in thin sensors depending on the dynamic range required. Also we present an edge-illuminated module that decouples the geometric constraints of the charge collection from that of the dynamic range. Our future work includes the development of plug in four-side buttable PHD-based detector module with vertically integrated higher function FPGAs to allow the creation of large flat panel array with smart pixels.

## Acknowledgments

The research for the development of the spectroscopic diffraction PHA camera was funded in part by the U.S. Department of Homeland Security (DHS), Science and Technology (S&T) Directorate through contract 70RSAT18CB0000002. The research for the development of the CT PHD detector was funded in part by DHS, S&T through contract 70RSAT18CB0000047. The research for the development of the DEXA detector was funded in part by the National Institutes of Health (NIH), National Institute of Biomedical Imaging and Bioengineering (NIBIB) through grant 1R44EB008612. The research for the development of the CT PHD detector was funded in part by the NIH, NIBIB through grant 1R44EB012379.

# 4

# *Photon-Integrating Hybrid Pixel Array Detectors for High-Energy X-Ray Applications*

**Katherine S. Shanks**

**CONTENTS**



## 4.1 Introduction: Synchrotron Radiation Science in the Hard X-Ray Regime

Synchrotron x-ray light sources offer the ability to probe matter at the atomic scale using diverse experimental methods, including absorption radiography, phase contrast imaging, diffraction, small-angle x-ray scattering, and a

DOI: 10.1201/9781003219446-4

multitude of spectroscopic techniques. This chapter will focus primarily on applications in the imaging and scattering realms, where an area detector is used to record a series of two-dimensional images produced by the interaction of x-rays with the sample under investigation. These methods, when applied in the hard x-ray regime (defined here as involving x-ray energies ≥20 keV), allow for the investigation of the material properties and structure of even thick and/or dense samples (e.g., structural materials such as metals and cements). Such studies are of great utility in the fields of materials science, industry, and engineering.

### 4.1.1 Sources

Synchrotron light sources comprise both storage ring sources, in which a beam of charged particles is directed through a circular accelerator, resulting in the radiation of x-rays as magnetic devices (i.e., bend magnets, wigglers, and undulators) cause the beam to move in a curved trajectory, and x-ray free electron lasers (XFELs), linear accelerators that use long undulators and either self-amplified spontaneous emission or self-seeding lasing processes to produce extremely bright, short pulses of coherent x-rays. The x-ray beams produced by these two types of light sources have different spectral and temporal characteristics, which impacts both the types of experiments possible and detector design considerations.

#### *4.1.1.1 Storage Rings*

The total power radiated as well as the spectrum of photon energies emitted by a storage ring source is dictated by the energy $E$ of the stored particle beam, with the power scaling as $E^4$ and the critical energy (defined as the energy at which half of the total emitted power is below and half is above) scaling as $E^3$ [1]. At the time of this writing, there are five storage ring sources in the world that operate with stored beam energy >5 GeV and that are generally considered "high-energy" rings: the Cornell High Energy Synchrotron Source (CHESS) (Ithaca, NY, USA), the Advanced Photon Source at Argonne National Laboratory (Lemont, IL, USA), the European Synchrotron Radiation Facility (ESRF) (Grenoble, France), PETRA-III (Hamburg, Germany), and SPring-8 (Hyogo Prefecture, Japan).

When the trajectory of the stored beam is bent by a magnetic device, x-rays are produced in a direction tangential to the ring and directed via x-ray optics into a beamline endstation, where scattering, imaging or spectroscopy measurements are performed. The range of energies over which a particular endstation operates is determined by the upstream x-ray optics and by the characteristics of the magnetic device used to produce the x-rays. The simplest magnetic device used in this context is a dipole or "bending magnet", which produces a continuous spectrum of x-rays radiated in a fan into the forward direction. An array of closely-spaced bending magnets can be assembled into a so-called "insertion device", i.e., a wiggler or an undulator, in order to produce brighter x-ray beams owing to the tighter radius over which the particle beam is bent. A wiggler produces a continuous spectrum with more power than a bending magnet and a tighter spatial distribution. An undulator produces a spectrum with distinct peaks at a fundamental energy, determined by the peak magnetic field and the period of the magnetic array, and its harmonics. In any of these three cases, a monochromator may be used to isolate a desired x-ray energy with bandwidth down to ~$10^{-4}$, or the full spectrum may be used, especially if more flux is required. As an example of the high-energy flux available at a storage ring endstation, following an upgrade in 2018 CHESS features six endstations fed by undulators expected to provide >$10^{13}$ ph/s/mm$^2$/0.1% bandwidth up to ~50 keV, and two endstations fed by a wiggler capable of providing >$10^{12}$ ph/s/mm$^2$/0.1% bandwidth at 100 keV [2].

Beam size influences many experimental parameters such as the size of diffraction spots, the area or volume of an extended object that can be probed simultaneously, and the maximum field of view in an imaging experiment. The simplest way to control the size of the beam is through the use of slits that are sufficiently absorbing at the x-ray energy or energies used. In practice, the beam is often slit down to hundreds of microns to millimeters on a side. Additional x-ray optics, such as mirrors, capillaries, zone plates, and compound refractive lenses, can be used to focus the beam to concentrate its intensity while reducing its footprint on the sample. Focusing optics have been developed that can produce beam spot sizes of <20 nm (FWHM) at >20 keV [3].

Although sometimes treated as continuous light sources, storage rings are inherently pulsed light sources, due to the fact that particles circulate the ring in discrete bunches with typical bunch length of order 10–100 ps. The x-ray pulse structure is determined by the fill pattern of the ring, i.e., the specific pattern of particle bunches that are stored. A variety of fill patterns are possible in a given storage ring; examples of some common fill patterns are shown in Figure 4.1. The minimum inter-bunch spacing for a given storage ring is determined by the resonant frequency of the RF system that supplies the accelerating field. Minimum inter-bunch spacings of the order 1–10 ns are often achievable, although many fill patterns do not use the minimum inter-bunch spacing throughout the ring but rather multiples thereof; bunch spacings of up to 100s of ns are common. Experiments that exploit the pulsed nature of storage ring sources to achieve picosecond time resolution of dynamic processes will be discussed further in Section 4.1.2.2.

#### 4.1.1.2 XFELs

The pulses produced by XFELs are considerably shorter (~10–100 fs) and more intense (~$10^{12}$ ph/pulse) than those produced by a storage ring. Pulse repetition rates vary widely, with the Linac Coherent Light Source (LCLS)

**FIGURE 4.1**
Schematic representation of three common storage ring fill patterns. Left: the ring is populated with evenly-spaced single bunches. Center: an example of a "camshaft" mode, with a train of bunches accompanied by a larger, isolated bunch. Right: the ring is populated with evenly-spaced trains of more closely spaced bunches. Variations on and combinations of these fill patterns are also possible.

(Stanford, CA, USA) running at 120 Hz, whereas the European XFEL (Hamburg, Germany) produces trains of many hundreds of pulses at 4.5 MHz. Since the goal of an XFEL experiment is generally to leverage the pulsed structure of the source, the ability to match the detector frame rate to the source repetition rate is desired.

The use of x-ray energies ≥20 keV has been more limited at XFELs than at storage ring sources, although some experiments have made use of the higher undulator harmonics in this range (e.g., [4]). The MaRIE XFEL [5] has been proposed by Los Alamos National Laboratory (Los Alamos, NM, USA) as a dedicated high-energy x-ray facility, with a first phase design focused on production of 42 keV x-rays, with possibilities of utilizing higher harmonics at >100 keV. In preparation for facility developments and upgrades that will expand the use of hard x-rays, the use of high-Z sensors in high-flux environments has been explored [6, 7].

XFELs present unique challenges to detector design, in particular the effects of extremely high instantaneous flux on the sensor, which can result in radiaiton damage and rapid material polarization, as well as stringent timing requirements if every pulse is to be captured, thus in some cases requiring frame rates in the MHz range. This puts significant demands not only on the readout electronics, but also on the sensor response, which can be relatively slow in some high-Z materials (e.g., CdTe) owing to low carrier mobility, and which generally trades off with efficiency for a given sensor if a planar sensor geometry is used, as will be discussed in Section 4.3.1.

### 4.1.2 Example Applications

This section provides an overview of two applications of high-energy x-rays: high energy diffraction microscopy (HEDM) and the time-resolved study of dynamic compression events. In both cases, high x-ray energy is required in order to achieve reasonable penetration through dense macroscopic samples.

#### *4.1.2.1 High Energy Diffraction Microscopy*

Many materials used in engineering applications are polycrystalline, i.e., they consist of many grains with different crystallographic orientations. As such, the microstructure is heterogeneous, and the behavior of the material or component depends on physics at the grain and subgrain level. A topic of particular interest is the response of materials and components to mechanical loading. Gaining a more complete and predictive understanding of material response and, in particular, failure under loading would allow for improved manufacturing processes resulting in lighter and cheaper components. Predictive models could also enhance safety by allowing for better estimates of component lifetime. HEDM is a techinque that allows one to extract the center of mass, crystallographic orientation, and stress and strain tensors for every grain in a bulk sample [8]. High-energy x-rays are an ideal probe for this application due to their large penetration depth in the materials of interest (allowing samples of millimeter to centimeter thickness to be studied) combined with their ability to resolve structures on the atomic scale.

HEDM experiments often consist of a suite of related measurements. In far-field HEDM, an area detector placed of order 1 meter away from the sample is used to measure the diffraction from the sample at a range of rotation angles. From this measurement, the crystallographic orientation and stress and strain tensor of each illuminated grain can be derived. In near-field HEDM, an area detector placed much closer (of order 10 mm) to the sample is used to detect diffraction spots from illuminated grains with much lower angular resolution but in such a way that the grain center of mass and shape can be determined. Finally, computed tomography measurements are often incorporated, relying on radiographic images of sample absorption collected at many sample orientations to reconstruct the 3D real-space image of the sample. To study the effect of loading, each of these measurements can be repeated at various load steps as the sample is placed under compression or tension via a load frame. This may be done in a quasi-static mode, where a load is applied and then held while images are collected, or in an in-situ mode, where images are acquired during loading.

Desired detector characteristics include a high-frame rate, to accomodate in-situ scans and facilitate high-throughout quasi-static measurements; reasonably high efficiency for high-energy x-rays; and for far-field HEDM, a sufficiently large area to capture many diffraction rings. The pixel size for the near-field or tomography detector must be sufficiently small to resolve spatial features on the order of microns, whereas the pixel size constraint is relaxed for the far-field camera, since in the far field diffraction spot size is generally much larger than microns. In many HEDM setups, an optical CCD coupled to a scintillator is used for the near-field and tomography measurements, while a larger-area detector is used in the far-field. The experiment described in [9] follows this format and is offered as a representative example of typical pixel sizes for HEDM: the tomography detector had an effective pixel size of 1.48 μm × 1.48 μm, whereas the far-field detector had a pixel size of 74.8 μm × 74.8 μm.

#### *4.1.2.2 Time-Resolved Studies of Dynamic Compression Events*

High-energy x-ray techniques have also been applied to the time-resolved study of dynamic compression events, which involve material dynamics on shorter timescales than typical HEDM and address topics such as the physics of shock waves, detonation, and related phenomena. Dynamic compression experiments often probe the sub-microsecond timescale, in some cases utitlizing the pulsed nature of synchrotron radiation sources to access the <100 ps timescale, as mentioned in Section 4.1.1.1; an example is the study of dynamics of shock-compressed metal alloys [10]. While it is possible to isolate the signal from individual x-ray pulses using a slower-framing detector in conjunction with a sufficiently fast shutter (e.g., the timing scheme used in [11]), a detector capable of capturing images from *neighboring* single pulses is highly desirable in order to increase throughput and allow for investigation of non-repeatable phenomena. In practice, this has required the use of detectors capable of burst-mode frame rates in the MHz regime [12]. This mode of operation can also preclude the use of photon-counting detectors, since many photons may arrive in a given pixel during a single x-ray pulse, especially in the case of bright diffraction spots.

As with HEDM, dynamic compression experiments have involved both the imaging and diffraction modalities. On the imaging side, propagation-based phase contrast imaging [13] has been used to obtain real-space images of the sample as a compression event takes place. Diffraction measurements have typically been polychromatic, which allows for larger flux, which can be critical to getting sufficient scattered signal in high-frame-rate experiments. Here again, the imaging modality benefits from a smaller pixel size of order 10 μm or less, whereas larger pixels are generally acceptable for diffraction.

## 4.2 Detectors

A range of area detectors with different sensor and readout architectures have been applied to the measurement of x-rays with energy ≥20 keV. Two major divides bear examination here, as they inform some of the design philosophy behind the specific detector systems described in Sections 4.2.3 and 4.2.4.

### 4.2.1 Indirect vs. Direct Conversion of X-Rays

Indirect conversion refers to detector systems in which x-rays are stopped in a medium such as a phosphor or scintillator where they are converted into visible light, which in turn is recorded by an optical camera. There are a number of area detector architectures that rely on indirect conversion of x-rays and that are suitable for detection of high-energy x-rays. Image plates offer large areas and good stopping power up to ~100 keV, and have been common in high-energy x-ray experiments (e.g., [14]). In these systems x-rays are stopped in a phosphor, where they generate photoelectrons that become trapped in a metastable state. After an exposure is complete, the image plate is scanned with an optical laser that excites the trapped electrons. This excitation process produces visible light that is detected by a photomultiplier tube and digitized into a final 2D image. Other widely used indirect conversion detectors involve a phosphor or scintillator coupled to a visible-light detector, either through an optical element such as a lens or fiber optic, or deposited directly onto the visible-light detector. For example, large-area flat-panel detectors incorporating scintillators coupled to CMOS active pixel sensors (e.g., [15]), originally developed for medical and industrial imaging, have been adopted at many synchrotron beamlines, where a large area is especially useful for collecting diffraction patterns spanning a wide swath of reciprocal space. One particular advantage of optical coupling is that, since visible light is more easily manipulated by conventional optics than x-rays are, it opens the possibility of magnifying, demagnifying, redirecting, and even splitting signal between mutliple cameras, as has been done in a multi-camera system used for dynamic compression experiments with a single scintillator screen optically coupled to a series of intensified CCDs [12].

In contrast, direct conversion refers to detector systems in which x-rays are stopped in a thick (typically 300 μm–1 mm) semiconducting material where they generate charge carriers that are drifted toward pixel collection nodes via application of an external bias. Germanium CCDs, under development for use in mainly astronomical applications, are one example [16]. Otherwise, the majority of direct-conversion systems for hard x-rays have been hybrid pixel array detectors (PADs) [17, 18], in which a sensor is bump-bonded pixel-by-pixel to a separate, customized readout ASIC. While originally developed primarily with Si sensors, their hybrid construction offers the advantage that the sensor and readout electronics can be optimized separately, and a given readout ASIC can be mated to, in principle, any sensor material amenable to either bump bonding or some form of deposition across the ASIC area. PADs have become prominent in synchrotron science since the late 2000s, largely due to the success of the photon-counting

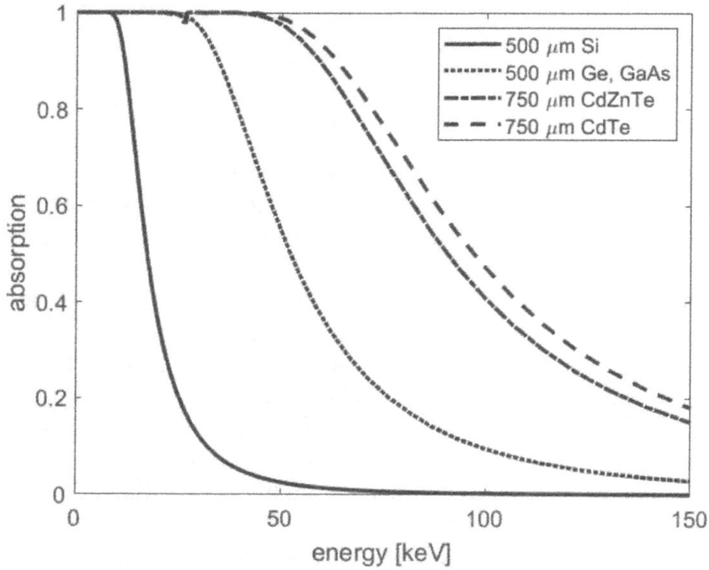


**FIGURE 4.2**
Absorption as a function of photon energy for Si, Ge, GaAs, CdTe, and CdZnTe.

PADs, such as the Pilatus [19], at storage ring sources, and due to the prominence of photon-integrating PADs, such as the CS-PAD [20] and AGIPD [21] at XFELs. In recent years, PADs with CdTe [22, 23], Ge [24], CdZnTe [7], and GaAs [25] have been developed and characterized, including those described in Sections 4.2.3 and 4.2.4. The absorption of these four sensor materials, along with Si for reference, is shown in Figure 4.2.

Direct conversion of x-rays offers many benefits. The gain in terms of signal per stopped x-ray is typically orders of magnitude larger; this particularly aids in the detection of single-photon events and otherwise weak or sparse signals. This also has the benefit of providing headroom for running readout amplifiers at high speeds in order to achieve high frame rates, since amplifier readout noise scales as the square root of frequency and without sufficient gain, single- or few-photon events are easily drowned out by read noise. The sensor point spread function is generally tigher in direct-conversion systems, being dictated by the diffusion of photo-generated charge through the length of the sensor as opposed to optical scattering processes.

### 4.2.2 Photon-Counting vs. Photon-Integrating Architectures

As noted earlier, two main classes of PADs have been developed for x-ray applications: photon-counting and photon-integrating. The distinction arises in how photogenerated signal from the sensor is processed by the pixel front end. In a photon-counting PAD, incoming signal is fed through a pulse shaper and single-photon events are individually identified and counted depending on whether a preset pulse height threshold is crossed. This allows for suppression of sensor dark current and of pixel read noise. The threshold can also be used to reject photons below a given energy, for example background sample fluorescence; photon counters have also been designed with multiple energy thresholds per pixel, allowing for "multi-color" x-ray imaging (e.g., [23]). However, pulse discrimination imposes a photon rate limit above which additional incoming signal is lost. Sustained photon rate limits are typically ~$10^6$–$10^7$ ph/pix/s [26], and in a single storage ring or XFEL pulse, a maximum of 1 photon per pixel per pulse can be measured. If the sustained photon rate is not exceeded, a maximum signal measureable by a photon-counting pixel is dictated by the counter depth, so that a pixel with a 14-bit counter can measure >$10^4$ ph/pix/frame; 17 bits are required to reach $10^5$ ph/pix/frame.

In contrast, in a photon-integrating PAD the front-end amplifier is configured as an integrator with feedback capacitance $C_F$, and incoming signal is simply accumulated without filtering or discrimination. As a result, sensor dark current is integrated along with the x-ray signal of interest, so that it is advantageous to run photon-integrating detectors at lower temperatures (≤0°C is typical), and the resulting signal inherently includes the pixel read noise. However, in a purely integrating pixel there is no inherent photon rate limit. Sustained photon rates up to the pixel well depth divided by the minimum frame time can be measured with fidelity. In particular, arbitrarily high instantaneous photon rates can be accepted up to the limit imposed by physical damage to the detector. This makes photon-integrating detectors the only option for XFEL applications in particular, where many photons can arrive at a given pixel during a <100 fs pulse.

The maximum signal per pixel measureable by a photon-integrating PAD is limited by the front-end feedback capacitance $C_F$, which is often implemented as a metal-insulator-metal (MiM) capacitor. In most of the CMOS processes used for these detectors, the MiM capacitance is 1–2 fF/μm$^2$. In a photon-integrating PAD with a CdTe sensor and a voltage swing of ~1 V at the integrator output, a feedback capacitor capable of storing the equivalent of $10^4$ 20 keV ph/pix/frame would have to be 60 μm × 60 μm. To measure $10^5$ 20 keV ph/pix/frame under the same conditions would require a feedback capacitor with area 190 μm × 190 μm. This tradeoff between pixel area and maximum measureable signal has lead to investigations into well-depth extension techqniues such as the one described in the following section.

Sections 4.2.3 and 4.2.4 describe two photon-integrating PADs developed at Cornell University geared toward synchrotron experiments requiring temporal resolution in the 100s of ns to millisecond range, and toward accommodating high-flux conditions that preclude the use of photon-counting PADs. Both were originally designed and characterized with silicon sensors, but have since been bonded to 750 μm In-Schottky CdTe sensors from Acrorad (Okinawa, Japan) as well.

### 4.2.3 The CdTe Mixed-Mode Pixel Array Detector

The Mixed Mode PAD (MM-PAD) [27, 28], originally designed with a 500-μm Si sensor, is a wide-dynamic-range imager that frames continuously at 1 kHz and uses a charge removal pixel architecture to achieve a dynamic range of ~$10^8$. A high-level single-pixel schematic is shown in Figure 4.3. The photon-integrating front end has a modest well depth, allowing for the integration of >$10^3$ keV per pixel before the integrator output reaches the externally-set global threshold voltage $V_{th}$. When this occurs, in-pixel control logic triggers a charge removal circuit that removes a fixed amount of charge from the integration capacitor $C_F$. An in-pixel 18-bit counter records the number of charge removal operations executed in a given frame. At the end of a frame, the counter value and the residual voltage at the integrator output are read out for every pixel in the detector array. The analog residuals are digitized off-chip, and then combined and scaled with the counter values to recover the total integrated intensity for each pixel. This mixed-mode readout allows for the use of a relatively small $C_F$, which helps preserve good signal-to-noise ratio at the single-photon level (charge-equivalent read noise is ~350 e-), while achieving a large pixel well depth and thus a wide dynamic range. A 256 × 384 pixel version of the detector with a Si sensor has been used in ptychographic experiments, where the wide dynamic range is critical to achieving high-spatial resolution images of extended samples [29, 30], and in time-resolved studies of material dynamics [31, 32].

It should be noted that the charge removal process incurs no dead time; incoming signal from the sensor continues to be integrated during charge removal operations. The only condition under which signal is lost is if it

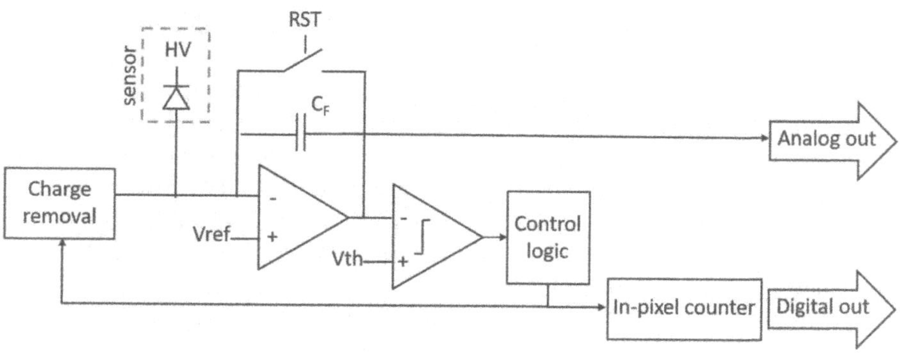


**FIGURE 4.3**
Single-pixel schematic of the MM-PAD.

arrives at a rate sufficient to saturate the integrating front-end before a charge removal is complete. In the first-generation MM-PAD, the sustained rate limit is $>3 \times 10^9$ keV/pix/s.

As noted earlier, a CdTe sensor has been used to extend the useful energy range of the MM-PAD. The CdTe-sensor MM-PAD [22] system has been used in synchrotron experiments probing the deformation mechanisms in metal alloys under mechanical loading [33, 34]. It has also been used to explore the polarization behavior of the sensor material, under both static [35] and dynamic conditions [36]. In these studies, the MM-PAD's high continuous frame rate has proven useful for characterization of sensor behavior with millisecond resolution over long time windows, allowing for studies of effects such as image persistence.

To further extend the utility of the MM-PAD detector platform, an upgraded version, the MM-PAD-2.1, has been developed [37]. A single-pixel schematic is shown in Figure 4.4. Key modifications to the pixel architecture are the implementation of two-stage adaptive gain on the front-end, a 50× increase in the maximum rate at which charge removal can operate, and the use of dual in-pixel counters and a dual track-and-hold circuit for the analog residual in order to allow for expose-while-readout operation with >90% duty cycle at maximum frame rate. The readout chain has been updated to allow for ≥10 kHz continuous framing of a $128 \times 128$ pixel array.

The MM-PAD-2.1 will extend the sustained photon rate limit of the MM-PAD by an order of magnitude, and is intended for operation with

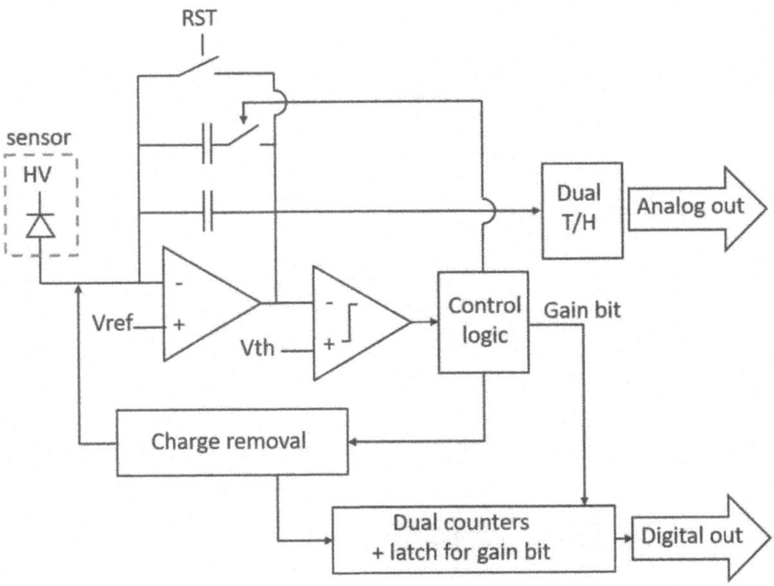


**FIGURE 4.4**
Single-pixel schematic of the MM-PAD-2.1.

a CdTe sensor, targeting applications at 20 keV and above. The explicit shift in orientation toward higher x-ray energies and a high-Z sensor has informed certain key design choices, e.g., regarding the energy-equivalent read noise, which is specified to be up to twice that of the first-generation MM-PAD, and what level of sensitivity to radiation damage is acceptable, since the CdTe sensor will provide significantly greater shielding to the readout ASIC compared to a Si sensor.

### 4.2.4 The CdTe Keck-PAD

The Keck-PAD [38] uses in-pixel frame storage to achieve a burst-mode frame rate approaching ~10 MHz for eight consecutive frames. As with the MM-PAD, 256 × 384 pixel versions of the detector have been assembled using both 500 μm Si and 750 μm thick CdTe sensors. A single-pixel schematic is shown in Figure 4.5. The front-end has a selectable (but not adaptable) gain (capacitors $C_{F1,...,4}$), allowing the user to choose a trade-off, to some extent, between read noise and maximum measureable signal. Burst-mode imaging is accomplished by switching which of the eight sampling capacitors $C_{S1,...,8}$ is connected to the front-end output during an exposure. When the selected sampling capacitor is disconnected from the front-end at the end of an exposure, it stores the sampled voltage until all eight frames are acquired and the full array is read out. The burst-mode frame rate is

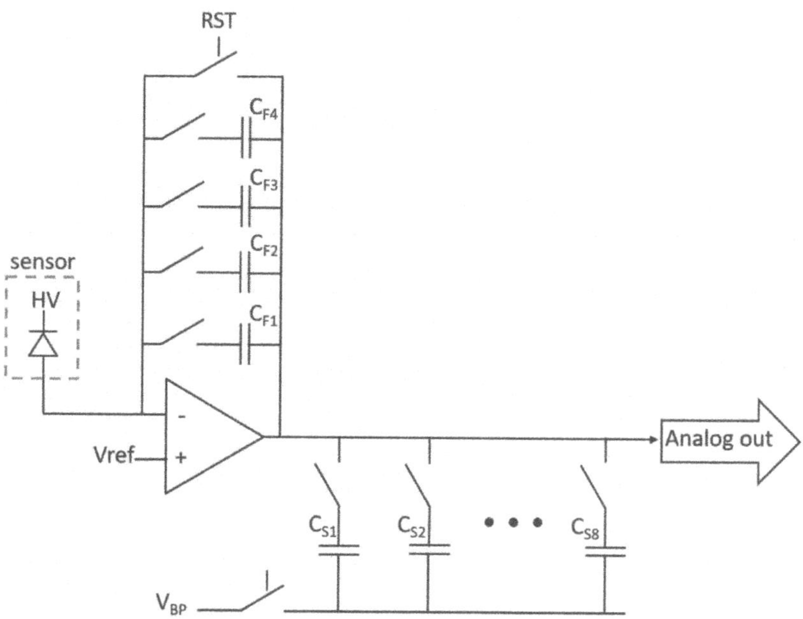


**FIGURE 4.5**
Single-pixel schematic of the Keck-PAD.

well-matched to single-bunch studies at many storage ring sources, as described in Section 4.1.1.1. Despite the relatively low charge carrier mobility in CdTe, which will be discussed further in Section 4.3.1, in [36] it was demonstrated that the CdTe-Keck-PAD is capable of resolving the passage of individual storage ring bunches with a bunch spacing of 153 ns.

The high burst-mode frame rate of the Keck-PAD also allows for characterization of sensor behavior down to the timescale of 10s–100s of ns. This has been useful for studying the polarization behavior of CdTe, as well as its charge transport properties, as will be discussed in the following section.

## 4.3 Specific Challenges

This chapter has given examples of direct-conversion, photon-integrating detectors that have been succesfully used to conduct measurements involving high-energy x-rays. Development on these systems, as well as others along similar lines by the broader detector community, continues. There remain some specific challenges inherent to the efficient detection of high-energy x-rays and to the use of currently-available high-Z sensors. These include the trade-off between sensor efficiency and speed, material defects that have been common in compound sensor materials, and sensor polarization.

### 4.3.1 Speed vs. Efficiency

For experiments relying on very high frame rates (~MHz and above), the speed of the sensor response to x-rays can become a limiting factor. The issue is exacerbated in the high-energy regime, because the need to efficiently detect high-energy x-rays suggests the need for thicker sensors. As the thickness of a planar sensor is increased, so is the time required to collect all of the charge generated by a stopped x-ray. [39] contains a useful discussion of the interplay between sensor thickness, efficiency, and collection time for Si, Ge, GaAs, and CdTe.

The speed with which photo-generated charge carriers are collected needs to be compared not only to the temporal requirements of the experiment, but also to the charge carrier lifetime; to ensure good charge collection efficiency, the average charge collection time should be considerably shorter than the charge carrier lifetime. Table 4.1 summarizes some of the carrier transport properties of several of the sensor materials discussed in this chapter. With the exception of Ge, the three main direct-conversion sensors that have been investigated for x-ray applications have significantly lower charge carrier lifetime, mobility, or both compared to Si. The issue of poor carrier transport has been found to be common in compound semiconductors [40].

**TABLE 4.1**

Properties, Including Charge Carrier Mobilities and Lifetimes, for a Subset of Sensor Materials

| | $\rho$ [g/cm³] | $E_{gap}$ [eV] | $E_{e-h}$ [eV] | $\mu_e$ [cm²/Vs] | $\mu_h$ [cm²/Vs] | $\tau_e$ | $\tau_h$ | Refs |
|---|---|---|---|---|---|---|---|---|
| Si | 2.33 | 1.12 | 3.6 | 1500 | 400 | >1 ms | >1 ms | [41] |
| Ge | 5.33 | 0.67 | 2.9 | 3900 | 1900 | >1 ms | >1 ms | [41] |
| GaAs:Cr | 5.32 | 1.42 | 4.3 | 2500 | 165 | ~40 ns | ~1 ns | [25] |
| CdTe | 5.85 | 1.44 | 4.4 | 1100 | 90 | ~3 µs | ~2 µs | [35] |
| CdZnTe* | 5.8 | 1.44 | 4.6 | 1100 | 88 | 11 µs | 0.2 µs | [42] |
| CdZnTe† | 5.8 | 1.44 | 4.6 | 940 | 114 | 1.2 µs | 2.5 µs | [42] |

$E_{gap}$ is the band gap energy; $E_{e-h}$ is the energy required to create one electron-hole pair. *GaAs:*Cr refers to chromium-compensated GaAs. CdZnTe* and CdZnTe† refer, respectively, to the spectroscopic-grade and high-flux materials vended by Redlen Technologies (Saanichton, BC, Canada).

When determining optimal sensor thickness, in the case of a semiconducting sensor, the voltage required to fully deplete (or, ideally, over-deplete) the sensor should also be considered. For example, while a ~120 mm thick Si sensor would be 50% efficient at 100 keV, a bias of >4 MV would be required to fully deplete the sensor (assuming a resistivity of at least 10 kΩ-cm). This is clearly impractical. In contrast, a 750-µm CdTe sensor offers the same efficiency at 100 keV and requires significantly lower bias during operation (<1 kV in all of the measurements conducted with the CdTe MM-PAD and Keck-PAD described earlier).

### 4.3.2 Material Defects

Many of the materials that have been investigated as sensors for high-energy x-rays are compound materials. This leads to the increased probability of encountering defects such as inclusions, lattice dislocations, and so on that are not as commonly found in high-purity, single-crystal materials such as Si and Ge. Defects in the sensor material can lead to inhomoegenieties in the sensor response, visible (for example) as spatial features in a uniform flood field exposure. If these nonuniformities are not stable with time, dose, temperature, etc. they can be difficult or impossible to correct in image post-processing. Sensor material defects can also lead to trapping of photogenerated charge, which can manifest as signal persistence, delayed photocurrent in response to stimulation, and incomplete charge collection. It should be noted that these phenomena are distinct from polarization effects, which will be summaries in the following section; that is, they may be present even before the onset of polarization.

### 4.3.3 Material Polarization

Material polarization is a term that has been used to cover a variety of deleterious effects on image quality after a sensor has been irradiated or biased for some amount of time. It has been particularly associated with CdTe, but has also been observed in other high-Z sensors to varying degrees, including CdZnTe [43], Cr-compensated GaAs [25], and amorphous selenium [44]. It should be noted that polarization can manifest differently in photon-integrating and photon-counting detectors. In photon-counting detectors, polarization is often discussed in terms of a count-rate defiicit, where increased charge trapping leads to a situation where the signal collected due to the absorption of an x-ray is insufficient to trigger the pixel circuitry to count the photon. In a photon-integrating scheme, the same root cause manifests as decreased charge collection efficiency.

Polarization in CdTe has been observed under irradiation as well as being induced by sensor dark current alone while the material is reverse-biased, and in a photon-integrating detector manifests as an increase in flat-field nonuniformity, a reduction of integrated signal with dose, the deterioration of image uniformity with time, and the lateral displacement of signal in the presence of differential dosing [35]. Fortunately, polarization can generally be reversed by periodic sensor "resets", often involving a change to the sensor bias. For the detectors described in Sections 4.2.3 and 4.2.4, it was found that the sensor should undergo a reset cycle, consisting of the application of a 5V-forward bias for at least 1 minute before returning to reverse bias, at least once every few hours or when exposed to a net flux of $\sim 10^{10}$ ph/mm$^2$. This reset cycle has been found to be extremely effective in restoring the sensor performance to its unpolarized state.

Polarization behavior has been observed to depend not only on the accumulated dose but also on the dose rate. In [36] the CdTe Keck-PAD was used to investigate the polarization behavior of CdTe under high-flux conditions ($1\text{–}2 \times 10^{10}$ ph/mm$^2$/s). It was found that allowing dose to accumulate at these rates without resetting the sensor lead to an increase in the delayed signal following the arrival of an x-ray pulse compared to a freshly-reset, unpolarized sensor. For reasons that remain unclear, the onset of polarization occurred more quickly at high dose rates than was anticipated from investigations at lower dose rates.

## 4.4 Summary

In spite of these complications, advances continue to be made, both in the range and quality of available sensor materials and in the understanding of their characteristics and nuances. The experiences described in this chapter with the MM-PAD and Keck-PAD demonstrate that photon-integrating PADs with CdTe sensors can be successfully used in synchrotron experiments requiring time resolution ranging from milliseconds down to ~150 ns. Additionally, photon-integrating detectors offer a different, complementary view into sensor behavior compared to photon-counting detectors, and can be valueable tools for interrogating sensor behavior.

## References

1. K. Wille, *The Physics of Particle Accelerators*. Oxford: Oxford University Press, 2005.
2. J. Shanks *et al.* "Accelerator design for the Cornell High Energy Synchrotron Source upgrade." *Phys. Rev. Accel. Beams*, vol. 22, p. 021602, 2019.
3. J. C. da Silva *et al.* "Efficient concentration of high-energy x-rays for diffraction-limited imaging resolution." *Optica*, vol. 4, pp. 492–495, 2017.
4. F. Seiboth *et al.* "Simultaneous 8.2 keV phase-contrast imaging and 24.6 keV X-ray diffraction from shock-compressed matter at the LCLS." *Appl. Phys. Lett.*, vol. 112, p. 221907, 2018.
5. R. L. Sheffield *et al.*, "Matter-Radiation Interactions in Extremes (MARIE): project overview,"*38th International Free Electron Laser Conference*, 2017.
6. M. Veale *et al.*. "Investigating the suitability of GaAs:Cr material for high flux X-ray imaging." *JINST*, vol. 9, pp. C12047, 2014.
7. M. Veale *et al.* "Cadmium zinc telluride pixel detectors for high-intensity x-ray imaging at free electron lasers." *J. Phys. D: Appl. Phys.*, vol. 52, p. 085106, 2019.
8. J. Schuren *et al.* "New opportunities for quantitative tracking of polycrystal responses in three dimensions." *Curr. Opin. Solid State Mater. Sci.*, vol. 19, pp. 235–244, 2015.
9. S. Nair *et al.* "Micromechanical Response of crystalline phases in Alternate cementitious Materials using 3-Dimensional X-ray Techniques." *Nature Scientific Reports*, vol. 9, 2019.
10. C. L. Williams *et al.* "Real-time observation of twinning-detwinning in shock-compressed magnesium via time-resolved in situ synchrotron XRD experiments." *Phys. Rev. Materials*, vol. 4, p. 083603, 2020.
11. F. Schotte *et al.* "Watching a protein as it functions with 150-ps time-resolved x-ray crystallography." *Science*, vol. 300, pp. 1944–1947, 2003.
12. S. Luo *et al.* "Gas gun shock experiments with single-pulse x-ray phase contrast imaging and diffraction at the Advanced Photon Source." *Rev. Sci. Instrum.*, vol. 83, p. 073903, 2012.
13. S. Wilkins *et al.* "Phase-contrast imaging using polychromatic hard X-rays." *Nature*, vol. 384, pp. 335–338, 1996.
14. L. Wcislak *et al.* "Hard X-ray texture measurements with an on-line image plate detector." *Nucl. Instr. Meth. Phys Res. A.*, vol. 467–468, pp. 1257–1260, 2001.
15. A. C. Konstantinidis *et al.* "The Dexela 2923 CMOS X-ray detector: A flat panel detector based on CMOS active pixel sensors for medical imaging applications." *Nucl. Instr. Meth. Phys Res. A.*, vol. 689, pp. 12–21, 2012.

16. C. Leitz *et al.* "Towards megapixel-class germanium charge-coupled devices for broadband x-ray detectors." *Proc. SPIE 11118, UV, X-Ray, and Gamma-Ray Space Instrumentation for Astronomy*, vol. XXI, p. 1111802, 2019.
17. H. Graafsma *et al.*, "Integrating hybrid area detectors for storage ring and free-electron laser applications," in *Synchrotron Light Sources and Free-Electron Lasers*, E. Jaeschke *et al.*, Eds. Springer, Cham, 2018.
18. C. Bronnimann and P. Trueb, "Hybrid pixel photon counting x-ray detectors for synchrotron radiation," in *Synchrotron Light Sources and Free-Electron Lasers*, E. Jaeschke *et al.*, Eds. Springer, Cham, 2020.
19. C. Bronnimann *et al.* "The PILATUS 1M detector." *J. Synchrotron Rad.*, vol. 13, pp. 120–130, 2006.
20. H. Philipp *et al.* "Pixel array detector for x-ray free electron laser experiments." *Nucl. Instr. Meth. Phys Res. A.*, vol. 649, pp. 67–69, 2011.
21. A. Allahgholi *et al.* "Megapixels @ Megahertz – The AGIPD high-speed cameras for the European XFEL." *Nucl. Instr. Meth. Phys Res. A.*, vol. 942, p. 162324, 2019.
22. H. Philipp *et al.* "High dynamic range CdTe mixed-mode pixel array detector (MM-PAD) for kilohertz imaging of hard x-rays." *JINST*, vol. 15, p. P06025, 2020.
23. R. Bellazzini *et al.* "Chromatic X-ray imaging with a fine pitch CdTe sensor coupled to a large area photon counting pixel ASIC." *JINST*, vol. 8, p. C02028, 2013.
24. D. Pennicard *et al.* "A germanium hybrid pixel detector with 55μm pixel size and 65,000 channels." *JINST*, vol. 9, p. P12003, 2014.
25. J. Becker *et al.* "Characterization of chromium compensated GaAs as an X-ray sensor material for charge-integrating pixel array detectors." *JINST*, vol. 13, p. P01007, 2018.
26. P. Trueb *et al.* "Improved count rate corrections for highest data quality with PILATUS detectors." *J. Synchrotron Rad.*, vol. 19, pp. 347–351, 2012.
27. D. R. Schuette, "A mixed analog and digital pixel array detector for synchrotron x-ray imaging," Ph.D. dissertation, Cornell University, USA, 2008.
28. M. Tate *et al.* "A medium-format, mixed-mode pixel array detector for kilohertz x-ray imaging." *J. Phys.: Conf. Ser.*, vol. 425, p. 062004, 2012.
29. K. Giewekemeyer *et al.* "High Dynamic Range Coherent Diffractive Imaging: Ptychography using the Mixed-Mode Pixel Array Detector." *J. Synchrotron Radiation*, vol. 21, pp. 1167–1174, 2014.
30. K. Giewekemeyer *et al.* "Experimental 3D Coherent Diffractive Imaging from photon-sparse random projections." *IUCrJ*, vol. 6, pp. 357–365, 2019.
31. J. Liu *et al.* "X-ray reflectivity measurement of interdiffusion in metallic multilayers during rapid heating." *J. Synchrotron Radiation*, vol. 24, pp. 796–801, 2017.
32. K. Overdeep *et al.* "Mechanisms of oxide growth during the combustion of Al:Zr nanolaminate foils." *Combusition and Flame*, vol. 102, pp. 442–452, 2018.
33. K. Chatterjee *et al.* "Study of residual stresses in Ti-7Al using theory and experiments." *J. Mech. Phys. Solids*, vol. 109, pp. 95–116, 2017.
34. K. Chatterjee *et al.* "Intermittent plasticity in individual grains – a study using high energy x-ray diffraction." *Structural Dynamics*, vol. 6, p. 014501, 2019.
35. J. Becker *et al.* "Characterization of CdTe Sensors with Schottky Contacts Coupled to Charge-Integrating Pixel Array Detectors for X-Ray Science." *JINST*, vol. 11, p. P12013, 2016.
36. J. Becker *et al.* "Sub-Microsecond X-Ray Imaging Using Hole-Collecting Schottky type CdTe with Charge-Integrating Pixel Array Detectors." *JINST*, vol. 12, p. P06022, 2017.

37. K. Shanks *et al.* "Characterization of a Fast-Framing X-Ray Camera With Wide Dynamic Range for High-Energy Imaging." *Nuclear Science Symposium and Medical Imaging Conference Proceedings (NSS/MIC), 2019 IEEE*, 2020.
38. L. Koerner, "X-ray analog pixel array detector for single synchrotron bunch time-resolved imaging," Ph.D. dissertation, Cornell University, USA, 2010.
39. H. Philipp *et al.* "Practical experience with high-speed x-ray pixel array detectors and x-ray sensing materials." *Nucl. Instr. Meth. Phys Res. A.*, vol. 925, pp. 18–23, 2019.
40. A. Owens and A. Peacock. "Compound semiconductor radiation detectors." *Nucl. Instr. Meth. Phys Res. A.*, vol. 531, pp. 18–37, 2004.
41. "Ioffe Physical-Technical Institute of the Russian Academy of Sciences," http://www.ioffe.ru/SVA/NSM/Semicond/.
42. B. Thomas *et al.* "Characterisation of Redlen high-flux CdZnTe." *JINST*, vol. 12, pp. C12 045–C12 045, 2017.
43. K. Iniewski. "CZT sensors for Computed Tomography: from crystal growth to image quality." *JINST*, vol. 11, p. C12034, 2016.
44. H. Huang and S. Abbaszadeh. "Recent Developments of Amorphous Selenium-Based X-Ray Detectors: A Review." *IEEE Sensors Journal*, vol. 20, pp. 1694–1704, 2020.

# 5

# *Position-Sensitive Virtual Frisch-Grid Detectors for Imaging and Spectroscopy of Gamma Rays*

**Aleksey E. Bolotnikov and Ralph B. James**

**CONTENTS**



## 5.1 Introduction

Most of the radiation detectors operating in charge collection mode are single polarity carrier devices, meaning that only one polarity carrier, usually electrons, has substantially higher mobility and lifetime compared to the

DOI: 10.1201/9781003219446-5

other carrier type, and therefore generates larger signals in the devices [1]. A drawback of this approach is that the signal amplitude exhibits a strong dependency on the specific location of interaction, which is called the induction effect. Unless special detector designs are employed, the induction effect degrades the spectral resolution of such gamma-ray detectors since the locations of interaction sites are typically randomly distributed over the active volume of the device. A virtual Frisch-grid (VFG) is one of the designs proposed to overcome this effect in detectors using different detecting media, including CdZnTe (CZT). The term Frisch grid was derived from gas ionization chambers in which case an electron-transparent metal grid is used to electrostatically shield the anode from the slow-moving positive ions [2]. For semiconductor detectors such embedded grids are practically impossible, but a shielding effect can be achieved by placing the electrodes outside of the detector volume, e.g., on the crystal's side surfaces, which produces the same Frisch-grid effect as if a real grid had been placed inside the detector.

There is a certain degree of confusion in relation to the name of the VFG detectors described here because its operating principle is used in many types of radiation detectors. It would not be an overstatement to say that majority of the semiconductor detectors operate as the VFG devices. However, historically this name refers to a particular kind of detectors that employs crystals with large geometrical aspect ratios (the ratio between the thickness and the width) and should not be confused with the three-terminal, hemispherical or pixelated detectors, even though they rely on the same operation principle. Following the tradition, we will use this name for the large aspect ratio CZT detectors (bars) described here.

Large aspect-ratio VFG detectors are particularly beneficial for integrating into large-area arrays. They are proposed to overcome the size limitations of individual CZT detectors, while retaining the excellent performance offered by relatively small monolithic crystals. This approach allows for making large area, yet affordable, instruments using commercial CZT crystals. A good example of CZT arrays is the gamma-ray telescope developed by NASA for the Swift space mission [3]. The instrument's detector plane employed 32,768 pieces of $4 \times 4 \times 2$ mm$^3$ CZT crystals arranged into sixteen arrays, which made up the entire detector area of ~5 m$^2$. Tailing together relatively thin planar detectors work well for low-energy gamma rays. For high-energy gamma rays, however, much thicker CZT crystals are required to gain high detection efficiency, and the spectral performance of planar detectors is unacceptable. Here the use of CZT bars with thicknesses up to several centimeters is particularly beneficial.

Since the performance of CZT detectors strongly depends on the material quality, there is always a tradeoff between cost, effective area, and additional functionality that determines the choice for the type and geometry of individual detectors selected for an array. Large-volume 3D pixelated and coplanar-grid (CPG) detectors, available today with volumes up 6 cm$^3$, represent suitable building blocks for arrays. The pixelated detectors offer the best spectroscopic performance and are much more advantageous compared to other CZT detectors [4], but they have not yet been integrated into large-area arrays. The CZT crystals have too high cost and too low availability for the required quality and efficiency of such detector applications. In contrast, CPG detectors have less demanding crystal quality requirements and thus are less expensive; however, the energy and position resolutions are inherently limited by the detector design. The lower cost and simplicity of design were crucial factors that helped the LBNL team build a very successful instrument based on large-area arrays of CPG detectors [5]. However, because of the CPG performance limitations, the energy and position resolutions of this instrument were not as good as would be expected from today's multi-pixel CZT detectors. In contrast, the VFG detectors with cross sections up to $10 \times 10$ mm$^2$ and thicknesses up to 40 mm, represent an economical approach to making large effective area position-sensitive detectors. As we describe later, such detectors provide performance similar to that achievable with 3D pixelated detectors but with higher detection efficiency due to the possibility of using crystals with greater thickness. In addition, crystals with such geometry are less expensive and easier to fabricate than, for example, $20 \times 20 \times 15$ mm$^3$ crystals used in large-volume 3D pixelated and CPG detectors.

The VFG detector design has been applied to different detector media: $HgI_2$ [6, 7], CZT [8, 9], TlBr [10], and even noble gases [11–13]. The designs went through several modifications and resulted in a steady increase in the availability of commercial CZT crystal. In the earlier devices [14–19], such as a CAPture device [14], the monolithic cathode contact was extended to cover approximately one half of the crystal side surfaces to form a shielding cap above the anode. The detectors employing this design were made of relatively small and thin crystals but worked well for low-energy gamma rays, which normally interact close to the cathode. However, since the side surfaces were not fully covered by the metal, the area near the anode was not 100% shielded, causing low-energy tailing in the pulse-height spectra. In addition, with this design, the extended portion of the cathode (placed directly on the crystal surface) affected the electric field inside the crystal. To avoid interference of the shielding electrode with the electric field and support longer crystals (the efficiency of shielding is enhanced by a large aspect ratio), McGregor and Rojeski [7] and independently Montemont et al. [8] separated the shielding electrode from the crystal side surfaces using a thin layer of insulating material (a shielding electrode called a noncontacting or capacitive Frisch-ring, respectively). In this case, the static electric field-lines that originated on the shielding electrode did not penetrate inside the semiconducting material, while the field lines generated by the fast-moving carriers escaped the crystal and were terminated by the shielding electrode. McGregor et al. [20, 21], McCane et al. [22], Kargar et al. [23], and Bolotnikov et al. [24] further optimized the geometry and investigated the performance of these detectors. It was shown that such devices could attain a good energy resolution of < 1% FWHM at 662 keV and a high peak-to-Compton ratio of ~5. In the earlier-mentioned devices, the noncontacting electrodes were an extension of the cathodes, meaning that the same cathode bias is also applied to the shielding electrode. This requires leaving a gap of several millimeters between the Frisch-ring edge and the anode's contact to prevent high leakage current or even possible electrical discharge between the two electrodes. On the other hand, the bare surface near the anode lowers the efficiency of shielding and results in a low-energy tailing effect in the pulse-height spectra. In addition, it is not possible to use a charge loss correction scheme in such devices. To improve the efficiency of shielding, one can reverse the detectors geometry by extending the shielding electron down to the anode and keeping it at the same potential as the anode. In this case, additional insulation will be required near the cathode side.

The BNL team introduced several improvements to the VFG design; the most important was the reduction of the shielding electrode down to a 5-mm-wide band positioned near the anode [25]. The electrode width was carefully optimized to provide a good electrostatic shielding of the anode (for crystals greater than three-aspect ratio) while leaving the cathode unshielded. This means that the cathode becomes sensitive to the interaction depth which allows for the use of the cathode signals to correct for the charge losses as the electron's cloud drifts from the origin point toward the anode. We called this a conventional VFG design. In addition, we proposed to group together a small sub-array (up to 4 × 4 detectors) under a single common cathode, which would mimic a pixelated detector. Using this design, the BNL team developed the first 6 × 6 array prototype of 15-mm-long detectors with 5-mm-wide shielding electrodes [26]. With these arrays, after accounting for drift-time corrections, we achieved < 1.3% FWHM at 662 keV using high homogeneity crystals. To increase the acceptance rate of CZT crystals and reduce instrument costs, we proposed using position-sensitive VFG detectors, which allowed us to correct response nonuniformity by virtually dividing the detector volumes into small voxels and equalizing the responses from each voxel. Additionally, the shielding electrode can be comprised of four separate pads placed on the four sides of the bar. The signals from the pads can be used to measure the locations of interaction points in the *X-Y* plane. Combined with data of the cathode-to-anode ratio (*C/A*) (or drift time) for measuring locations along the *Z* coordinate, the device operates as a mini time-projection chamber [27].

The detector design allows for integrating arrays of different geometries and thicknesses with easily replaceable crystals. At the same time, the 3D correction technique ensures good spectral resolution of the arrays over a wide dynamic range: < 2% at 200 keV and < 1% at 662 keV at 25°C. The position resolution, which is important for the instrument's imaging capability, is typically about 0.5 mm or less, and is determined by the local variations of the electric field. However, the intrinsic X-Y position resolution (within small areas of the detector) can be as good as 100 μm at above 400 keV, which allows correcting for response nonuniformity caused by device geometry, readout electronics, and crystal defects present inside even the highest quality CZT crystals. The ability to use the majority of commercially available CZT crystals brings down the cost significantly, making this design useful in the mass manufacturing of commercial detectors.

In this chapter, we summarize the development of position-sensitive VFG detectors to advance this technology for a broad range of practical applications, including basic science, medical and industrial imaging, nuclear security, non-proliferation, and safeguards.

## 5.2 VFG Detector Designs

The main distinguishing feature of VFG detectors is their geometry. We use bar-shaped CZT crystals with an aspect ratio greater that 3 and a relatively small, up to $10 \times 10$ mm$^2$, cross-section with a thickness up to 40 mm. This geometry offers two benefits for making the VFG. The first benefit comes from the enhanced shielding efficiency in large aspect-ratio crystals, as well as their ease of integration into large-area arrays. The second is their cost, as such crystals are relatively inexpensive to produce.

CZT detectors operate as single type charge carrier devices for which only fast-moving electrons are collected and used to measure the energies deposited by ionizing photons and particles. The main drawback of such devices is that the output signals depend on the locations of the interaction points inside the detector. This dependence is common among ionization detectors operating in the charge-collection mode. The classic example is a gas ionization chamber, which can be used as a good model for CZT detectors. As a result of the induction effect, single charge type detectors show no spectral response unless a special detector design is implemented to minimize the induction effect.

In the electron-only ionization detector, the amplitude of the charge signal, $Q_{out}$, generated on the anode, has two components:

$$Q_{out} = Q_{col} + Q_{ind} \tag{5.1}$$

where $Q_{col}$ is the total collected charge (electrons) and $Q_{ind}$ is the total charge induced on this electrode by the uncollected (trapped) holes and electrons. Equation 5.1 can be understood from very basic principles: $Q_{col}$ reflects the fact that for every charge that actually reaches the electrode, an identical opposite polarity charge appears on the electrode from the external circuitry. The term $Q_{ind}$ is the image-charge induced by the uncollected charges and its value depends on the position of the source charge. For every location of the source charge (trapped electrons or holes) the induced charge $Q_{ind}$ can be calculated by solving a Dirichlet problem with certain boundary conditions following from a Green's reciprocity formula, which is equivalent to the weighting potential concept well described in the literature [1, 28, 29]. Equation 5.1 is equivalent to the Shockley-Ramo equation [30, 31] integrated over the carriers' drift time. Equation 5.1 suggests two ways of reducing the contribution of the induced component: (1) signals subtraction (used for CPG-type devices) and (2) electrostatic shielding (implemented in VFG-type detectors). However, in practical devices, it is impossible to fully eliminate the induced charge contribution for the entire volume of the detector. Using the terminology originally applied to the classic Frisch-grid gas ionization chamber [32], the residual effect of the uncollected holes in VFG detectors can be described as the shielding inefficiency of the VFG. The shielding inefficiency, measured as a fraction of the induced charge, directly contributes to the total width of photopeaks, setting an intrinsic limit on the energy resolution. Furthermore, small "unshielded" regions always remain (their sizes depend on the device's geometry) near the collecting electrodes where the induction effect cannot be reduced. Thus, the detector's volume can be divided in two unequal parts: the larger one, called the drift region, where the device response weakly depends on the location of the interaction points, and the smaller second one, called the induction region, in which the amplitude of the output signal changes from the maximum value to zero. In the classic Frisch-grid ionization chambers [32], the amplitude of the output signal changes relative to the Z coordinate as a linear function in both regions. In this case, the shielding inefficiency of the Frisch-grid is defined as the total change of the output signal over the drift region. Assuming that the interaction points are uniformly distributed inside the detector, then the geometrical width of the photopeak will be equal to the shielding inefficiency. Events interacting in the induction region generate a uniform background in the pulse-height spectra.

The electron trapping reduces the amplitude of the output signal proportionally to the drift distance, which compensates for the shielding inefficiency effect. The actual variation of the output signals in the drift region is a result of the interplay between the shielding inefficiency and the electron trapping, both of which depend on the device geometry. From here we can formulate two important criteria for detector designs: (1) how small are the variations of the output signals in the drift region and (2) how small is the ratio between the volumes of the drift and induction regions. The former determines the "geometrical" width of the photopeak, i.e., the width of the photopeak attributed to the device geometry when all other factors are neglected, and the latter is related to the photopeak efficiency and background continuum in the pulse-height spectra. We note that the gradual response changes due to the device geometry (inefficiency of shielding) can be corrected altogether with the charge losses, while the interaction events from the induction region can be rejected if the coordinates of the interaction points are known.

### 5.2.1 Conventional VFG Detectors

A conventional VFG detector is a simple, inexpensive and easy-to-fabricate device that employs a bar-shaped crystal with two monolithic contacts. Several examples of such designs are shown in Figure 5.1: (a) CAPture [14, 15], (b) Frisch-ring [7] or capacitive Frisch-grid [8], and (c and d) two designs that combine the elements of the previous two [24]. In the CAPture design the cathode is extended halfway toward the anode. Such geometry causes very nonuniform charge collection due to the weakening of the electric field in the detector's corners, near the cathode, and poor shielding near the anode. Although the main advantage of the capacitive Frisch-grid design is that it provides a uniform electric field, it does not allow for charge loss corrections, which are required in long drift-time detectors. As will be explained next, the variations in charge collection efficiency within the drift time (or interaction depth) may not be perfectly compensated by the shielding inefficiency, which degrades detector performance. The third design offers a few minor improvements. The cathode is extended 2–3 mm down to the side surfaces like in the CAPture device. The shielding electrode is disconnected from the cathode and extended to the very edge of the anode. A negative voltage bias is applied to the cathode, while the shielding electrode is kept at the ground potential. A gap of a few millimeters of the insulator is enough to block the leakage current between the cathode and shielding electrodes. This design

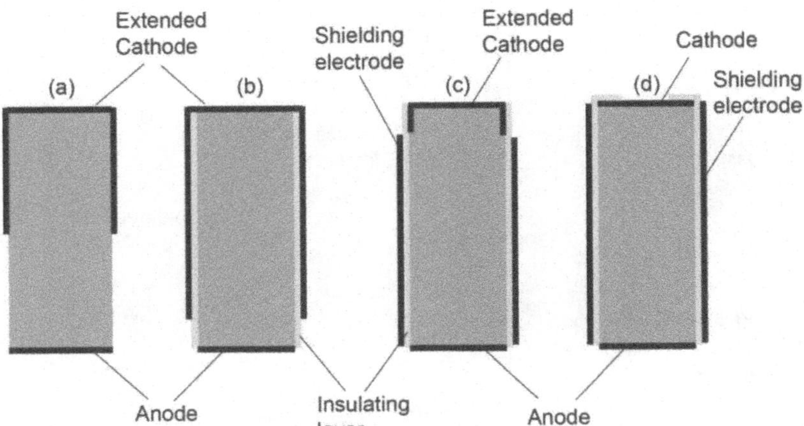


**FIGURE 5.1**
Schematics of the designs used for bar-shaped virtual Frisch-grid CZT detectors: (a) CAPture [14], (b) Frisch-ring [7] or Capacitive Frisch-grid [8], and (c and d) modified designs which include elements of earlier designs [24]. In the case (c and d), a shielding electrode is decoupled from the cathode and extended up to the anode.

provides a means for decoupling the shielding electrode from the cathode while at the same time ensuring that the detector remains properly shielded. It also keeps the shielding electrode at the ground, which is important for assembling arrays of such detectors. Neither of these designs allows for charge-loss corrections, which would be possible only by reading the cathode signals.

The next improvement to VFG detectors involved reducing the shielding electrode, originally covering the entire side surfaces of the detector, with an arrow band placed at the anode [25] and then grouping several detectors together under a common cathode. For 2D arrays, the detectors were mounded side-by-side with minimal gaps between them and grouped into $2 \times 2$ or $3 \times 3$ sub-arrays with a common cathode [33] for each sub-array module. Because of the shared shielding of the detector's anodes, the height of the shielding electrode could be reduced to 4–5 mm without compromising the efficiency of the shielding. Figure 5.2 shows a photograph of a conventional VFG detector (a) and its spectral response to an uncollimated $^{137}$Cs source before and after grounding the shielding electrode placed near the anode. Furthermore, by leaving the remaining surface of the detectors unshielded, we make the common cathode sensitive to the depth positions of the interaction sites. The common cathode signals provide independent measures of the interaction depths, and the electron cloud drift times can be used for charge-loss corrections. This has a great practical importance because it allows use of thicker CZT crystals or material with lower $\mu\tau$-product. The drift time (or interaction depth) correction is a standard technique for improving the performance of CZT detectors.

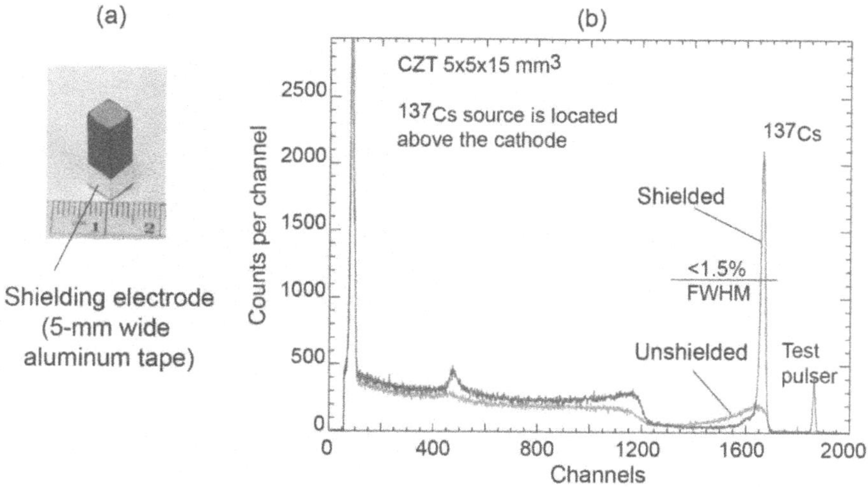


**FIGURE 5.2**
Photograph of a conventional VFG detector (a) and its spectral response to an uncollimated $^{137}$Cs source before and after placing the shielding electrode near the anode (b).

Using conventional 5 × 5 × 15 mm$^3$ VFG detectors, researchers from the BNL team integrated and tested several 6 × 6 array modules and readout system based on a 36-channel AVG1 ASIC [26]. The energy resolution achieved with these arrays was < 1.5% FWHM at 662 keV after drift-time corrections. Using carefully selected CZT crystals, the energy resolution could be improved to < 1%. The energy resolution for all the detectors was in a range of 0.7–1.3% FWHM at 662 keV. However, this performance improvement came at a higher detector cost. To increase the acceptance rate of CZT crystals and reduce instrument costs, we proposed use of 3D position-sensitive VFG detectors, which allowed us to correct response nonuniformity by virtually dividing the detector volumes into small voxels and equalizing the responses from each voxel. Position-sensitive VFG detectors, described next, allowed for further improvement of the energy resolution by correcting the detector response nonuniformity in the lateral directions.

### 5.2.2 Position-Sensitive VFG Detectors

As was mentioned previously, drift-time (or 1D) corrections are applied to correct continuous charge loss due to trapping of carriers by point and other nanoscale defects, which are assumed to be uniformly distributed inside the detector volume. Unfortunately, these are not the only reasons for the response variations of CZT detectors; the extended defects (particularly sub-grain boundaries and dislocation networks) cause spatial variations in the distributions of the trapping centers. These extended defects cause a nonuniform response of CZT detectors, which degrades their spectral performance. Moreover, the extended defects cause local variations in the electric field, which in turn cause additional variations in the collected charge. To overcome these problems, detector developers are forced to use more carefully selected, and therefore more expensive CZT crystals, which may significantly increase the cost of the actual instruments. A more economical approach would be to implement detector designs with the ability to correct for response nonuniformities caused by crystal defects. This approach is justified by the fact that the defect locations inside crystals are fixed, and the output signal variations are drawn from random distributions of interaction points. As such, the nonuniformities can be corrected by virtually segmenting the detector's active volume into small voxels and equalizing their responses.

Position-sensitive VFG detectors employ a 5-mm-wide shielding electrode as in the conventional design described earlier, but it is comprised of four separated pads, each on one side of the detector (Figure 5.3(a)). The transient signals captured from the cathode and position-sensitive pads are processed to evaluate the locations of interaction points. The amplitudes of the signals captured from the pads are used to evaluate the *X*-*Y* coordinates, while the measured drift time or the *C*/*A* ratio is used to independently evaluate the *Z* coordinate for each event. In other words, the device operates as a miniature

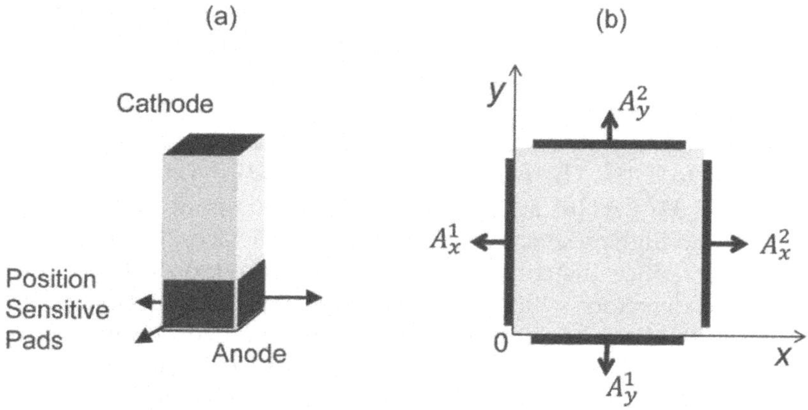


**FIGURE 5.3**
Schematic of a position-sensitive VFG detector (a) and its coordinate system (b).

time-projection chamber that provides two lateral coordinates and drift times for event reconstruction. We note that it is more convenient to use normalized pad signals, taken as the pad-to-anode ratios $P/A$, to make the measurements of coordinates independent of the photon energies.

Figure 5.4 depicts typical charge-signal waveforms captured from a position-sensitive VFG detector [34, 35]. The plots (a), (b) show the inverted cathode, $C$, and the sum of the pad signals, $P$, while the plot (c), (d), (f), and (g) represent the pad signals. The plot (e) shows the anode, $A$, signal. As seen from the plots (a) and (b), the leading edge of the sum of the pad signals is almost identical to the leading edge of the cathode signal, which means that the latter can be used to substitute for the cathode signals. The solid vertical lines in the plots indicate the arrival time of the event being identified using the cathode signal or the sum of the pad signals (whichever has the strongest amplitude). The cathode and anode amplitudes have different polarities, but the same range of amplitudes. They can be measured using either a shaping amplifier or sampled waveforms. However, the pads' signals can have positive and negative slope components: the positive slope component represents the electron cloud drift before it reaches the pads, while the negative slope occurs when the cloud crosses the pads' level and approaches the anode. This negative swing contains the information needed for position reconstructions. For both the anode and the cathode, either a shaping amplifier or sampled waveforms can be used to measure the amplitude of the swing (the signal maximum with respect to the negative base); however, the shaping amplifier does not preserve a proportionality between the negative amplitude of the shaped signals and the actual charge signals induced on the pads (we will illustrate this in the next paragraph). The sum of the positive peaks of the pad signals, measured with the respect to the baseline level, is equivalent to

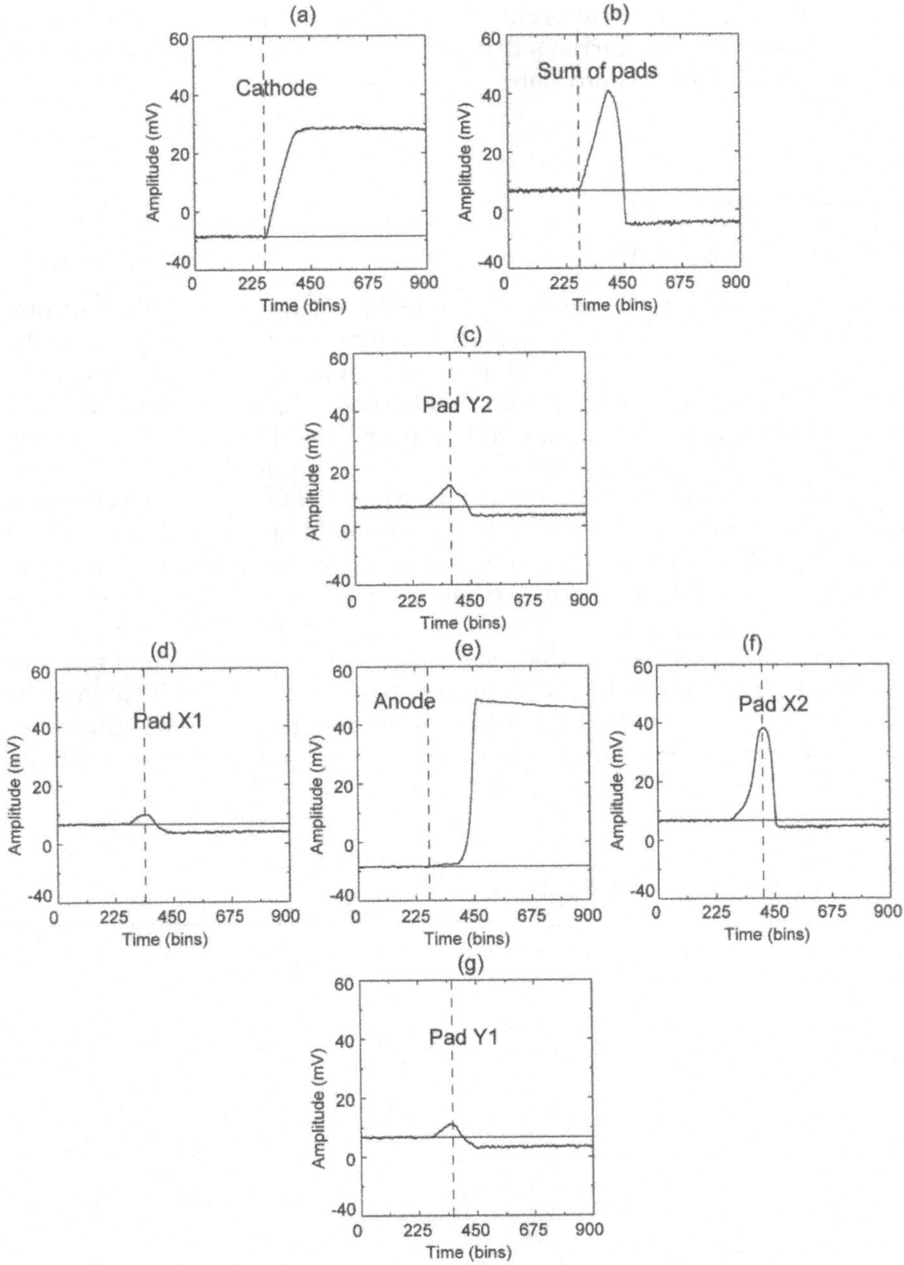


**FIGURE 5.4**
Waveforms captured from the cathode (a), the sum of the pad signals (b), and four pads (c, d, f, and g). Plot (e) represents the cathode. The dashed vertical lines in (a) and (b) indicate the moment when the interaction occurs, which is evaluated using the cathode signal or the sum of the pad signals. The lines in the pad plots indicate the pulse maxima, which corresponds to the different times. Each waveform contains 1,000 samples recorded every 8 ns.

the cathode signal. Therefore, either the *C/A* or *P/A* can be used to evaluate the *Z* coordinate. The carrier's drift time can be used as well to independently estimate the *Z* coordinate.

## 5.3 Detector Modeling

For the purpose of modeling, the VFG detector can be viewed as a rectangular metal box, of which the top and bottom surfaces correspond to the anode and cathode, while the side surfaces represent the shielding electrodes. Here we replaced the 5-mm-wide shielding strip located near the anode with a solid electrode extending over the entire side surface of the detector, which is a good approximation to illustrate the operating principle of the VFG device. In this case, the problem of finding the electrostatic field from a point-like charge inside the grounded metal box can be solved analytically [36, 37]. The total charge induced on the anode is then calculated by integrating the normal component of the electric field strength over the anode area. Figure 5.5(a) shows the dependence of the induced charge on the anode versus distance from the anode for different ratios of anode width to detector length. A source charge is in the middle of the box. The same curves represent the temporal dependencies of the signal from a charge-sensitive preamplifier as the charge drifts between the cathode

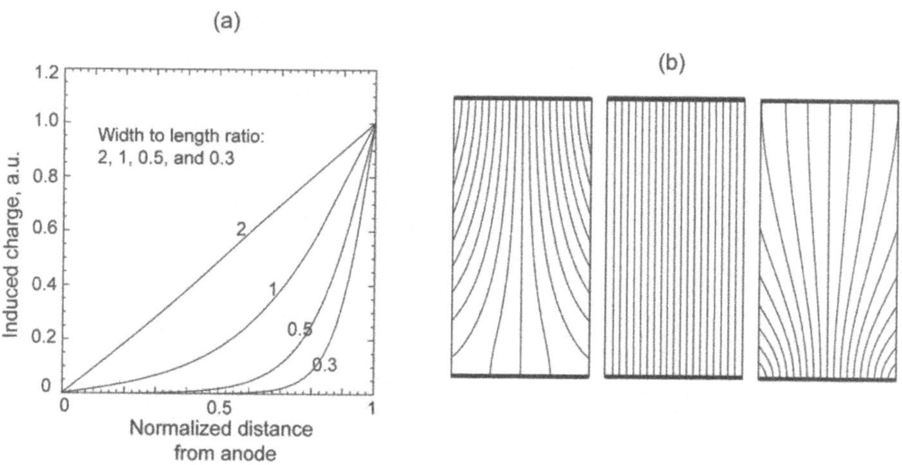


**FIGURE 5.5**
Dependence of the induced charge on the anode versus the distance from the anode for different ratios of the anode width to the detector thickness (a); three possible types of drift-field inside a bar-shaped CZT crystal (b): defocusing (left), uniform (middle), and focusing (right). The cathodes are on the top, the anodes are on the bottom.

and the anode. As one can see, for small aspect-ratio detectors, most of the signal is induced near the anode.

The position of the VFG, which virtually divides the detector volume into the drift and induction regions, is defined as a surface beyond which the output signal decreases below a certain fraction of its maxima. The volume fraction and the magnitude of the signal variations in the drift region are the two main parameters that determine the quality of the radiation detector. The first determines the photopeak area relative to the continuum in the pulse-height spectra. The second determines the geometrical width of the photopeak, which affects the total energy resolution. The shape and location of the VFG is practically independent of device thickness, while the volume fraction of the drift regions, which contributes to the photopeak efficiency, increases proportionally with device thickness, meaning that the device with greater thicknesses give better performance. It is worth mentioning that in multielectrode devices like pixelated, strip, and CPG detectors, a characteristic distance at which the weighing function becomes nearly uniform (location of the VFG) is determined by the pitch size.

The electric field distribution inside long bar-shaped detectors is determined by the electrostatic potential distribution on the side surfaces (the boundary conditions), which depends on the surface resistivity. Field simulations predict that depending on the surface potential distributions, there could be three outcomes for the electric field distribution inside the device as illustrated in Figure 5.5(b). If the surface potential decreases from the cathode value toward the anode faster than the potential along the device axes, a defocusing field (left) is generated inside the device from which a poor response is expected. If the opposite occurs, a focusing drift-field (right) will be formed inside the device, which steers the electrons toward the anode. Since the surface conductance is much higher than the bulk, the surface can be treated as a conductive skin with a uniform surface resistivity of ~$10^{11}$ Ohm/square. Thus, we can expect a linear decrease of the potential between the cathode and anode levels and a uniform electric field inside the detectors. However, because of the effect of charging at deep levels, which results in the formation of space charge, the electric field in thick CZT detectors slightly decreases toward the anode, which means that the surface potential decreases slower than the potential in the center of the detector. Even a small difference in potential distributions results in a focusing field that repels the electrons away from the side surfaces. The focusing effect is small in CZT crystals with moderate thicknesses, < 5 cm, but sufficient to avoid charge losses near the surface while preserving an early uniform field distribution. The latter is important for achieving good position resolution. In the first approximation, the electric field can be considered uniform inside VFG detectors. However, everything can change if the crystal has large extended defects, such as twins, sub-grain boundaries or dislocation networks (prismatic defects) that can significantly perturb the electric field.

The distribution of the amplitude of the output signal versus the drift time (interaction depth) is the most important characteristic of a VFG device. At a certain distance from the anode, the distribution converges into a line that represents the correlation between the device response and the interaction depth. For events located close to the detector edges, the shielding effect is even stronger. The slope of this line is determined by the variation of the weighting potential inside the drift region and electron loss due to trapping. These two factors, which can be approximated by linear functions with opposite slopes, partially cancel each other. Kargar et al. [38] investigated the uniformity of the gamma-ray response for a Frisch collar device by probing the detector experimentally with a collimated 662-keV gamma-ray source. Their results confirmed the charge collection efficiency predicted by simulations.

## 5.4 Assembling Arrays of VFG Devices

The detectors are encapsulated inside high dielectric-strength, ultra-thin, 100–200 μm thick, polyester shrink tubes [36]. Use of the thin insulating layer is important for achieving efficient shielding and good position reconstruction. The ultrathin polyester shrink tubes have proven to be a good choice for insulating material. In addition to excellent dielectric properties, it provides good mechanical protection of the crystals during the handling of detectors for integration.

The encapsulation process includes two steps. First, the rectangular polyester shells are preformed using an aluminum mandrel and a heat treatment. Then a detector is inserted inside the shell, and the open ends are sealed by two aluminum mandrels with small grooves to ensure tighten capsulation near the ends of the crystal. The above assembly is self-supported and self-aligning. Finally, heating up to 85°C is applied to form a thin seamless encapsulating layer that follows the exact shape of the crystal without damaging the surface or adversely affecting the device's electrical or mechanical properties. The mandrels are pulled away, and the tube edges are accurately and carefully trimmed ~1 mm above the crystal's top and bottom surfaces. After that, two flat spring CuBe contacts are placed on the cathode and anode sides (with intermediate protection layers to avoid metal contacts scratching) and secured by applying additional heat to the polyester shell edges that firmly envelops the contacts around the edges. To protect the anode and cathode surfaces from any potential scratches during the encapsulation, a thin layer of conductive rubber or aluminum foil is placed underneath the spring contacts. Finally, a solid metal strip or separate pads, cut from copper or aluminum tape, are glued to the side surfaces of the detector. For array integration, we developed special device holders optimized for particular

**FIGURE 5.6**
Encapsulated $6 \times 6 \times 20$ mm$^3$ detectors with spring contacts inserted inside a cell formed by the array of the finger-spring contacts soldered to the detector board.

instruments and tasks. As an example, Figure 5.6 shows a photograph of a partially integrated array of $6 \times 6 \times 20$ mm$^3$ detectors encapsulated in the polyester shrink tubes with spring contacts inserted inside the cells formed by the four finger-spring contacts soldered to the backside of the detector board. The spring contacts on the detector cathodes are clearly seen at the top. The identical contacts are attached to the anode side of the detectors.

## 5.5 Testing VFG Detectors

We tested our detectors using two experimental setups: one is based on AVG ASICs [39, 40] that employs the conventional analog signal shaping and the second uses a set of charge-sensitive preamplifiers and a waveform digitizer (digital oscilloscope) [41]. During the measurements, the assembled detectors were mounted individually (or as arrays inside the detector holding crates) on the fan out board and gently pressed with the cathode board on top to engage the spring contacts [35]. The cathode board carries a decoupling capacitor and a 100-MΩ resistor to read signals from the detector's cathode. The backside of the board has multi-pin connectors and can be plugged into the readout system's motherboard.

The AVG1ASIC [9] was optimized for conventional VFG detectors. It had 36 anodes and 9 cathode inputs, and its architectural design was adopted from its predecessor, the H3D ASIC, developed for 3D pixelated detectors by BNL's Instrumentation Division in collaboration with the University of Michigan [42–44]. A full description of this powerful and multifunctional ASIC is given by De Geronimo et al. [42]. The ASIC implements a conventional processing chain of analog signals including a low-noise charge-sensitive amplifier with continuous reset, a baseline stabilizer, a fifth-order unipolar shaping amplifier with an adjustable peaking time, and two peak and timing detectors for measuring the amplitude and timing for both the positive and negative pulses. Several gain settings (20, 40, 60, and 120 fC/mV) are available with the maximum dynamic range up to 3 MeV at the lowest gain. The cathode channel also has a charge-sensitive amplifier with continuous reset and a baseline stabilizer. It is followed by two parallel filtering circuitries – one (with a long shaping time) for measuring the amplitude and the other (with a short shaping time) for measuring the timing. The timing assigned to the anode signals is measured at their peaking time and is, to a first order, independent of the signal amplitude.

In the case of the waveform sampling readout, the detector outputs were directly connected to a digital oscilloscope. The anode signals are used (after fast shaping) to trigger the oscilloscope. The signals are typically sampled in the sequential mode and stored in the memory of the oscilloscope for further analysis. Typically, each waveform contains 1,000 amplitudes sampled over 10-ns time intervals. The event arrival time and the time when the carriers are collected to the anode are evaluated after the waveform fittings [40]. The first 100 samples are used to calculate and subtract the base lines. The starting point of the cathode signal, $t_1$ is evaluated as an intersection of the line that results from fitting the first 5–7 samples of the cathode's front edge and zero level. Accordingly, an intersection of the line fitting the last 10 samples of the anode's front edge and the line fitting the saturated level of the anode signal gives the time, $t_2$ when the electron cloud reaches the anode. The solid lines in the plots of the cathode and anode signals indicate the times $t_1$ and $t_2$. The difference between $t_2$ and $t_1$ gives the electron drift time. The amplitudes of the cathode and the anode signals are calculated by averaging 30–40 samples after $t_2$, which is equivalent to the correlated double sampling [45] with the "flat top" adjusted according to the drift time.

Since the pad signals do not reach their maxima at the same time, this introduces errors in the coordinates determined using the pad amplitudes. A better approach is to use the synchronized magnitudes of the pad waveforms at a fixed time to ensure the highest accuracy. For example, we use the pad samples at the time corresponding to the earliest maximum among the different pad signals. The pad amplitudes can be calculated with respect to the negative pad levels.

## 5.6 Performance of VFG Detectors

### 5.6.1 Leakage Current Measurements

Although VFG detectors have a simpler design and are easier to fabricate, they inherently suffer from consequences of using much larger area anodes in comparison to pixelated detectors. The large area anodes increase electronic noise (due to a higher leakage current and a large anode capacitance), which ultimately limits the energy resolution of the VFG detectors. The surface leakage is the dominant dark current component in VFG detectors. Because of this, the total leakage current scales linearly with the detector lateral dimension (width). A difference between the bulk and surface currents can be significant for VFG detectors. The surface leakage current strongly depends on the surface treatment and surface encapsulation. The highest surface resistivity reported for CZT detectors was ~$2 \times 10^{12}$ Ohm/square [46], which corresponds to a surface leakage current of less than 3 nA for the $10 \times 10 \times 32$ mm$^3$ detector at 4,000 V. Typically, the total leakage current is < 10 nA at 4,000 V for the $10 \times 10$ mm$^2$ are a detector, which is sufficiently low for achieving good spectroscopic performance.

### 5.6.2 Electron Lifetime Measurements

The electron lifetime is another factor limiting the performance of CZT detectors and should be long to ensure good spectral performance. Traditionally, the electron lifetime measurements rely on using the Hecht equation [47, 48, 49]. This method works well for thin samples with relatively low mu-tau product, < $10^{-2}$ cm$^2$/V. However, it assumes a constant electric field over the entire detector volume, which is not always justified, especially at low biases. This inevitably leads to a significant underestimation of the measured electron lifetimes. A more reliable approach is to directly measure the dependence of the collected charge on the electron cloud drift time (an exponential decay) to evaluate the lifetime. In this case, the measurements are independent of the electric field distribution inside the sample, which makes this technique very accurate for measuring the electron lifetimes in high mu-tau product, > $10^{-2}$ cm$^2$/V, materials [49].

The geometry of bar-shaped VFG detectors makes it very convenient for evaluating electron lifetimes by measuring the amplitude of the anode signals generated by 662-keV photons interacting near the cathode and plotting them against the corresponding drift times (Figure 5.7(a)). To make these plots we digitally processed the sampled cathode and anode charge signals and measured the drift times and anode amplitudes. This method allows us to accurately evaluate electron lifetimes, which were found to be in the range 80–90 μs for high-quality commercial CZT. This method provides high

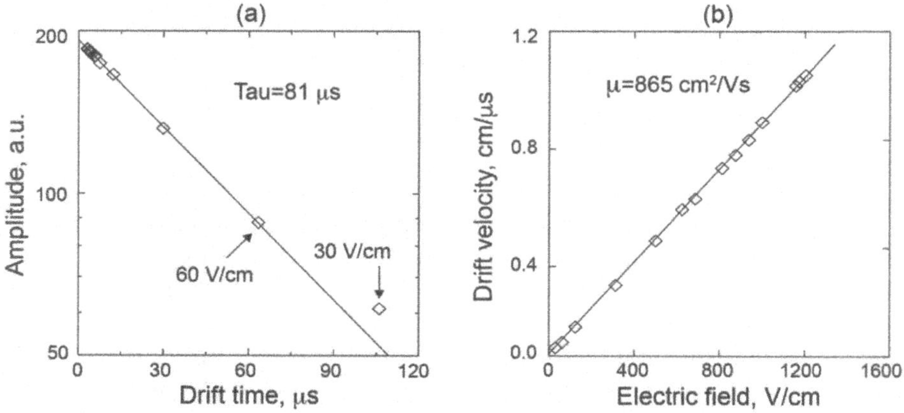


**FIGURE 5.7**
Dependences of the amplitude of the signals generated by 662-keV photons interacting near the cathode on the drift time measured at different biases (a) and the electron drift velocity on the electric field strength (b). From the fitting of the above curves, we determined the electron lifetime and mobility to be 81 μs and 865 $cm^2$/V-s, correspondingly. From here, the mu-tau product was found to be $7.0 \times 10^{-2}$ $cm^2$/V.

statistical accuracy of +/− 2 μs and is not sensitive to the systematic errors related to the electric field nonuniformity and the presence of holes. Using the same data, we also evaluated the electron mobility as the slope of the plot in Figure 5.7(b) representing the dependence of drift velocity on electric field strength. The electron mobility was found to be ~865 $cm^2$/V-s [50]. This value is slightly smaller than the values measured for thin detectors, which can be explained by a nonuniform electric field inside the long drift-length detector. Taking these numbers into account, we found the mu-tau product to be (7–8) × $10^{-2}$ $cm^2$/V, which is higher than the values typically measured using the Hecht equation [9].

### 5.6.3 Spectral Responses from a Position-Sensitive VFG Detector Measured Using Charge Signals Sampling

Output signal processing based on waveform sampling has many advantages for achieving the best performance in comparison to that based on signal shaping. It allows for more accurate position reconstructions and 3D corrections. To illustrate how it works, we applied the waveform sampling approach to evaluate signal amplitudes and timing, measured from a $10 \times 10 \times 32$ $mm^3$ position-sensitive detector irradiated with an uncollimated $^{137}$Cs source, and used the data to generate correlation plots (Figure 5.8) specific for these types of devices [35]. As seen from Figure 5.8, the dependence of the anode signals on the *C*/*A* ratio (a) and the sum of the pads to

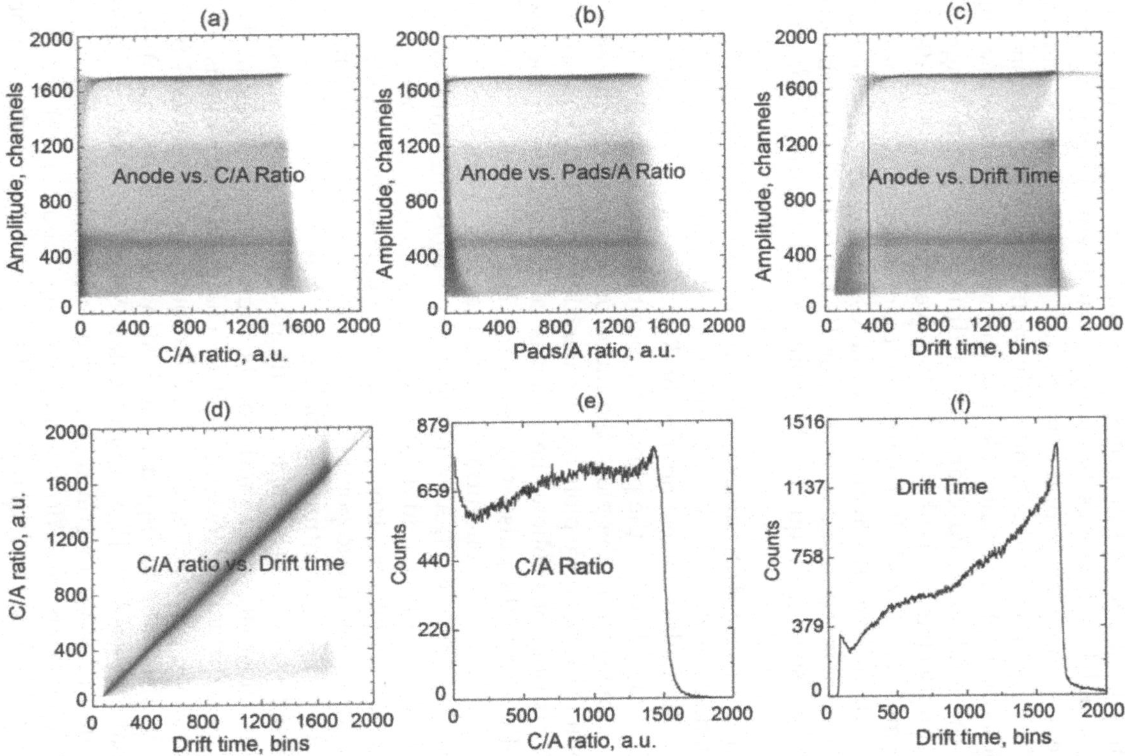


**FIGURE 5.8**
The dependences of the anode amplitudes vs. the drift time (c) can also be used to correct the charge loss along the Z direction. The first line in the anode vs. drift-time distribution (c) indicates the approximate position of the virtual Frisch-grid. The second line indicates the edge from the events interacting near the cathode corresponding to the maximum drift time when electrons drift between the cathode and the anode [35]. The detector under test was biased at 3,500 V.

anode, *P/A*, (b) ratios look very similar, demonstrating that the sum of the pad signals can be substituted for the cathode signals. Both dependences can be used to apply the interaction depth (1D) corrections. The dependence of the anode amplitudes on the drift time (c) can also be used to correct the charge loss along the *Z* direction. The first line in the anode vs. drift-time distribution (c) indicates approximately the position of the VFG. The second line indicates the edge of the events interacting near the cathode corresponding to the maximum drift time, as electrons drift from the cathode to the anode. The anode amplitudes from the 662-keV photo absorption events taken at the maximum drift time correspond to the total collected charge (after drifting the entire length of the crystal) unaffected by the holes (for the events near the cathode the holes are completely shielded). Finally, the correlation between the *C/A* ratio and the drift time (*d*) illustrate that the *C/A* (or *P/A*) ratios and the drift time give two independent estimates for the *Z* coordinate. To complete the picture, we also show frequency distributions of *C/A* ratio (*e*) and drift times (*f*).

The amplitudes and drift times in Figure 5.8 are given in channels. The dependencies of *A* vs. *T*, *A* vs. *C*, and *C* vs. *T* were measured before applying the charge-loss correction. For the *R* vs. *T*, we used the corrected charge signal normalized to the photopeak position at a channel of 1,300. To keep the same scale in the plots, the ratios of *C/A* were multiplied by 1,300. The points in the above plots represent individual interaction events. The points' distributions follow the patterns corresponding to the features normally seen in the pulse-height spectra. The photo absorption and Compton edge events are concentrated along the lines clearly seen in the plots presented in the *A* vs. *T*, *A* vs. *C*, and *C* vs. *T* plots (first three rows). The dependencies of *A* vs. *C* and *A* vs. *T* are used to correct the continuous charge loss in CZT detectors. However, one should bear in mind that the continuous charge loss due to point defects (impurities) is proportional to the drift time, while the continuous charge loss due to extended defects (e.g., Te inclusions) is proportional to the distance, and that the measured drift times may fluctuate significantly for the same drift distances. In this analysis we used the dependencies *A* vs. *T* for charge loss corrections. The dependencies *A* vs. *T* and *C* vs. *T* indicate the positions of the VFG from the anodes; in the cathode-to-grid region (also called drift region), the cathode signal changes linearly with the distance measured from the grid position, while the anode-to-grid region (also called collection region) is shielded from the cathode. Correspondingly, the anode signal remains practically unchanged in the drift region and rises linearly in the collection region. This is a full analog of the classic ionization chamber with a Frisch-grid. The dependence *C/A* vs. *T* relates three correlated variables measured for each interaction event in each detector: the anode and cathode signals and the drift times.

### 5.6.4 Using Position Information to Correct Response Nonuniformities

As we mentioned earlier, the response nonuniformities depend on the locations of interaction points. This is due to either detector material nonuniformities (defects) or detector geometry (variations of the shielding inefficiency as previously described). If the detector provides high position resolution, such correlations can be employed to correct the response nonuniformities and improve the detector performances. Using position information, the detector can be virtually segmented into small voxels, so small that responses within individual voxels can be considered uniform. Then, to correct the response of the entire detector, one just needs to equalize the "gains" of the individual voxels. The corresponding scaling coefficients, which must be generated during detector calibration, are stored into a 3D-look-up matrix and used to adjust the measured amplitudes on-the-fly. We call this procedure the 3D corrections versus the 1D corrections used for drift-time (or interaction-depth) corrections. Figure 5.9 illustrates the improvement of energy resolution after applying 1D and 3D corrections for a very large $10 \times 10 \times 32$ mm$^3$ detector biased at 2,200 V [35]. The energy resolution improves from 3.6% (raw data) to 0.9% FWHM at 662 keV (after 3D corrections).

The 3D correction technique was tested for detectors with thicknesses up to 50 mm and cross-section areas up to $10 \times 10$ mm$^2$ [34, 40]. Figure 5.10 shows the representative spectra measured for 20-, 30-, 40-, and 50-mm-long detectors before (first column from left) and after (second column from left) corrections [50]. The two spectra representing each group of detectors are plotted on the same scale, so improvements in the energy resolution are also seen as increases in the photopeak height. After the corrections we achieved the corresponding energy resolutions of 0.76%, 1.06%, 1.10%, and 1.6% FWHM at 662 keV for the 20-, 30-, 40-, and 50-mm-long detectors, respectively.

To illustrate the excellent performance of VFG detectors over a broad energy range, we measured the pulse-height spectra from the largest CZT bars available today: $8 \times 8 \times 32$ mm$^3$ and a $10 \times 10 \times 32$ mm$^3$ [35]. Figure 5.11 shows representative spectra after 3D corrections, acquired from $^{133}$Ba, $^{60}$Co, and $^{232}$U sources. The $^{232}$U spectrum (in linear and logarithmic scales) was measured using a 5-mm lead shield to reduce low-energy counts. The cathode bias was 3,200 V for all the detectors. For the low energy 356-keV line, we obtained 1.5% and 0.5% for the high-energy gamma at 2.6 MeV. We note that two $10 \times 10 \times 32$ mm$^3$ detectors have the same volume as one $20 \times 20 \times 15$ mm$^3$ H3D detector with comparable spectroscopic performance. The measurements were taken at the temperature of ~23°C. The typical electronic noise, measured at ~3,000 V bias on the cathode using the test-pulse signals, was found to be in the range of 4–5 keV FWHM (relatively high due to the long wires used to connect the detector contacts), which is equivalent to 0.8% FWHM.

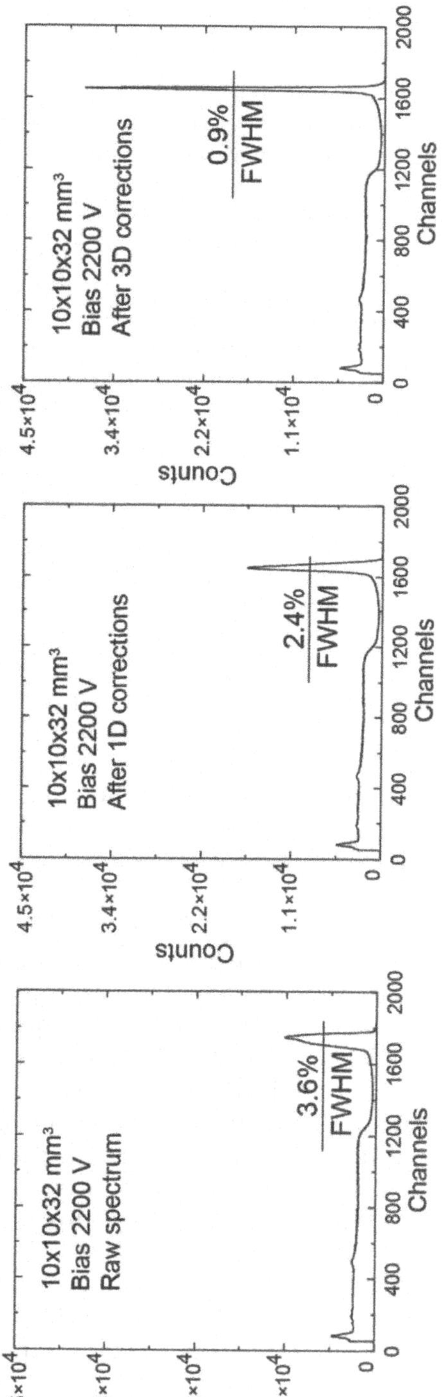


**FIGURE 5.9**
The $^{137}Cs$ spectra after 1D and after 3D corrections for a $10 \times 10 \times 32$ mm$^3$ detector biased at 2,200 V.

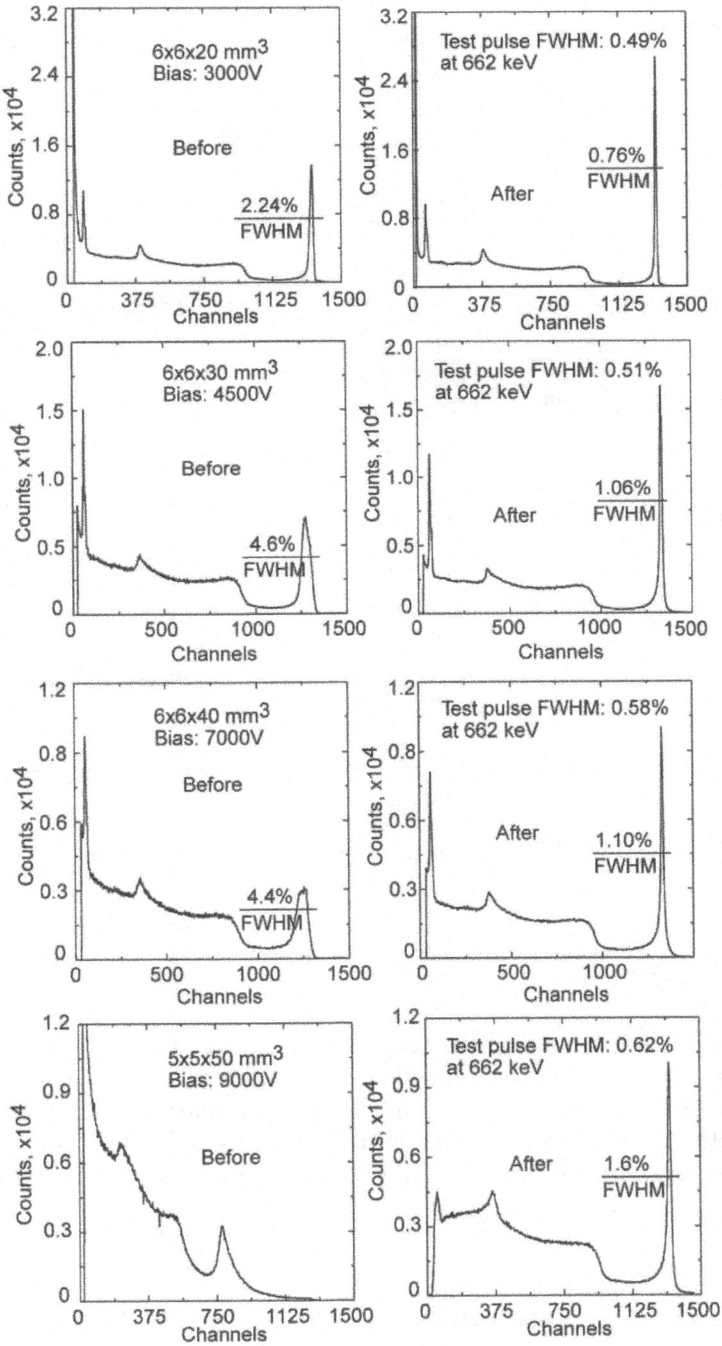


**FIGURE 5.10**
Representative pulse-height spectra and response maps measured for the best 20-, 30-, 40-, and 50-mm long detectors. The first and second columns show the all-event spectra before (first column from left) and after (second column from left) corrections.

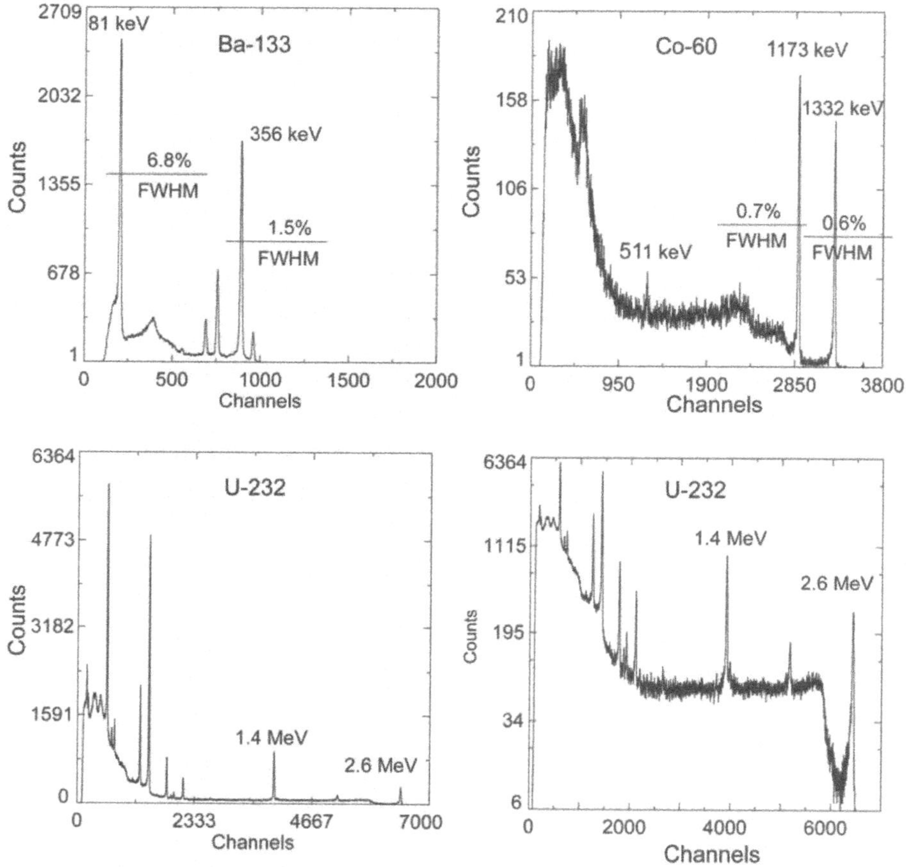


**FIGURE 5.11**
The $^{133}Ba$, $^{60}Co$, and $^{232}U$ spectra measured from the 8 × 8 × 32 mm³ (top) and 10 × 10 × 32 mm³ (bottom) detectors biased at 3,200 V at the temperature of ~23°C. The energy resolution (% FWHM) is 6.8% at 81 keV, 1.5% at 356 keV, and 0.5% at 2.6 MeV [35].

The main factor limiting the energy resolution of position-sensitive VFG detectors is electronic noise related to the leakage current, 5–20 nA, and the large anode capacitance, ~3 pF. As an example, the detector's energy resolution stays below 1% FWHM at 662 keV for temperatures up to 35°C. At room temperature, an electronic noise as low as 2.5 keV is possible for VFG detectors.

## 5.7 *X-Y* Position Resolution

The position-sensitive VFG detectors operate as a mini time-projection chambers, in which the amplitudes of the pads' signals are used to evaluate *X-Y* coordinates for each event, while the drift time or the *C/A* ratio is used to evaluate the Z coordinate, assuming no lateral movement of the electron cloud.

The signal amplitudes from the pairs orthogonal pads form four amplitude domains denoted as $\left(A_x^1, A_y^1\right)$, $\left(A_x^2, A_y^2\right)$, $\left(A_x^1, A_y^2\right)$, and $\left(A_x^2, A_y^1\right)$, as illustrated in Figure 5.3(b). We define the pad response functions as the dependences of the pad amplitudes on $(x,y)$ coordinates: $A_x^i = R_x^i(x,y)$ and $A_y^i = R_y^i(x,y)$, where $i = 1, 2$. Each point from the real space domain $(x,y)$ corresponds to the point $\left(A_x^i, A_y^i\right)$ from one of the amplitude domains, which means that four estimates can be obtained for X-Y coordinates using the signals readout from the orthogonal pads:

$$\begin{pmatrix} x \\ y \end{pmatrix} = \begin{pmatrix} R_x^{-1}\left(A_x^1, A_y^1\right) \\ R_y^{-1}\left(A_x^1, A_y^1\right) \end{pmatrix} \tag{5.2}$$

where $R_x^{-1}(A_x, A_y)$ and $R_y^{-1}(A_x, A_y)$ are the inverse response functions $R_x(x,y)$ and $R_y(x,y)$. However, only two estimates corresponding to the pads from the opposite corners are independent. We note that the possibility of using the signals from only two orthogonal pads is important when detectors are arranged in tightly packed arrays in which case the pads facing each other share the same readout input. Combing two uncorrelated estimates corresponding to pairs of pads on the opposite corners, we can write:

$$\begin{pmatrix} x \\ y \end{pmatrix} = \sum_{i=1}^{2} \begin{pmatrix} x_i \\ y_i \end{pmatrix} w_i \tag{5.3}$$

where $w_i$ are the weights. There are several ways for finding the weights; for example, one can choose $w_i$ to be proportional to the pad amplitudes. The discussed equations can also be applied when signals from one pair of pads are used. Using a linear approximation for the response functions, we arrive to the center of gravity formulas: $x' = \frac{A_x^1}{A_x^1 + A_x^2}$ and $y' = \frac{A_y^1}{A_y^1 + A_y^2}$. The center of gravity formula is a simple and robust estimator that works well for correcting the detector response nonuniformities but resulted in geometrical distortions near the detector edges. To avoid such distortions, one must use the actual response functions, which can either be experimentally measured or calculated. However, it is more practical to start with the center of gravity formula, as the first approximations, to calculate $x'$ and $y'$ and then apply a conformal transformation to convert $x'$ and $y'$ into the true $x$ and $y$ coordinates: $x = T_x(x', y')$ and $y = T_y(x', y')$. In practice, the transformation matrixes are evaluated during the calibration. We note that a set of matrixes can be generated for different Z segments to improve the accuracy.

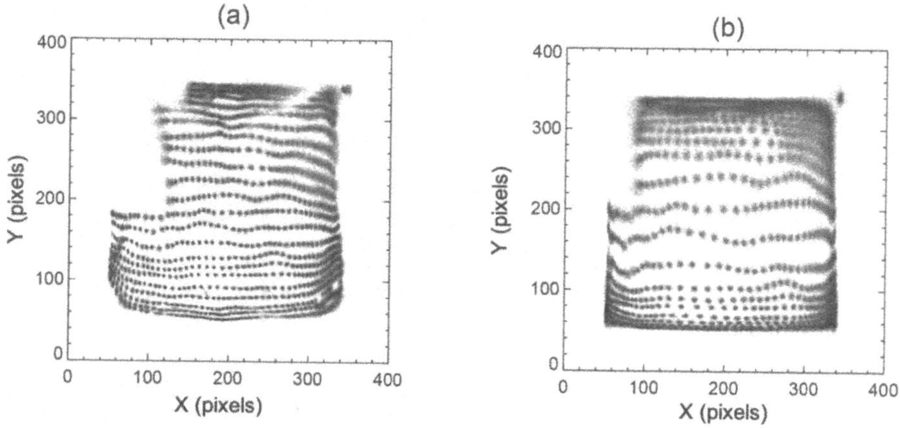


**FIGURE 5.12**
Response maps evaluated from the same raster scan using two signal processing approaches based on: the digitized waveforms (a) and the shaped signals (b). The scan consisted of 30 × 30 beam positions evenly distributed over the 6 × 6 mm² cathode area of the 20-mm-long detector. The missing portions in the response maps in the upper left corners correspond to the location of the metal strip used to apply the bias to the cathode.

To investigate position sensitivity of the VFG detectors, a 550-nm pulse laser beam focused down to a 20-μm diameter was employed to carry out raster scans over the detector cathode areas. Figure 5.12 represents a scan consisting of 30 × 30 beam positions evenly distributed over the 6 × 6 mm² cathode area of a 20-mm-long detector. For each location of the beam, we generated 200 pulses. Each pulse produced an electron cloud near the surface equivalent to the photon deposited energy of ~400 keV. The locations of the electron clouds in the response maps were calculated using the center-of-gravity formula. To emphasize the importance of using the synchronized pad signals, we show two maps from the same scan evaluated by two readout approaches: sampling the charge waveforms (a) and shaping the charge signals (b). In the first approach, we used the time correlated samples of the pad waveforms, while in the second approach, the peaks of the shaped signals were not time correlated. The missing portions in the response maps in the upper left corners correspond to the location of the metal strip used to apply the bias to the cathode.

The maps show the patterns of the dot clusters, each corresponding to a beam position in the raster scan. Each cluster consists of 200 dots whose distribution can be used to evaluate the local position resolutions, which were found to be < 50 μm FWHM at ~400 keV. Geometrical distortions, resulting from using the center-of-gravity formulas, are clearly seen in both response maps. In both cases, the measured beam positions are compressed as the beams move from the center of the detector toward the edges. However, the distortions are notably stronger in the case of shaped signals (b), because the peak amplitudes used are not time-correlated, meaning that they do not correspond to the same location of the electron cloud, which introduces additional systematic errors. The position resolutions in position-sensitive VFG detectors can be as good as 50 μm, which is the main reason why the 3D corrections work so well. However, the geometrical resolution is limited by the local nonuniformities of the electric field, which is a common problem for CZT detectors. The lines in Figure 5.12, which are supposed to be straight, are curved due to the local variation of the electric field. This effect may degrade the actual position resolution of CZT detectors up to > 0.5 mm depending on the crystal quality.

Figure 5.13 shows the results of the position resolution measurements for a 10 × 10 × 32 mm³ detector irradiated with an uncollimated $^{137}$Cs source [35]. In these measurements, a 0.8-mm-wide tungsten slit collimator was placed 65 mm above the detector cathode. The thickness of the collimator was 26 mm. The $^{137}$Cs source was placed ~25 mm above the collimator as shown in the figure. It shows the image of the slits for the single-point photo absorption events distributed over the entire thickness of the detector. The pixel size is 0.2 × 0.2 mm². The integrated event distribution with respect to the line representing the center of gravity of the image is shown at the bottom. Its FWHM was found to be 14 pixels or 2.8 mm. By comparing a Monte-Carlo simulated image for the same geometry with the measured one, we estimated the spatial resolution to be < 1 mm.

The position resolution in the Z-direction was evaluated by measuring coincidences between the signals of the CZT detector and a small $BaF_2$

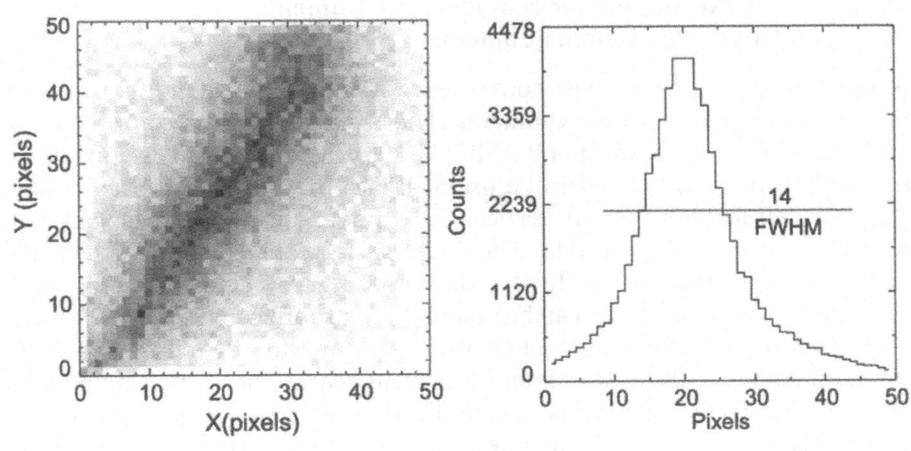


**FIGURE 5.13**
A schematic of the experimental setup used to evaluate the position resolution of one of the 10 × 10 × 32 mm³ detectors biased at 3,700 V (left); an image of the slits after combining all the single point photo absorption events over the entire length of the 32-mm-thick detector (top-right); an event distribution with respect to a line representing the center of gravity of the image (bottom-right). The FWHM is 14 pixels or 2.8 mm. The pixel size is 0.2 mm [35].

scintillator, generated with two back-to-back 511-keV photons emitted from the $^{68}Ge$ source placed between the detectors. Fast signals generated in the $BaF_2$ detector provide triggers for the incident photons, while the leading edges of the digitized cathode and anode signals in the CZT detector were used to identify the photon interaction and electron cloud arrival times. The time resolution was found to be ~10 ns, which corresponds to less than 0.5-mm spatial resolution in the Z-direction.

## 5.8 Arrays of Position-Sensitive VFG Detectors

In this section, we describe two examples of using position-sensitive VFG detectors in practical instruments. Arrays of VFG bars with thicknesses up to 4 cm offer an economical way to produce large area, high detection efficiency, and high energy resolution gamma cameras. With the capability to correct response nonuniformity caused by crystal defects, they allow standard grade (unselected) CZT crystals to be used, which reduces instrument cost. The arrays provide performance and functionality comparable to 3D pixelated detectors but at a lesser cost. The potential applications of the arrays, which can be configured into detection planes with different geometrical form factors and dimensions, include nonproliferation, safeguards, gamma-ray astronomy, and other areas that require spectroscopy and imaging of gamma-ray sources over a wide dynamic range, from ~10 keV up to several MeV.

### 5.8.1 4 × 4 Array Module for Handhelds Instruments and Large Area Gamma Cameras

A modular approach is most convenient for integrating handheld instruments and large-area gamma cameras. The basic module is a compact 4 × 4 array coupled with the front-end ASIC [34, 39]. The mechanical design of our first prototype is illustrated in Figure 5.14. Each detector is inserted inside a square cell formed by four vertical CuBe finger-spring contacts soldered directly to the detector board (a). The finger-spring contacts connect the pads to the ASIC readout inputs. In this design, the gaps between the physical edges of the detector crystal can be potentially minimized below 0.5–0.7 mm. To reduce the number of readout channels and simplify the design, we connected the pads of the adjacent detectors facing each other to the same readout channel (b). This may cause a signal ambiguity when interactions occur in two adjacent detectors. In such cases, we cannot use signals captured from the common pads and instead must rely on signals from the other two or three available pads to evaluate the positions of the interaction points. The signals from just two orthogonal pads are sufficient to measure the coordinates; this is further explained in Section 5.7. The anode board is bolted to a 5-mm aluminum frame housing, and the detectors are attached on top with a 15-mm-thick insulating frame made from G10 fiberglass (c). After inserting

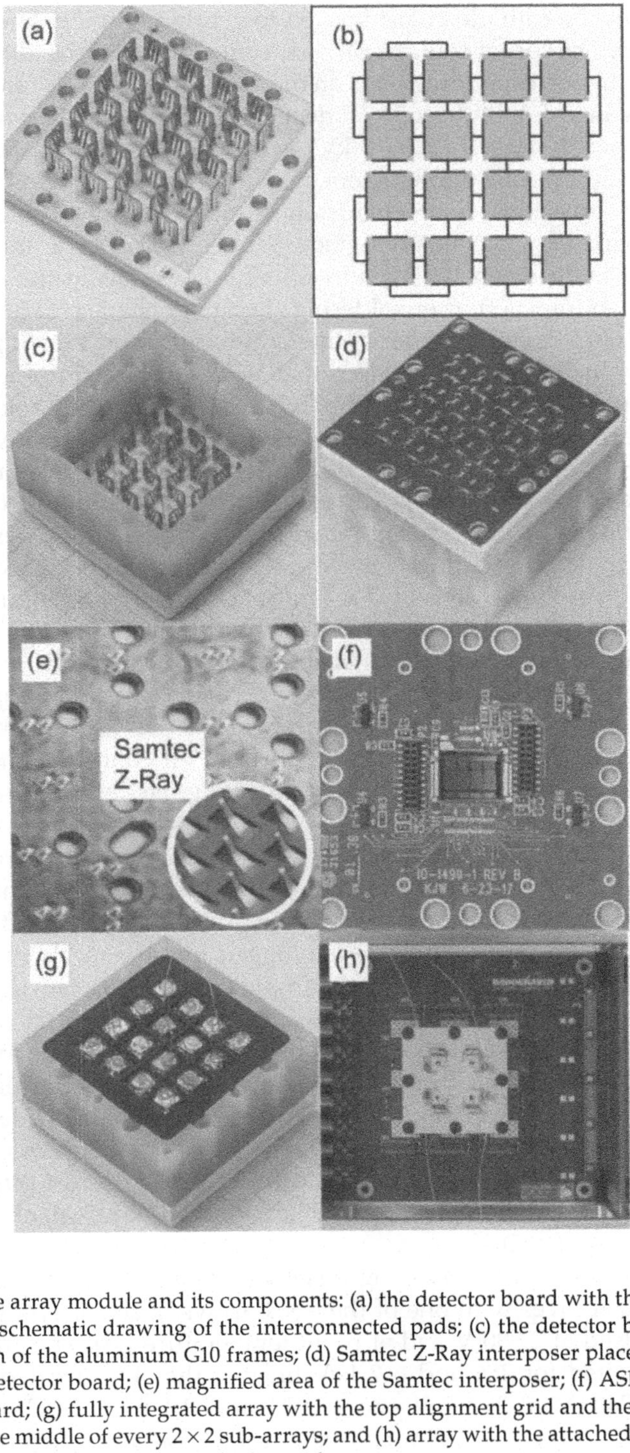


**FIGURE 5.14**
Design of the array module and its components: (a) the detector board with the CuBe finger-contacts; (b) schematic drawing of the interconnected pads; (c) the detector board bolted to the sandwich of the aluminum G10 frames; (d) Samtec Z-Ray interposer placed on the back-side of the detector board; (e) magnified area of the Samtec interposer; (f) ASIC mounted on the ASIC board; (g) fully integrated array with the top alignment grid and the cathode wires passing in the middle of every 2 × 2 sub-arrays; and (h) array with the attached cathode board plugged into the motherboard inside the test box.

the detectors, the cathode board is used to gently press the detectors to the anode board. The spring contacts provide connections for the anodes and the cathodes to the corresponding pads on the anode and cathode fan out boards. The cathode board, with four 50 pF decoupling capacitors and four 100 MΩ resistors (typically biased to 2.5–3 kV), is bolted to the top of the insulating frame with eight plastic screws to ensure a uniform distribution of the load.

The ASIC board (f) is mechanically attached to the backside of the detector board (a) via a custom-made Samtec Z-Ray type interposer (e). The adapter is designed with a number of holes to avoid the interference of the solder bumps left after soldering the finger contacts. The anode and ASIC boards are bolted together to the aluminum frame housing. The cathode board, which is typically biased to 2.5–3 kV, is bolted to the insulating frame with plastic screws. Thin cathode wires (f) are directly soldered to the detector board and inserted between each group of four detectors and then through the four holes of the cathode board outside the array housing. They are soldered to the decoupling capacitors mounted on top of the cathode board as a final step of the array integration. The discussed design allows us to easily replace individual detectors if necessary and can accommodate detectors with cross-sections up $7 \times 7$ mm$^2$ and up 40-mm lengths.

The backside of the ASIC board has two multi-pin connectors and can be plugged into the motherboard (h) of the readout system. The motherboard carries the ultra-stable low-voltage passive converters supplying power to the ASIC chips, two analog-to-digital converters (ADCs) for digitizing the peak amplitudes and timing information from all channels, and the Field Programmable Gate Array (FPGA) for processing the data and communicating with the ASICs and the USB port.

For testing the arrays, we acquired $6 \times 6 \times 20$ mm$^3$ standard detector-grade (unselected) CZT crystals from Redlen Technologies and Kromek. The only requirement specified was a maximum leakage current of < 25 nA at 3,000 V and 25°C. The arrays were plugged into the motherboard of the readout system and enclosed inside an aluminum box. During the measurements, we placed the test box inside an environmental chamber to maintain the detectors at a specified constant temperature during the measurements. We used uncollimated $^{137}$Cs and $^{232}$U sources normally placed 2–3 cm above the array cathode board to generate signals inside the detectors. The cathode bias was set at 2,500–3,000 V. The detectors' electronic noise, measured at high voltage on the cathode using the test-pulse signals, was found to be in the range of 2.5–3.5 keV, which is equivalent to 0.5–0.6% FWHM at 662 keV at 26°C.

Figure 5.15 shows the pulse-height spectra measured from $^{232}$U and $^{137}$Cs uncollimated sources with one of the $4 \times 4$ arrays. It shows spectra from 16 detectors biased at 2,800 V and at room temperature [39]. The energy resolutions evaluated for 286 and 662 keV lines of the 3D corrected spectrum were 1.8% FWHM at 200 keV and < 0.9% at 662 keV at 23°C. In the case of just 1D correction, we obtained 2.8% and 2.5%, respectively. The raw

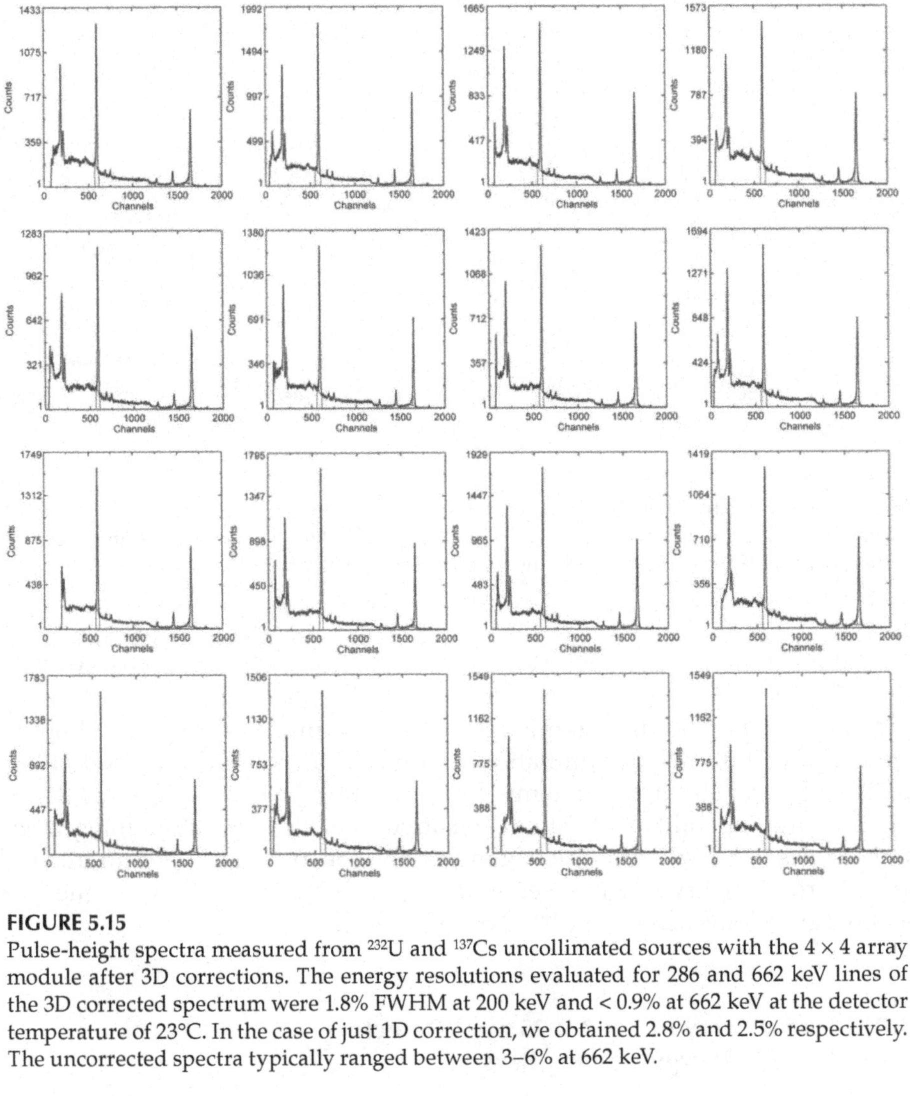


**FIGURE 5.15**
Pulse-height spectra measured from $^{232}U$ and $^{137}Cs$ uncollimated sources with the 4 × 4 array module after 3D corrections. The energy resolutions evaluated for 286 and 662 keV lines of the 3D corrected spectrum were 1.8% FWHM at 200 keV and < 0.9% at 662 keV at the detector temperature of 23°C. In the case of just 1D correction, we obtained 2.8% and 2.5% respectively. The uncorrected spectra typically ranged between 3–6% at 662 keV.

(uncorrected) spectra (not shown) typically resulted in 3–6% energy resolution range at 662 keV. The main factor limiting the energy resolution of the VFG detectors, in comparison to H3D pixelated detectors, is electronic noise primarily related to the high leakage current, 5–25 nA, and large anode capacitance, ~3 pF, on the anodes.

The sensitivity of the array was measured using a standard 9.6 μCi $^{137}Cs$ source placed at several distances from the detector's cathode board, 1.0, 1.3, 1.6, and 2.0 meters in the lab environment. The averaged minimum time to detect the 10 μCi source at 1 minute was found to be (8 +/−1) s at 3 sigma (or

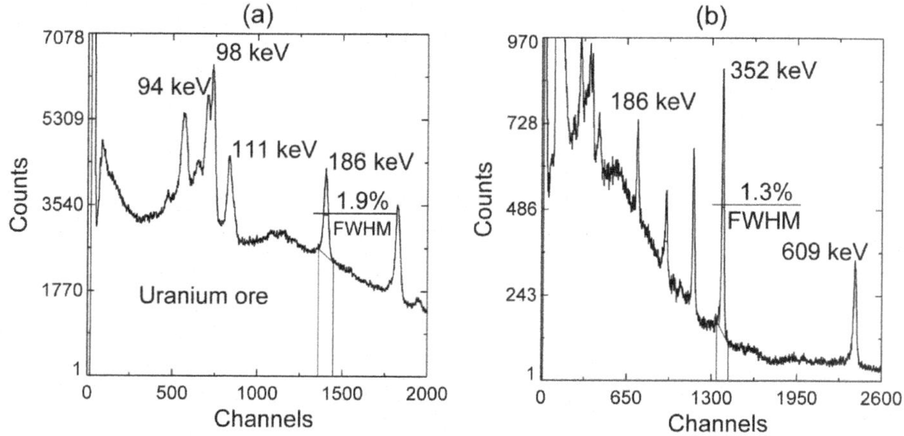


**FIGURE 5.16**
Spectra measured from ~1 kg of U ore placed ~10 cm from the top surface of the array: (a) gain is 120 fC/mV, accumulation time is 7 h; (b) gain is 60 fC/mV, accumulation time is 25 min. The cathode bias is 2800 V, and the detector temperature is 23 C.

32 s at 6 sigma) confidence level. Figure 5.16 shows the combined spectra measured from ~1 kg of U ore bag placed ~10 cm from the top surface of the array: (a) preamplifier gainis 120 fC/mV, accumulation time is 7 hours; (b) gain is 60 fC/mV, accumulation time is 25 minutes. The cathode bias is 2,800 V, and the detector temperature is 23°C [39]. These results demonstrate the capability of position-sensitive VFG CZT detectors integrated into arrays for high resolution gamma-ray spectroscopic measurements over a broad energy range. After applying the 3D corrections, we achieved an energy resolution below 0.9% FWHM at 662 keV.

### 5.8.2 Linear Array of 5 × 7 × 25 mm³ Detectors for Low Energy Gamma Rays

A linear array consisting of six position-sensitive VFG detectors fabricated from 5 × 7 × 25 mm³CZT crystals is shown in Figure 5.17 [51]. The detectors have a simple design. The geometry of the array was optimized to fit inside a future handheld detector proposed for uranium enrichment measurements. During prototype testing, the detectors were placed vertically on the detector board and gently pressed from the top using the cathode board holding the decoupling capacitors and resistors. The anode spring contacts touch the designated anode pads on the board, while the charge-sensing pads were soldered to the board contacts. The signals generated by the incident photons on the anodes, cathodes and four position-sensing pads were routed to the corresponding front AVG1 ASIC inputs. Decoupling

**FIGURE 5.17**
Frontal view of the linear array with one side of pads connected to the PCB board. The detector board, with the detectors on the front and the multi-pin connectors on the opposite side, was plugged into the motherboard inside the test box.

circuitries are required for reading the signals from cathodes, which were biased at 2,500–3,000 V.

Figure 5.18 shows three rows of the spectra: after applying 3D corrections (top), 1D correction (middle), and the raw data (bottom) measured from a fresh fuel rod, ~93% $^{235}U$, located ~40 cm from the detector plane. The measurements were taken at a temperature of ~19°C. The same cathode bias of 2,750 V was applied to all detectors. The energy resolution was found to be 1.9% FWHM at 186 keV, which is very good for such big detectors with an area of $5 \times 7$ mm$^2$. If we only apply a 1D correction, the energy resolution is only ~3%.

This design was proposed to minimize the number of readout channels while retaining the detector's large effective area. With this in mind, we investigated the possibility of interconnecting the corresponding pads from all the detectors, i.e., only four readout channels are used to read signals from all pads. In this case, we will need only 10 channels to read signals from four pads and six anodes. To test this approach, we took measurements of spectra from a $^{232}U$ source. We also connected their cathodes together without increasing the electronic noise. We note that the cathode signals are needed to evaluate the *Z* coordinates of the interaction points. However, if the waveform sampling readout is used, then one can substitute the cathode signals with the sum of the pad signals. In this case no readout channels will be required at all. Thus, the total number of channels can be reduced to 10 for the 10 cm$^2$ area detector. This will also simplify the detector

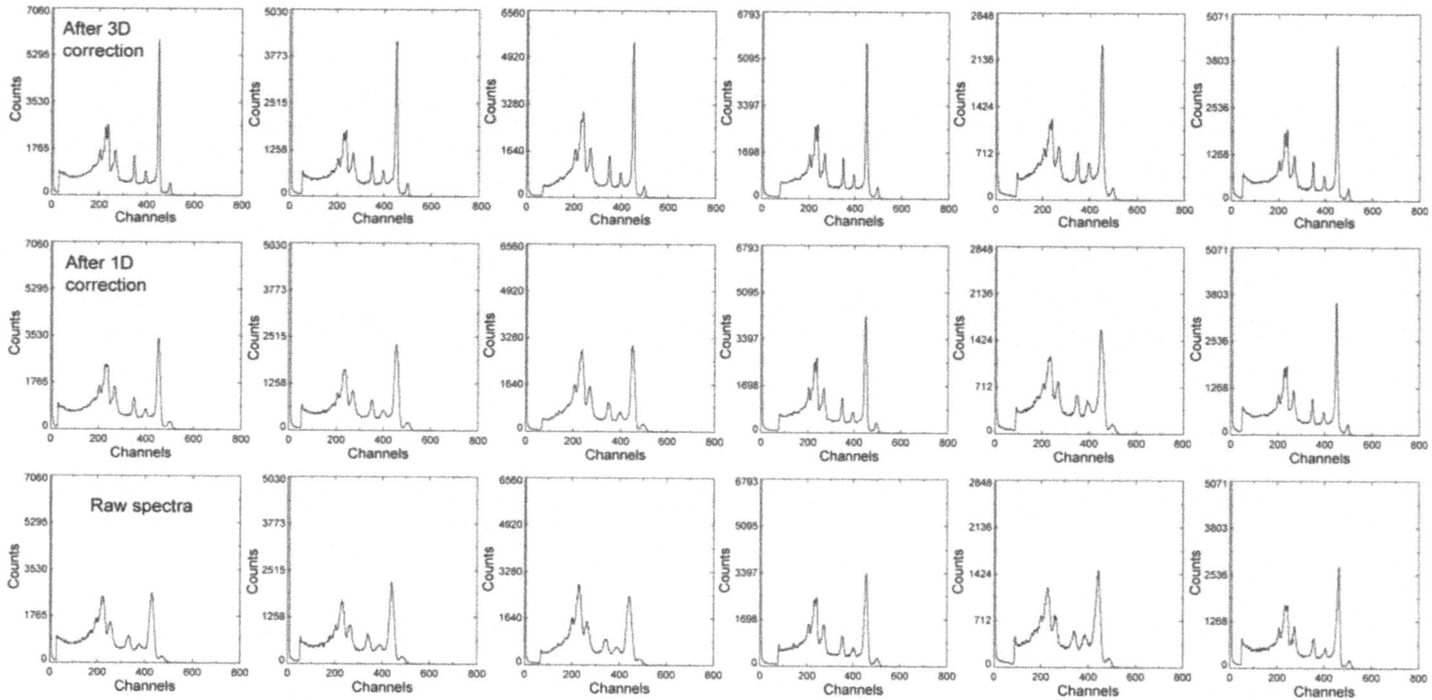


**FIGURE 5.18**

The pulse-height spectra measured from a fresh fuel rod, ~93% $^{235}U$, located ~40 cm from the detector plane at 19°C and corrected using calibration data measured at the same temperature.

design by eliminating decoupling capacitors and the cathode signals wires. Furthermore, one can avoid using an ASIC by replacing it with hybrid preamplifiers and a 10-channel digitizer.

## References

1. G. F. Knoll, *Radiation Detection and Measurement*, 3rd Ed. (Wiley, New York, 2000).
2. O. Frisch, British Atomic Energy Report, BR-149 (1944).
3. NASA Swift Mission, http://www.nasa.gov/mission_pages/swift/main/index.html.
4. Z. He, W. Li, G. F. Knoll, D. K. Wehe, J. Berry, C. M. Stahle, "3-D position sensitive CdZnTe gamma-ray spectrometers", *Nucl. Instr. and Meth.* A. 422, pp. 173–178, 1999.
5. P. N. Luke, "Single-polarity charge sensing in ionization detectors using coplanar electrodes", *Appl. Phys. Lett.* 65(22), pp. 2884–2886, 1995.
6. A. Kargar, E. Ariesanti, D. S. McGregor, "Charge collection efficiency characterization of a $HgI_2$ Frisch collar spectrometer with collimated high energy gamma rays", *Nucl. Instrum. and Meth.* A. 652, pp. 186–192, 2011.
7. A. E. Bolotnikov, J. Baker, R. DeVito, J. Sandoval, L. Szurbart, "HgI2 detector with a virtual Frisch ring", *IEEE Trans. Nucl. Sci.* 52, pp. 468–472, 2005.
8. D. S. McGregor, R. A. Rojeski, *U.S. Patent*. 6, 175, 120 (16 January 2001).
9. G. Montemont, M. Arques, L. Verger, J. Rustique, "A capacitive Frisch grid structure for CdZnTe detectors", *IEEE Trans. Nucl. Sci.* pp. 278–281, 2001.
10. A. Kargar, J. Christian, H. Kim, L. Cirignano, M. Squillante, M. S. Squillante, E. Weststrate, A. Bolotnikov, G. Carini, A. Dellapenna, J. Fried, M. B. Smith, E. M. Johnston, M. R. Koslowsky, A. L. Miller, K. S. Shah, "Design of Capacitive Frisch Grid TlBr Detectors for Radionuclide Identification", Nuclear Science Symposium and Medical Imaging Conference (NSS/MIC), 2019 IEEE, Pages: 1–5.
11. V. V. Dmitrenko, A. C. Romanuk, Z. M. Uteshev, V. K. Chernyatin, "Spectrometric applications of an ionization-type drift chamber", *Pribory I Tekhnika Eksperimenta*. 1, pp. 51–53, 1982.
12. A. E. Bolotnikov, R. Austin, A. I. Bolozdynya, J. D. Richards, "Virtual Frisch-grid ionization chambers filled with high-pressure Xe", Proc. SPIE 5540, Hard X-Ray and Gamma-Ray Detector Physics VI (21 October 2004), pp. 216–224.
13. R. Austin, "Gamma ray detection with a 3×3 virtual Frisch grid array", Nuclear Science Symposium Conference Record (NSS/MIC), 2008 IEEE, Pages: 887–893.
14. K. Parnham, J. B. Glick, Cs. Szeles, K. G. Lynn, "Performance improvement of CdZnTe detectors using modified two-terminal electrode geometry", *Journal of Crystal Growth*. 214–215, pp. 1152–1154, 2000.
15. Cs. Szeles, D. Bale, J. Grosholz, Jr., G. L. Smith, M. Blostein, and J. Eger, "Fabrication of high performance CdZnTe quasi-hemispherical gamma-ray CAPture™ plus detectors", Hard X-Ray and Gamma-Ray Detector Physics VIII,

edited by Larry A. Franks, Arnold Burger, and Ralph B. James, *Proceedings of SPIE*, Vol. 6319 (SPIE, Bellingham, WA, 2006).
16. C. L. Lingren, J. F. Butler, B. Apotovsky, R. L. Conwell, F. P. Doty, S. J. Friesenhahn, "Semiconductor radiation detector with enhanced charge collection", *U.S. Pat.* 5, 677, 539, 1997.
17. K. Parnham, C. Szeles, K. G. Lynn, and R. Tjossem, "Performance improvement of CdZnTe detectors using modified two-terminal electrode geometry", in *Hard X-ray, Gamma-Ray and Neutron Detector Physics, Proceedings of SPIE*. 3786, pp. 49–54, 1999.
18. D. Bale and Cs. Szeles, "Design of high performance CdZnTe quasi-hemispherical gamma-ray CAPture (TM) plus detectors", Hard X-Ray and Gamma-Ray Detector Physics VIII, edited by Larry A. Franks, Arnold Burger, and Ralph B. James, *Proceedings of SPIE*, Vol. 6319 (SPIE, Bellingham, WA, 2006).
19. P. Dorogov, V. Ivanov, A. Loutchanski, L. Grigorjeva, D. Miller, "Improving the performance of quasi-hemispherical CdZnTe detectors using infrared stimulation", *IEEE Trans. Nucl. Sci.* 59(5), pp. 2375–2382, 2012.
20. D. S McGregor, Z. He, H. A. Seifert, D. K. Wehe, R. A. Rojeski, "Single charge carrier type sensing with a parallel strip pseudo-Frisch-grid CdZnTe semiconductor radiation detector", *Appl. Phys. Lett.* 72, pp. 792–794, 1998.
21. D. S. McGregor, R. A. Rojeski, "Performance of geometrically weighted semiconductor Frisch grid radiation spectrometers", *IEEE Trans. Nuclear Science.* 46, pp. 250–259, 1999.
22. W. J. McNeil, D. S. McGregor, A. E. Bolotnikov, G. W. Wright, R. B. James, "Single-charge-carrier-type sensing with an insulated Frisch ring CdZnTe semiconductor radiation detector", *Appl. Phys. Lett.* 84, pp. 1988–1990, 2004.
23. A. Kargar, A. M. Jones, W. J. McNeil, D. S. McGregor, "CdZnTe Frisch ring detectors for low-energy gamma ray spectroscopy", *Nuclear Instruments and Methods.* A558, pp. 487–503, 2006.
24. A. E. Bolotnikov, G. S. Camarda, G. A. Carini, M. Fiederle, L. Li, D. S. McGregor, W. McNeil, G. W. Wright, R. B. James, "Performance characteristics of Frisch-ring CdZnTe detectors", *IEEE Trans. Nucl. Sci.* 53(2), pp. 607–614, 2006.
25. J. K. Polack, M. Hirt, J. Sturgess, N. D. Sferrazza, A. E. Bolotnikov, S. Babalola, G. S. Camarda, Y. Cui, S. U. Egarievwe, P. M. Fochuk, R. Gul, A. Hossain, K. Kim, O. V. Kopach, L. Marchini, G. Yang, L. Xu, R. B. James, "Variation of electric shielding on virtual Frisch-grid detectors", *Nucl. Instr. and Meth.* A. 621, pp. 424–430, 2010.
26. A. E. Bolotnikov, K. Ackley, G. S. Camarda, C. Cherches, Y. Cui, G. De Geronimo, J. Fried, D. Hodges, A. Hossain, W. Lee, G. Mahler, M. Maritato, M. Petryk, U. Roy, C. Salwen, E. Vernon, G. Yang, R. B. James, "An array of virtual Frisch-grid CdZnTe detectors and a front-end application-specific integrated circuit for large-area position-sensitive gamma-ray cameras", *Review of Scientific Instruments.* 86(7), 013114–013124, 2015.
27. D. R. Nygren, "A time projection chamber", presented at 1975 PEP Summer Study, PEP 198, 1975 and included in Proceedings.
28. J. D. Eskin, H. H. Barrett, H. B. Barber, "Signals induced in semiconductor gamma-ray imaging detectors", *J. Appl. Phys.* 85, pp. 647–659, 1999.
29. V. Radeka, "Low noise techniques in detectors", *Ann. Rev. Nucl. Part. Sci.* 38, pp. 217–277, 1988.

30. S. Ramo, "Currents induced by electron motion", *Proc. IRE.* 27, pp. 584–585, 1938.
31. W. Shockley, "Currents to conductors induced by a moving point charge", *J. of Appl. Phys.* 9(10), p. 635, 1938.
32. O. Buneman, T. E. Cranshaw, J. A. Harvey, "Design of grid ionization chambers", *Can. J. Res.* A27, pp. 191–206, 1949.
33. A. E. Bolotnikov, J. Butcher, G. S. Camarda, Y. Cui, G. De Geronimo, J. Fried, P. M. Fochuk, A. Hossain, K. H. Kim, O. V. Kopach, G. Mahler, M. Marshall, B. McCall, M. Petryk, E. Vernon, Ge Yang, R. B. James, "Design considerations and testing of virtual Frisch-grid CdZnTe detector arrays using the H3D ASIC", *IEEE Trans. Nucl. Sci.* 60(4), pp. 2875–2882, 2013.
34. L. A. Ocampo Giraldo, A. E. Bolotnikov, G. S. Camarda, S. Cheng, G. De Geronimo, A. McGilloway, J. Fried, D. Hodges, A. Hossain, K. Ünlü, M. Petryk, E. Vernon, V. Vidal, G. Yang, R. B. James, "Arrays of position-sensitive virtual Frisch-grid CdZnTe detectors: Results from a 4×4 array prototype", *IEEE Trans. Nucl. Sci.* 64(10), pp. 2698–2705, 2017.
35. A. E. Bolotnikov, J. MacKenzie, E. Chen, F. J. Kumar, S. Taherion, G. Carini, G. De Geronimo, J. Fried, Kihyun Kim L. Ocampo Girado, E. Vernon, R. B. James, "Performance of 8×8×32 and 10×10×32 $mm^3$ CdZnTe position-sensitive virtual Frisch-grid detectors for high-energy gamma-ray cameras", *Nucl. Instr. Meth.* A 969, p. 164005, 2020.
36. A. E. Bolotnikov, N. M. Abdul-Jabbar, S. Babalola, G. S. Camarda, Y. Cui, A. Hossain, E. Jackson, H. Jackson, J. R. James, A. L. Luryi, and R. B. James, "Optimization of virtual Frisch-grid CdZnTe detector designs for imaging and spectroscopy of gamma rays", in *Proceedings of SPIE Hard X-Ray and Gamma-Ray Detector Physics IX*, Vol. 6702, edited by R. B. James, A. Burger, and L. A. Franks, (SPIE, San Diego, CA, 2007), pp. 670603-1-14.
37. A. E. Bolotnikov, S. Babalola, G. S. Camarda, Y. Cui, S. U. Egarievwe P. M. Fochuk, R. Hawrami, A. Hossain, J. R. James, I. J. Nakonechnyj, Ge Yang, and R. B. James, "Spectral responses of virtual Frisch-grid CdZnTe detectors and their relation to IR microscopy and X-ray diffraction topography data", in *SPIE Conference on Hard X-Ray, Gamma-Ray and Neutron Detector Physics X*, Vol. 7079, edited by A. Burger, L. Franks, and R. B. James (SPIE, Bellingham, WA, 2008), pp. 707903–707912.
38. A. Kargar, A. M. Jones, W. J. McNeil, M. J. Harrison, and D. S. McGregor, "Angular response of a W-collimated room temperature-operated CdZnTe Frisch collar spectrometer", *Nuclear Instrum. and Meth.*, A562, pp. 262–271, 2006.
39. A. E. Bolotnikov, G. S. Camarda, G. De Geronimo, J. Fried, R. B. James, "A 4 × 4 array module of position-sensitive virtual Frisch-grid CdZnTe detectors for gamma-ray imaging spectrometers", *Nuclear Instruments and Methods in Physics Research Section A: Accelerators, Spectrometers, Detectors and Associated Equipment.* 954, p. 161036, 2020.
40. E. Vernon, G. De Geronimo, A. Bolotnikov, M. Stanacevic, J. Fried, L. Ocampo Giraldo, G. Smith, K. Wolniewicz, K. Ackley, C. Salwen, J. Triolo, D. Pinelli, K. Luong, "Front-end ASIC for spectroscopic readout of virtual Frisch-grid CZT bar sensors", *Nucl. Instr. and Meth.* A940, pp. 1–11, 2019.
41. A. E. Bolotnikov, G. S. Camarda, G. A. Carini, M. Fiederle, L. Li, G. W. Wright, and R. B. James, "Performance studies of CdZnTe detector by using a pulse-shape

analysis", in *Proc. SPIE, Hard X-Ray and Gamma-Ray Detector Physics VII* (SPIE, Bellingham, WA, 2005), pp. 59200K-1–59200K-12.
42. G. De Geronimo, E. Vernon, K. Ackley, A. Dragone, J. Fried, P. O'Connor, Z. He, C. Herman, F. Zhang, "Readout ASIC for 3D position-sensitive detectors", *IEEE Trans. Nucl. Sci.* 55(3), pp. 1593–1603, 2008.
43. E. Vernon, K. Ackley, G. De Geronimo, J. Fried, P. O'Connor, Z. He, C. Herman, F. Zhang, "ASIC for high rate 3-D position sensitive detectors", *IEEE Trans. Nucl. Sci.* 57(3), pp. 1536–1542, 2010.
44. F. Zhang, C. Herman, Z. He, G. De Geronimo, E. Vernon, J. Fried, "Characterization of the H3D ASIC readout system and 6.0 cm3 3-D position sensitive CdZnTe detectors", *IEEE Trans. Nucl. Sci.* 59, pp. 236–242, 2012.
45. V. Radeka, "Signal processing for particle detectors", in *Elementary Particles, Subvolume B: Detectors for Particles and Radiation*, edited by H. Schopper and C. Fabjan (Springer-Verlag Berlin Heidelberg, 2011).
46. A. E. Bolotnikov, C. M. Hubert Chen, W. R. Cook, F. A. Harrison, I.Kuvvetli, S. M. Schindler, "Effects of bulk and surface conductivity on the performance of CdZnTe pixel detectors", *IEEE Trans. Nucl. Sci.*, 49(4), pp. 1941–1949, 2002.
47. K. Hecht, "Zum mechanismus des lichtelektrischen Primärstromes in isolierenden kristallen", *Z. Phys.* 77, pp. 235–245, 1932.
48. Z. He, G. F. Knoll, D. K. Wehe, "Direct measurement of product of the electron mobility and mean free drift time of CdZnTe semiconductors using position sensitive single polarity charge sensing detectors", *J. Appl. Phys.* 84, pp. 5566–5569, 1998.
49. A. E. Bolotnikov, G. S. Camarda, E. Chen, R. Gul, V. Dedic, G. Geronimo, J. Fried, A. Hossain, J. MacKenzie, L. Ocampo, P. Sellin, S. Taherion, E. Vernon, Ge Yang, Uri El-hanany, R. B. James, "Use of the drift-time method to measure the electron lifetime in long-drift-length CdZnTe detectors", *J. of Appl. Phys.* 120, p. 104507, 2016.
50. A. E. Bolotnikov, G. S. Camarda, E. Chen, S. Cheng, Y. Cui, R. Gul, R. Gallagher, V. Dedic, G. De Geronimo, L. Ocampo Giraldo, J. Fried, A. Hossain, J. M. MacKenzie, P. Sellin, S. Taherion, E. Vernon, G. Yang, U. El-Hanany, R. B. James, "CdZnTe position-sensitive drift detectors with thicknesses up to 5 cm", *Appl. Phys. Lett.* 108, p. 093504, 2016.
51. L. Ocampo Giraldo, A. E. Bolotnikov, G. S. Camarda, G. De Geronimo, R. B. James, "A linear array of position-sensitive virtual Frisch-grid CdZnTe for low-energy gamma rays", *Nuclear Instruments and Methods in Physics Research Section A: Accelerators, Spectrometers, Detectors and Associated Equipment.* 903, pp. 204–214, 2018.

# 6

# *Detector Design for Surgical Guidance*

**Muhammed Emin Bedir and Bruce R. Thomadsen**

**CONTENTS**



DOI: 10.1201/9781003219446-6

## 6.1 Background and Setting

### 6.1.1 Conventional Breast Surgery Localization and Problems

Surgery to remove breast cancer is performed about 1,000 times a day in the United States. Most of these surgeries aim to remove the tumor with some margin, leaving the remaining breast tissue in place, in a procedure called a tylectomy, or in the vernacular, a lumpectomy. In many of these cases, the tumor presents as non-palpable, meaning that for a physical exam and during surgery, the physician cannot feel the difference between the tumor and the rest of the breast tissue. During the operation, the tumor also looks like the surrounding normal tissue. This makes identification of the boundary of the tumor, and where to cut to remove all the cancer, difficult to assess. The result of this ambiguity is that sometimes the margins of the surgery indicate that tumor was left behind requiring additional surgery. In an observational study of breast surgery in 2,206 patients between 2003–2008, 23% needed one re-excision, 2% received a second, and 0.3% only achieved clear margins with a third procedure [1]. Applying those values to the number of procedures performed in the United States alone means that each year this problem affects around 60,000 patients.

While the tumors cannot be delineated during surgery, they can be detected through mammography, ultrasound, or magnetic-resonance imaging. Three methods are in use to use the imaging information to guide the surgery. Hayes presents a good review of all these localization methods [2].

#### *6.1.1.1 Wire-Guided Localization*

During wire-guided localization (WGL), a wire is placed in the center of the tumor as estimated on the mammogram or ultrasound image. The wire is held in place in the patient as she goes from the imaging to surgery. This method presents several shortcomings:

- Coordination complexity. Imaging has to take place immediately before the procedure, requiring coordination between departments and opening the procedure to delays when either department has scheduling problems.
- Surgical approach. The surgical incision mostly needs to follow the wire. This may not be the approach that the surgeon would prefer.
- Patient comfort. During the time the wire is in place between imaging and surgery, the wire is not comfortable for the patient.
- Tumor definition. The wire only indicates the center of the tumor, giving the surgeon no guidance on the location of the boundary of the cancer.

#### 6.1.1.2 Radioactive Seed Localization

An improvement in localization methodology came with radioactive seed localization (RSL). With this technique, during the pre-procedure imaging, a radioactive seed is placed at the center of the tumor. The radioactive seed is the same type as used for brachytherapy implants, most typically for prostate cancers. The seeds are titanium capsules, approximately 5 mm long and 0.8 mm in diameter, containing about 3.5–5.6 MBq of $^{125}$I. In the operating room, the seed is located using a hand-held radiation detector to find the maximum signal at each of several orientations. The RSL eliminates some, but not all, of the problems with WGL.

- Coordination complexity. The imaging not long needs to be immediately before the surgery and can be several days ahead, simplifying coordination.
- Surgical approach. The surgeon is free to use any approach that leads to the center of the tumor as determined by measurement.
- Patient comfort. While inserting the seed can cause slight pain, it is usually performed with local anesthesia. Once inserted, it usually is not felt.
- Tumor definition. RSL, just as with WGL, only indicates the center of the tumor, giving the surgeon no guidance on the location of the boundary of the cancer. Some practitioners will implant a few additional seeds to mark particular edges of the tumor, but due to the radioactive, the number of seed that can be used is limited, and with that so also is the information on the boundary of the tumor.

#### 6.1.1.3 Beacons

Beacons are passive implants that respond to some stimulus signal. Three examples have found their way into breast-tumor localization. Each beacon is inserted into the center of the tumor similarly to the radioactive seeds. One shows under infrared light shined on the patient, although deeper than 4 cm in tissue the beacon may not be visible. Another contains iron particles that become magnetic through induction by an alternating magnetic field. As with the infrared system, beacons deeper than 4 cm in tissue may not provide an adequate signal for detection. The last system uses radiofrequency identification (RFID) tags as beacons, the implanted tag responding to a signal from a radiofrequency antenna. The RFID beacons can respond to depths of 6 cm. Compared with WGL or RSL:

- Coordination complexity, surgical approach, and patient comfort. Each of these facets is the same as with RSL.

- Tumor definition. Each of the systems has been used with multiples beacons to mark some points on the boundary of the tumor as well as the center. Regardless of the system, there are practical limits on how many beacons can be implanted. The use of multiple beacons seems to reduce the re-resection rate, but only to 15%.

#### 6.1.1.4 Volumetric Localization

An approach to provide information on the boundary of a tumor can use a radiopharmaceutical that localizes in breast cancers. The first report of this approach was by Audisio et al. in 2005 using $^{99m}$Tc-labeled colloidal human serum albumin [3]. The results of their study still had a re-excision rate of 24%. A possible cause for the failure to improve the complete removal rate may have been the failure of the colloidal albumin actually to attach to the tumor, rather than just remain in the blood pool. An example of a radiopharmaceutical that labels tumor cells is CLR1404 (Cellectar Biosciences Inc, Madison, WI), which can be labeled with iodine radionuclides [4]. While breast imaging could use $^{124}$I for positron-emission tomography or $^{131}$I for single-photon emission computerized tomography (SPECT), both present radiation hazards in the operating room because of their high-energy photons. With a long half-life (59 days) and low photon energy (average of 28 keV), $^{125}$I could be a health risk if it dissociates from the carrier molecule. A better choice for surgical localization would be $^{123}$I, with an effective half-life of around 12 hours and a dominant photon emission energy of 159 keV.

### 6.1.2 Localization of the Sentinel Lymph Node

Because breast cancer metastasizes via the lymphatics, beginning at the end of the 19th century, surgery for breast cancers also entailed removal of much of the axillar lymph chain to reduce the likelihood that cancer cells would use that route to spread to the rest of the body [5]. In the middle of the 20th century, Haagensen found that cancer cells invade the axillary lymph chain in an orderly manner, from the breast distally, and often only would be in the first node, and called it the sentinel node [6]. In 1994, Giuliano et al. reported injecting blue dye into the tumor and assessed how far along the lymph chain it went as an indicator of the lost of integrity of the lymph node resulting from tumor cells disrupting the normal node structure, and then removed those nodes [7]. The concept was that if the removed sentinel node contained no tumor cells according to pathological examination, the lymph chain did not need to be removed, saving the patient from a much prolonged recovery period and the potential for developing lymphedema, a condition where fluid accumulated in the arm without adequate lymph drainage, causing great pain. The blue dye was replaced with radiolabeled colloids by Krag et al. and localized during surgery using a gamma detector [8].

Currently, the most common approach to find the sentinel nodes uses $^{99m}$Tc-labeled nanoparticles injected into the breast tissue near the tumor a few hours before surgery. At the time of surgery, the extent of the activity along the lymph chain shows which nodes need to be removed and examined.

## 6.2 Characteristics for a Volumetric Detector

As with any detector, the conditions of use determine the necessary characteristics. In this case the requirements are: (1) to detect reliably low activities of $^{123}$I and $^{99m}$Tc; (2) to be able to distinguish the counts recorded to separate anatomy that contains one or the other radionuclide; and (3) to identify geometrically the margins of anatomy containing radionuclide and distinguish different anatomy in close proximity, each of which contains one of the nuclides. These requirements lead to the need for the characteristics of the detector system that follow. The discussion is focused toward providing an understanding for the comparison of the detectors in this chapter for this application. For a more complete explanation of detectors, the reader is directed to a textbook such as that by Knoll [9].

### 6.2.1 Resolution

Resolution, in this context, entails the ability to separate counts from the two radionuclides, which would be energy resolution, and to distinguish two objects in close proximity as separate, that is geometric resolution.

#### *6.2.1.1 Energy Resolution*

In concept, energy resolution of a detector is measured as the full width at half the value of the peak of the measured spectrum produced by a monoenergetic photon emission. The energy resolution determines whether two emissions with different energy can be distinguished as separate photons. This is illustrated in Figure 6.1, which shows the contributions to the recorded counts from two hypothetical emissions. In Figure 6.1(a), the energies of the photons are separated by a large amount compared with the full width at half the maximum of the peak (FWHM) and the curve from each can be clearly resolved. In Figure 6.1(b), the energies fall close together and the sum of the contributions appear as a single curve, albeit with a broadened total FWHM. Figure 6.1(c) shows the energies just fall such that the upper end of one FWHM just touches the lower end of the other FWHM. In this case, the two emission energies can just be resolved. For resolving the photons, the energy difference between the two photons must be separated by a

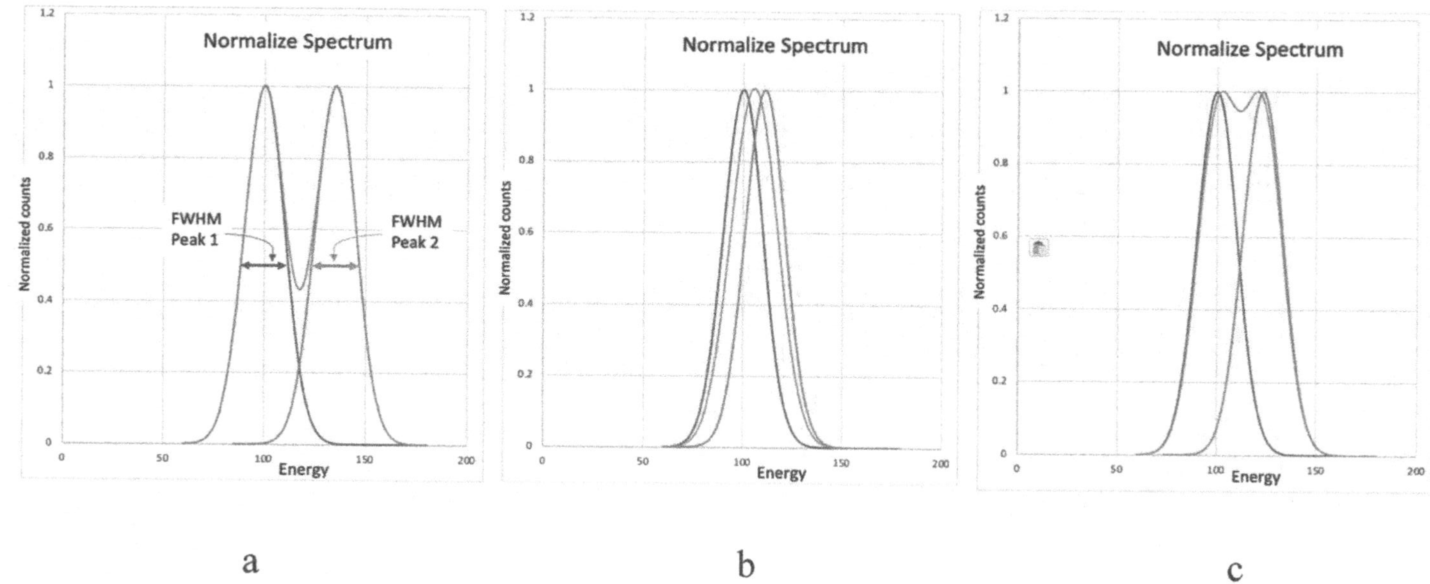


**FIGURE 6.1**
The resultant observed spectrum (gray) from two hypothetical photon emissions of equal frequency (blue and orange), all normalized to the maximum counts: (a) when the photon energies are separated by more than the FWHM, (b) when the energy separation is less than the FWHM, and (c) when the energies are separate by the FWHM. The two emissions can be easily resolved in (a) and just resolved in (b), but appear as a single, slightly broader peak between the two true peak energies in (c).

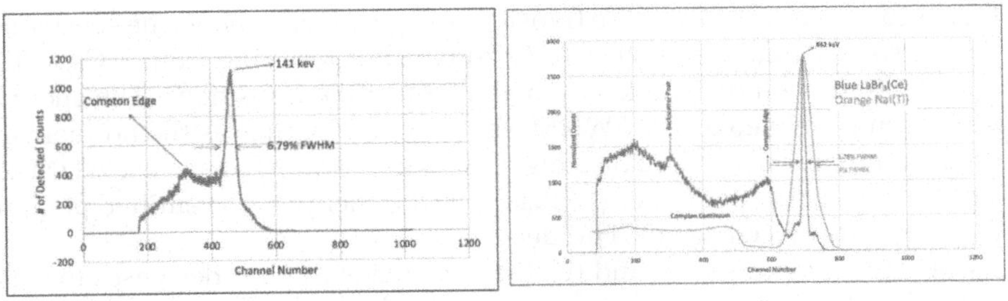


a b

**FIGURE 6.2**
Measured spectra: (a) for $^{99m}$Tc using a $LaBr_3$(Ce) crystal and a silicon photomultiplier; and (b) for $^{137}$Cs, using the same detector as in (a) in blue, and a conventional NaI-photomultiplier tube in orange.

difference equal to the value of the FWHM in keV. Interestingly, even though the percentage FWHM decreases as the photon energy increases, the absolute value increases, as discussed here.

Figure 6.2(a) shows a measured spectrum from $^{99m}$Tc in a specific detector system. While close to the ideal monoenergetic source, $^{99m}$Tc, with a half-life of 6.0 hours, emits a 140.5 keV gamma 89% of the deexcitations. Table 6.1 gives some of the photons emitted with their frequency for those

**TABLE 6.1**
Photon Emission Energies for $^{99m}$Tc, $^{123}$I, and $^{137}$Cs with a Frequency Greater than 0.1%

| $^{99m}$Tc | | $^{123}$I | | $^{137}$Cs | |
|---|---|---|---|---|---|
| Energy [keV] | Frequency [%] | Energy [keV] | Frequency [%] | Energy [keV] | Frequency [%] |
| 2.42 | 0.45 | 3.77 | 9.0 | 4.47 | 0.9 |
| 18.3 | 2.2 | 27.2 | 24.7 | 31.8 | 2.0 |
| 18.4 | 4.1 | 27.4 | 45.7 | 32.2 | 3.6 |
| 20.6 | 0.33 | 30.9 | 4.2 | 36.3 | 0.35 |
| $20.6_2$ | 0.64 | 31.0 | 8.1 | 36.4 | 0.67 |
| 21.0 | 0.15 | 31.7 | 2.3 | 37.3 | 0.21 |
| 140.5 | 89.0 | 159.0 | 83.3 | 661.7 | 85.1 |
| – | – | 346.4 | 0.13 | – | – |
| – | – | 440.0 | 0.43 | – | – |
| – | – | 505.3 | 0.32 | – | – |
| – | – | 529.0 | 1.4 | – | – |
| – | – | 538.5 | 0.38 | – | – |

Frequencies below 1% rounded to two decimal places while those greater than 1% rounded one decimal place [10].

that occur in at least 0.1% of the transitions. Figure 6.2(b) shows the spectrum for the same detector as Figure 6.2(a) in blue but resulting from $^{137}Cs$, with a photon emission of 661.6 keV in about 85% of the transitions. The energy resolution is defined as the FWHM, shown in the figures as the horizontal, doubled-headed arrow. Usually, this resolution is expressed as the FWHM as a percentage of the peak energy. Clearly, as the energy of the emitted photon increases, the percentage FWHM decreases.

Comparing Figure 6.2(a) and (b), the percentage FWHM decreases for the same detector as the energy increases, even though the actual size in keV remains almost the same (9.6 keV at 141 keV and 11.7 keV at 662 keV). Also as seen in Figure 6.2(b), the energy resolution of the $LaBr_3(Ce)$ detector is considerably better than that of the conventional NaI system.

The energy resolution depends on a number of characteristics of a detector, including the detection mechanism (e.g., scintillation, solid state); size; decay time for a signal from a detection event; and, if a scintillator, reflectivity of the container and speed of the photomultiplier, among a host of other factors. One of the oldest scintillation detector systems, and maybe still one of the most common, is a NaI crystal connected to a tube-type photomultiplier. Figure 6.2(b) shows the spectrum for $^{137}Cs$ using a NaI crystal and a conventional tube-type photomultiplier. Compared with Figure 6.1(b), the FWHM for the same input photon energy is much broader, reducing the resolving power of the detector.

The size of the crystal is important for the energy resolution because smaller crystals allow a higher percentage of the energy from a photon interaction to escape the crystal before being absorbed. This loss of energy results in the signal from the original photon being recorded erroneously low, and in the aggregate response curve, the peak associated with a photon energy broaden, becoming less well defined.

#### 6.2.1.2 Geometric Resolution

The geometric resolution of a detector is measured as the distance between two-point sources (or sometimes two-line sources) separated perpendicularly to the detector axis, where they can still be distinguished as two sources instead of one larger source. The dominant characteristic of the detector system in determining the geometric resolution is the collimation, which is a separate topic from the discussion in this chapter and will not be considered here.

### 6.2.2 Sensitivity

A very important aspect of a detector is its sensitivity that is the signal generated and measured from the radiation of interest. This can often be expressed as counts per unit of activity at a specified distance. This method of expressing the sensitivity is more practical for measurement sake, but essentially the same, as using the efficiency of detecting radiation. The efficiency would be defined as the signal (e.g., count rate) divided by the total flux from the source times the solid angle subtended by the detector as seen by the source, or

$$\varepsilon = \Phi \cdot \Delta\Omega = \Phi \frac{A_{dec}}{4\pi d^2} \tag{6.1}$$

The sensitivity is a function of many features of a detector, the most influential being: the inherent conversion from a photon interacting in the detector to a measurable quantity, the detecting volume, and the geometry of the detecting volume. Each type of detector has additional aspects that affect the sensitivity, for example the voltage used for a GM counter or the reflective coating around a scintillation crystal.

Sensitivity becomes very important in a clinical environment, particularly when a patient is on the table, possibly under anesthesia, and the procedure is waiting for the information from the detector.

### 6.2.3 Speed

The speed of a detector relates to how quickly the signal from a single event, that is usually a photon interacting with the detector's sensitive volume, completes the steps to produce a count. The time is called the decay time or relaxation time. The speed becomes important for both the energy resolution and the sensitivity. An accurate counting of a photon of a given energy depends on the absorption of the energy from the photon by the detector and the conversion of that energy into the measured signal. If the process is slow, a second photon may interact with the detector while the energy from the first photon is still being processed. That could confuse the detector system into adding the energy from both the photons and scoring the output as one photon with a large energy. This coincidence effect reduces the number of counts that should be recorded of a given energy. While inadvertent coincidences occur in all detectors, the faster the detector processes the signal from a photon, the smaller the coincident effect. Since this effect changes the spectrum recorded by the detector, it is deleterious to the energy resolution.

Speed also affects the sensitivity of a detector. Because the job the detector performs, for example building an image, usually requires counting photons of the specified energy, the coincident counts do not contribute to the job, and waste the precious counting time. Once a photon interacts in the sensing volume, the detector cannot produce a useful count from another photon until it finishes the first one. Essentially, the detector is insensitive during the processing time, which is why it is referred to as dead time. Dead time reduces the sensitivity of the detector. In some detectors in high radiation fields, continually have photons interacting in the sensitive volume and never can record a count, a situation referred to as paralysis.

For scintillation detector, the speed has two components. The first is the time for the scintillation in the crystal to be completed, while the second is for the photomultiplier to finish processing the electron multiplication process from the scintillation produced in the crystal. Speed in a semiconductor-based detector can be much faster than for a scintillator-based detector. The speed and dead time for a GM counter is more complicated and will not be discussed here (again the reader is referred to Knoll [9]).

### 6.2.4 Background, Extraneous Noise, and Dark Current

In the absence of a radiation source that is a target for measurement, counts can be recorded from background radiation (natural and man-made) and electrical leakage though the circuits (referred to as dark current). The signals from the background and dark current must be measured and removed from measurements so not to give erroneous values. The combination of the extraneous signals also reduces the difference between the true signal from the source in the patient and locations with no source, making the generation of an activity map more difficult. Thus, a good detector either minimizes the amount of extraneous signal produced or has efficient methods for removing it live time.

### 6.2.5 Physical Limitations

The intention for the detector is to provide live-time imaging during surgery. To be held by the surgeon, the detector needs to be compact and light. Moving it around the patient in tight quarters argues against fixed stands or arms, detectors the size of gamma-camera crystals and gantries. Power should be provided by batteries to avoid the constrictions of a power cord.

## 6.3 Review of Possible Detectors

More than 30 different semiconduction and scintillation crystals could be considered for surgical localization, including, but not limited to: NaI(TI), $LaBr_3$(Ce), CdZnTe, $CeBr_3$(Ce), $SrI_2$(Eu), BGO, LSO, GSO, YaP, LuAP, LPS, CsI(TI), CsI(NaI), $CaWO_4$, $YTaO_4$(Nb), $Gd_2O_2S$(Tb), $Gd_2O_2S$(Pr,Ce,F), $Y_{1.34}Gd_{0.6003}$(Eu,Pr), $Gd_3Ga_5O_{12}$(Cr,Ce), $CdWO_4$, $Lu_2O_3$(Eu,Tb), $CaHfO_3$(Ce), $SrHfO_3$(Ce), and $BaHfO_3$(Ce); only those listed in Table 6.2 can be considered the most promising based on their performance characteristics. The characteristics of the rest will not be discussed here, so as not to go beyond the aim of this chapter.

Apart from the Table 6.2, excluding YaP, semiconductors, such as BGO, LSO, GSO, LuAP, LPS, and CsI(Na), provide energy resolution values above

**TABLE 6.2**

Some Physical and Performance Characteristics of Several Detection Crystals

| Detector Material | Density (g/cm³) | Light Yield (Photons/MeV) | Decay Time (ns) | Hygroscopicity | Wavelength of max emission (nm) | Energy Res. % at 662 keV | Energy Res. % at 81 keV | Reference |
|---|---|---|---|---|---|---|---|---|
| NaI(TI) | 3.67 | 41,000 | 230 | Yes | 230 | 5.6 | 9.9 | [11–15] |
| $LaBr_3$(Ce) | 5.3 | 63,000 | 16 | Yes | 380 | 2.6 | 9.1 | [14, 16] |
| $LaCl_3$(Ce) | 3.86 | 46,000 | 25 | Yes | 330 | 3,3 | - | [15, 17] |
| CdZnTe (very small volume) | 6.2 | - | - | No | - | 1.8 | - | [18] |
| CdZnTe (large volume) | 6.2 | - | - | No | - | 5.1 | 18.7 | [14] |
| $Gd_2O_2S$(Tb) | 7.3 | 60,000 | $10^6$ | No | 545 | Integ. Mode | - | [15, 19, 20] |
| CsI(TI) | 4.51 | 66,000 | 800 | Slightly | 550 | 6.6 | - | [12, 13, 21] |
| $CeBr_3$(Ce) | 5.07 | 45,000 | 17.2 | Yes | 370 | 5.8 | 11 | [22, 23] |
| $SrI_2$(Eu) | 4.6 | 120,000 | 1200 | Yes | 435 | 3 | - | [24, 25] |

7% at 662 keV photon energy. Although YaP allows a good energy resolution value (4.3%) at the same energy, even the light yield of NaI(TI), which has the least light efficiency among the crystals listed in Table 6.2, has been found to be more than twice the light yield of YaP [13, 15, 26–32]. Therefore, these crystals were eliminated in the first stage without having to be included in Table 6.2.

$LaBr_3$(Ce) is an inorganic scintillator with very attractive properties when used in medical imaging for gamma spectrometry purposes. Several studies conducted with this crystal have proven that it provides excellent energy resolution, very fast decay, high temperature stability, high gamma detection efficiency, and excellent energy linearity [16, 33]. The following section will include a one-to-one comparison of the lanthanum bromide ($LaBr_3$) with the characteristic of other crystals listed in Table 6.2.

## 6.4 Characteristics of $LaBr_3$(Ce)

As the sensing core of a radiation detector, $LaBr_3$(Ce) has a number of advantages over other scintillation and semiconduction crystals.

### 6.4.1 Inherent Energy Resolution

Looking at the percentage energy resolution in Table 6.2, only two of the detectors have comparable energy resolution with $LaBr_3$(Ce): one is a smaller crystal of CdZnTe (CZT), but that advantage disappears for a crystal of similar size; and the other is $SrI_2$(Eu). This latter detector also produces about twice the light output of $LaBr_3$ and seems like it might be a better choice for a localization detector, but the decay time, of $SrI_2$(Eu) is 75 times as long as $LaBr_3$(Ce), with limits its use to very low interaction rates in the detector, wasting the increased sensitivity. The very small volume (< 1.0 $cm^3$) of CZT detectors provides excellent energy resolution (< 2% at 662 keV) [14, 18]. However, the sensitivity of the CZT is limited by the small size of the crystal, and they are hard to grow and become expensive for larger sizes. The smaller the active volume of the CZT crystal results in a lower the detection efficiency. Also, as the size of the crystal increases, the energy resolution becomes poorer [14]. A CZT crystal with 5.445 $cm^3$ volume (the largest volume available) and a $LaBr^3$(Ce) crystal with 12.9 $cm^3$ volume exhibits 5.1% and 3.2% resolution at 662 keV peak respectively [14]. It becomes even more marked at 81 keV, where the energy resolution for CZT is 18.7% and $LaBr_3$(Ce) is 9.1% [14]. Thus, at sizes large enough to provide a useful sensitivity, the CZT likely would no longer have any advantage over $LaBr_3$(Ce), particularly at the energies in surgical guidance.

### 6.4.2 Crystal Relaxation Time

While a few of the detectors in Table 6.2 have relaxation times approximately the same as $LaBr_3$(Ce), they pale when comparing the energy resolution or sensitivity.

### 6.4.3 Sensitivity

Compared with $LaBr_3$(Ce), the only crystals with comparable or better sensitivity are $SrI_2$(Eu) (considered above), CsI(TI), and $Gd_2O_2S$(Tb). CsI(TI) has a long decay time and poor percentage energy resolution, while $Gd_2O_2S$(Tb) has an extremely long decay time and can only be useful in an integrating mode, not distinguishing photons of different energies.

### 6.4.4 Background and Internal Radioactivity

While $LaBr_3$(Ce) has outstanding characteristics for a surgical-guidance detector, it has one troublesome property. The detecting crystal contains 99.91% $^{139}$La and 0.09% naturally radioactive $^{138}$La. Present in a small abundance and with an extremely long half-life of $10^{11}$ y, there is not a lot of radiation produced. However, the betas, conversion electrons, gammas, and X-rays that come from the decays can be measured easily. Fortunately, for measurements of spectra with a maximum energy less that 700 keV, the background seldom represents 10% of the signal in a channel, and it is well characterized and easily subtracted. Figure 6.3 shows the radioactive background as part of the signal from a source of mixed radionuclides.

## 6.5 Solid-State Photomultiplier (SiPM)

The conventional photomultiplier tube (PMT) starts with a photocathode, that is, a window to the tube that is coated with a material to convert the light from the scintillation crystal into photoelectrons. The electrons from the photocathode are attracted to a more positively charged electrode, called a dynode, which upon impact by the photoelectron, emits several electrons. These emitted electrons are attracted to the next dynode, which is more positive by about 100 V. This process continues through approximately a dozen dynodes until the electrons are collected by the final anode as about $10^6$ electrons from the initial photoelectron, giving a gain of $10^6$:1. Typically, PMTs require very stable voltage supplies of around 1,000 V and are bulky and heavy.

An alternative to the conventional PMT is the SiPM. The SiPM is a collection of solid-state microcell diodes operating in avalanche mode. A review paper by Gundacker and Heering provides a good discussion on the physics of SiPM [35].

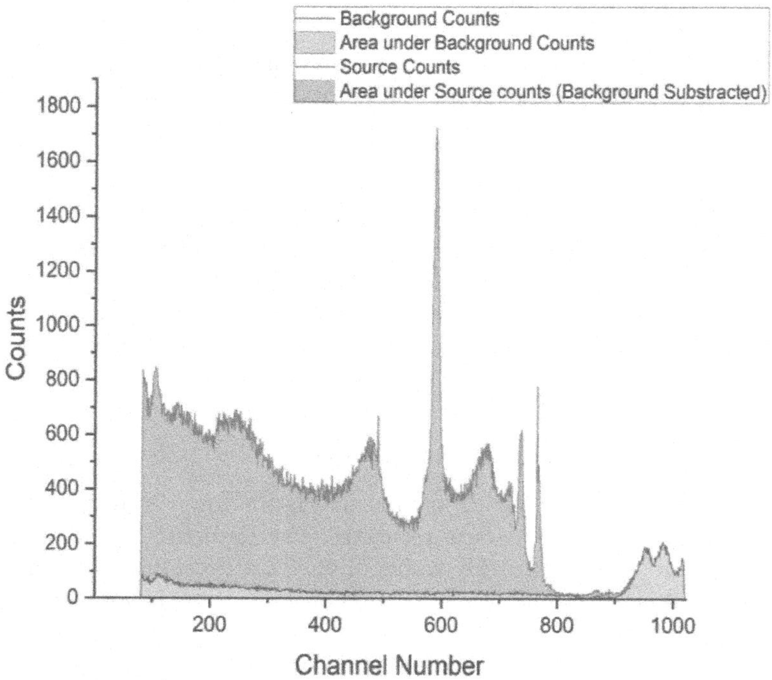


**FIGURE 6.3**
The spectrum of a multi-nuclide source taken with a $LaBr_3$(Ce) detector by Bedir et al. The background (in tan) is subtracted from the multi-nuclide counts (in red line) to obtain the net counts (in green). (Figure copied from [34] with permission.)

### 6.5.1 Speed and Sensitivity

Because the solid-state cells provide all the steps in the photomultiplication process, the distance traveled by source carriers is short and the readout is direct, the SiPM can respond very fast with little dead time. The gain approximates that of a PMT, with a comparable sensitivity.

### 6.5.2 Energy Resolution

Because of the fast response, and the stability of the multiplication gain, SiPM can produce sharper measured spectra from a scintillation crystal than PMT. While not as sharp as a complete solid-state detector system, the SiPM proved a good compromise for practical applications.

### 6.5.3 Practicality

The first practical advantage for a portable detector system of the SiPM compared to PMT is their small size and light weight. The lower voltage and the reduced stability need for the voltage source both reduce the weight and size of the power source for the photomultiplication. Altogether, the SiPM and power supply cost much less that would be the case with a conventional PMT.

## 6.6 Combined Detector System

### 6.6.1 Design Parameters and Integration

SiPM, along with a LaBr(Ce) crystal, could be considered to be an effective pair working on a radiation localization mission. The most important basis of this idea is that these two components have separately improved features as described earlier. But before building the detector with these two components and experimentally examining it, there are some aspects that can theoretically be investigated. One of the most important indicators of compatibility is that the wavelength of the maximum photon emission of both components to be harmony with each other as shown in Figure 6.4.

As it can be deduced from Figure 6.4, within the range of wavelength of the photons emitted from the $LaBr_3(Ce)$ crystal, the silicon photomultiplier provides a wide range of high photon detection efficiency that includes about 60% efficiency at a wavelength of 420 nm. That is one of the good evidences that of how efficient $LaBr_3(Ce)$ will be when it works together with the SiPM.

For easy and portable use, the materials used in the detector should be as small and light as possible. In this sense, and for the low cost of the detector, the dimensions of the crystal to be used are also important. Monte Carlo simulations were conducted in a study to determine the minimum required diameter of the $LaBr_3(Ce)$ crystal when an isotropic $^{123}I$ point source was placed at a source-crystal surface distance of 30 mm [38]. The thickness of the cylindrical crystal was set at 25 mm in the simulation. This was because the internal photon absorption efficiency of the 1-inch thick Brillance™–380 $Labr_3(Ce)$ produced by Saint-Gobain (Hiram, Ohio, USA) had been reported as almost 100% at the 159 keV photon-energy peak of $^{123}I$, based on the company's own internal studies [37]. Then, intrinsic percent photopeak efficiency of the detection crystal was reported as a function of radius of the crystal. The results were then plotted as seen on Figure 6.5 [38]. The rapid increase in photopeak efficiency continues up to a radius of 5 mm, and after, the increase tends to plateau.

Figure 6.6 shows the integrated detector (as a graphic on the left and picture on the right) containing a 10-mm diameter × 25-mm thickness $LaBr_3(Ce)$ cylindrical crystal, hermetically coupled to a 12.46 mm × 12.46 mm SensL Micro Array FJ60035-4P silicon photomultiplier (ON Semiconductors ISG, Cork, Ireland), along with all necessary electronics including built-in amplifier and Li-on rechargeable battery.

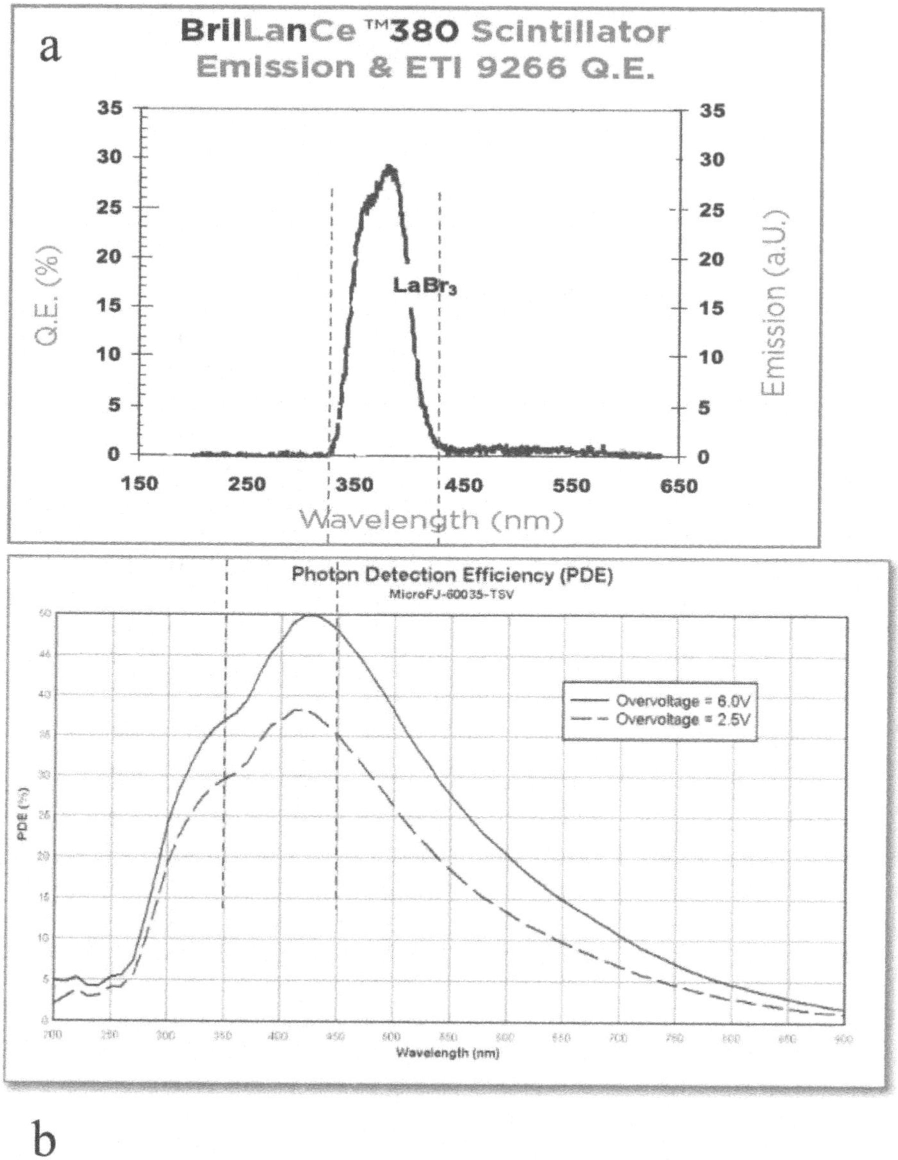


**FIGURE 6.4**
(a) plot of the photon detection efficiency of a SensL-MicroFJ 60035 silicon photomultiplier (*SensL-ON Semiconductors ISG, Cork, Ireland*) as a function of the wavelength of the photons absorbed. Taken from [36] with permission (b) shows the range of the wavelength of the photons emitted from Brillance™-380 $LaBr_3(Ce)$ crystal and their corresponding emission probabilities. (Reproduced from [37] with permission.)

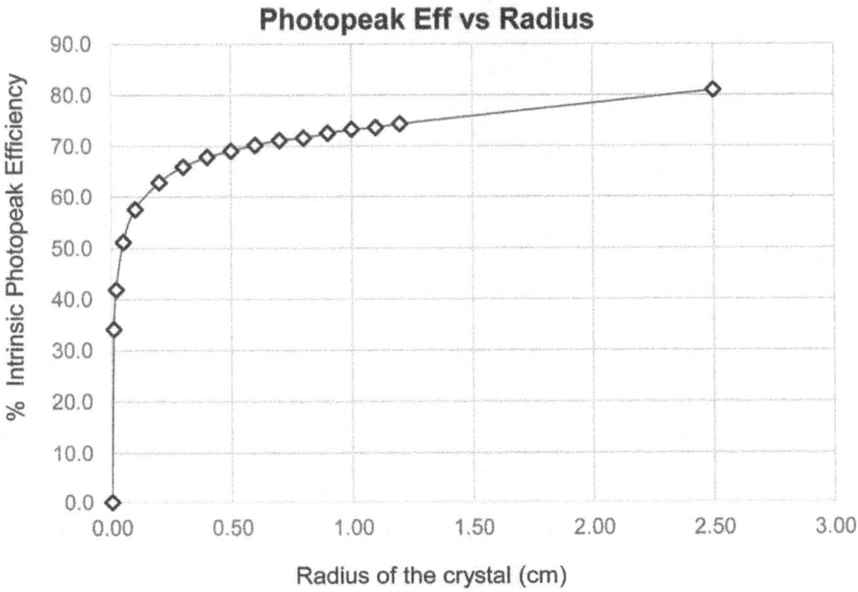


**FIGURE 6.5**
Plot by [38] shows the results of the MCNP simulations which provided percent intrinsic photopeak efficiency of the crystal as a function of the radius.

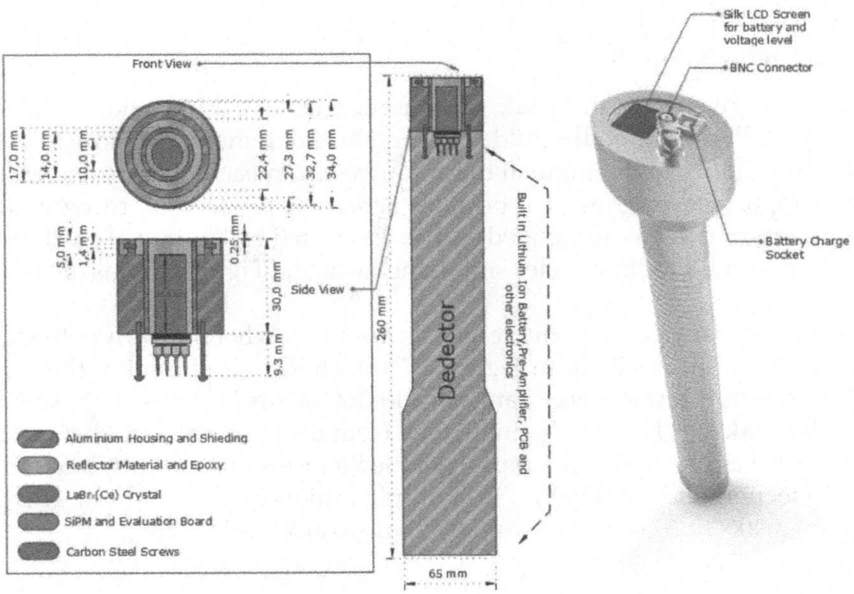


**FIGURE 6.6**
The integrated detector system: on the left, a graphical representation, and on the right, the picture of the integrated detector. (Reproduced from [34] with permission.)

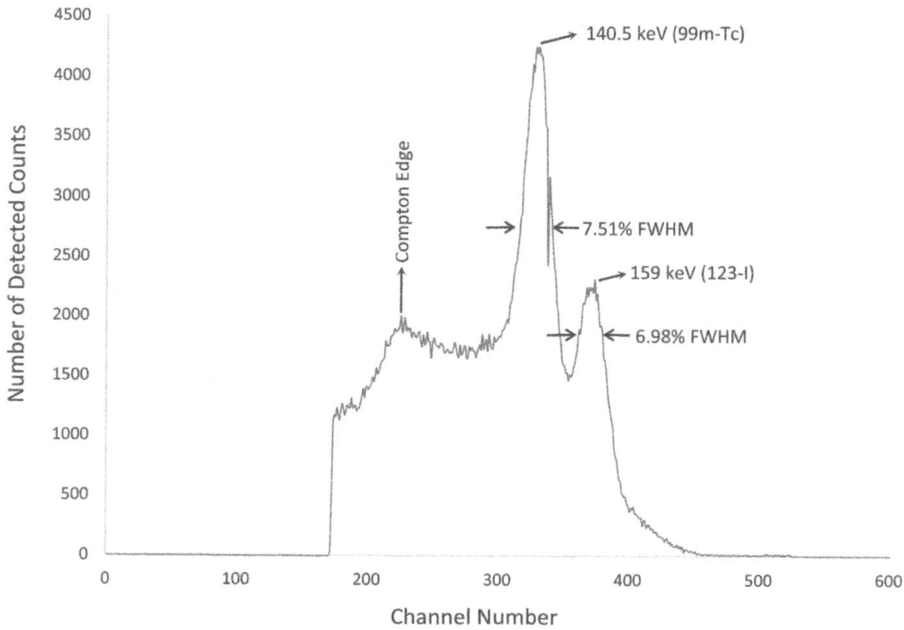


**FIGURE 6.7**
Pulse height MCA spectrum of $^{99m}$Tc and $^{123}$I taken with detector discussed Section 6.6.1 (Reproduced from [34, 38] with permission.)

### 6.6.2 Energy Resolution

Especially given that the peak photon energies of radionuclides such as $^{99m}$Tc and $^{123}$I used in radio-guided operations are remarkably close to each other, the ability to distinguish these sources from each other is one of the most important features of a portable radiation probe. Accordingly, this feature of the detector integrated earlier was tested by Bedir et al. by detecting $^{99m}$Tc and $^{123}$I radionuclides simultaneously and Figure 6.7 was obtained [34, 38].

Figure 6.7 is a good example of Figure 6.1(a), where two hypothetical peaks are easily resolved. In Figure 6.7, the energy resolution of the combined detector in the energy range of 140–160 keV is better than 20 keV, so that the peak of $^{123}$I can be distinguished from the peak of $^{99m}$Tc. That can be considered as one of the first reports in the literature that allows simultaneous detection and localization of multiple radionuclides with low-energy photon peaks (< 200 keV) and are used for surgical guidance.

### 6.6.3 Detection Efficiency

As already explained in Section 6.2.2 sensitivity of a detection system encompasses many functions of a detector. The most important among those is the inherent conversion from a photon interacting in the detector to a measurable quantity, which is defined as the intrinsic detection efficiency. Full energy peak efficiency or absolute photopeak efficiency of a detection crystal, on the other hand, can be determined by fraction between the number of net counts detected under a photopeak area and number of total photons emitted from a radiation source at that peak energy. It has been reported by Bedir et al. [34] that the integrated detector in Figure 6.6 has an absolute 122 keV photopeak efficiency of about 0.32% compared to the same energy absorption efficiency of about 0.07% obtained with a NaI(TI) detector [39]. Although the volume of the $LaBr_3$(Ce) crystal used in Figure 6.6 is only 1/77th of the crystalline volume used in the NaI(TI) detector, the photopeak efficiency obtained with $LaBr_3$(Ce) is much better compared to that obtained with NaI(TI). This, in turn, can be considered as another indicator of how good $LaBr_3$(Ce)-SiPM pair is when combined with a feasible plan.

## 6.7 Conclusion

Detectors used for source localization during surgery require high performance characteristics, namely high energy resolution, sensitivity, and response time, as well as being light and easily handled. With the current state of the art, one system that seems fit those specification the best is using a combination of a $LaBr_3$(Ce) crystal with a solid-state silicon photomultiplier.

## References


1. L. E. McCahill *et al.*, "Variability in reexcision following breast conservation surgery," *JAMA - J. Am. Med. Assoc.*, 2012, doi: 10.1001/jama.2012.43.
2. M. K. Hayes, "Update on preoperative breast localization," *Radiol. Clin. North Am.*, vol. 55, no. 3. W.B. Saunders, pp. 591–603, 01-May-2017, doi: 10.1016/j.rcl.2016.12.012.
3. R. A. Audisio, R. Nadeem, O. Harris, S. Desmond, R. Thind, and L. S. Chagla, "Radioguided occult lesion localisation (ROLL) is available in the UK for impalpable breast lesions," *Ann. R. Coll. Surg. Engl.*, 2005, doi: 10.1308/1478708051595.
4. J. P. Weichert *et al.*, "Alkylphosphocholine analogs for broad-spectrum cancer imaging and therapy," *Sci. Transl. Med.*, 2014, doi: 10.1126/scitranslmed.3007646.
5. W. S. Halsted, "The results of operations for the cure of cancer of the breast performed at the Johns Hopkins Hospital from June, 1889, to January, 1894," *Ann. Surg.*, vol. 20, no. 5, pp. 497–555, Jul. 1894, doi: 10.1097/00000658-189407000-00075.

6. C. Haagensen, "Lymphatics of the Breast," in *The Lymphatics in Cancer*, J. Haagensen, C. Feind, K. R. Herter, F. P. Slanetz, C. A. Weinberg, Ed. Philadelphia, PA: W.B. Saunders, 1972, pp. 300–87.
7. A. E. Giuliano, D. M. Kirgan, J. M. Guenther, and D. L. Morton, "Lymphatic mapping and sentinel lymphadenectomy for breast cancer," *Ann. Surg.*, vol. 220, no. 3, pp. 391–401, 1994, doi: 10.1097/00000658-199409000-00015.
8. D. N. Krag, D. L. Weaver, J. C. Alex, and J. T. Fairbank, "Surgical resection and radiolocalization of the sentinel lymph node in breast cancer using a gamma probe," *Surg. Oncol.*, 1993, doi: 10.1016/0960-7404(93)90064-6.
9. G. F. Knoll, *"'General Properties of Radiation Detectors,' in Radiation Detection and Measurement,"* Hoboken, NJ: John Wiley, 2010, pp. 103–128.
10. "Brookhaven National Labaratory. NNDCBN. Nudat 2.8 Interactive Chart of the Nuclides," 14-Dec-2020. Available: https://www.nndc.bnl.gov/nudat2/. [Accessed: 15-Dec-2020].
11. C. W. E. Van Eijk, "REVIEW To cite this article: Carel W E van Eijk," 2002.
12. C. W. E. Van Eijk, P. Dorenbos, E. V. D. Van Loef, K. Krämer, and H. U. Güdel, "Energy resolution of some new inorganic-scintillator gamma-ray detectors," *Radiat. Meas.*, 2001, doi: 10.1016/S1350-4487(01)00045-2.
13. P. Dorenbos, J. T. D. de Haas, and C. W. V. van Eijk, "Non-proportionality in the scintillation response and the energy resolution obtainable with scintillation crystals," *IEEE Trans. Nucl. Sci.*, 1995, doi: 10.1109/23.489415.
14. A. Syntfeld *et al.*, "Comparison of a $LaBr_3$(Ce) scintillation detector with a large volume CdZnTe detector," *IEEE Trans. Nucl. Sci.*, 2006, doi: 10.1109/TNS.2006.885385.
15. C. W. E. van Eijk, "Inorganic scintillators in medical imaging," *Phys. Med. Biol.*, 2002, doi: 10.1088/0031-9155/47/8/201.
16. E. I. Prosper, O. J. Abebe, and U. J. Ogri, "Characterisation of Cerium-Doped Lanthanum Bromide scintillation detector," *Am. J. Phys. Educ.*, vol. 6, no. 1, p. 162, 2012.
17. C. W. E. van, E. E. V. D. van Loef, and P. Dorenbos, "High-energy-resolution scintillator: $Ce^{3+}$ activated $LaBr_3$," *Appl. Phys. Lett.*, 2000, doi: 10.1063/1.1308053.
18. D. Alexiev, L. Mo, D. A. Prokopovich, M. L. Smith, and M. Matuchova, "Comparison of $LaBr_3$:Ce and $LaCl_3$:Ce with NaI(T1) and Cadmium Zinc Telluride (CZT) detectors," *IEEE Trans. Nucl. Sci.*, 2008, doi: 10.1109/TNS.2008.922837.
19. N. Miura, "Phosphors for X-ray and Ionizing Radiation," in *Phosphor Handbook*, Boca Raton, FL: CRC Press, 2006.
20. F. Rossner, W. Ostertag, M. Jermann, "Properties and applications of gadolinium oxysulfide based ceramic scintillators," *Electrochem. Soc. Proc.*, 1999, pp. 98–124, 187–194.
21. N. Garnier, C. Dujardin, A. N. Belsky, C. Pédrini, J. P. Moy, H. Wieczorek, P. Chevallier, A. Firsov, "Spectroscopy of CsI(Tl) layers," 2000.
22. F. G. A. Quarati, I. V. Khodyuk, C. W. E. Van Eijk, P. Quarati, and P. Dorenbos, "Study of $^{138}$La radioactive decays using $LaBr_3$ scintillators," *Nucl. Instruments Methods Phys. Res. Sect. A Accel. Spectrometers, Detect. Assoc. Equip.*, 2012, doi: 10.1016/j.nima.2012.04.066.
23. P. Guss, M. Reed, D. Yuan, M. Cutler, C. Contreras, and D. Beller, "Comparison of $CeBr_3$ with $LaBr_3$:Ce, $LaCl_3$:Ce, and NaI:Tl detectors," 2010, doi: 10.1117/12.862579.

24. L. J. Mitchell and B. Phlips, "Characterization of strontium iodide scintillators with silicon photomultipliers," *Nucl. Instruments Methods Phys. Res. Sect. A Accel. Spectrometers, Detect. Assoc. Equip.*, 2016, doi: 10.1016/j.nima.2016.02.057.
25. E. V. Van Loef *et al.*, "Crystal growth and scintillation properties of strontium iodide scintillators," *IEEE Trans. Nucl. Sci.*, 2009, doi: 10.1109/TNS.2009.2013947.
26. C. W. E. Van Eijk, "Inorganic-scintillator development," *Nucl. Instruments Methods Phys. Res. Sect. A Accel. Spectrometers, Detect. Assoc. Equip.*, 2001, doi: 10.1016/S0168-9002(00)01088-3.
27. M. E. Casey and R. Nutt, "A multicrystal two dimensional bgo detector system for positron emission tomography," *IEEE Trans. Nucl. Sci.*, 1986, doi: 10.1109/TNS.1986.4337143.
28. H. Ishibashi, K. Shimizu, K. Susa, and S. Kubota, "Cerium doped G S O scintillators and its application to position sensitive detectors," *IEEE Trans. Nucl. Sci.*, 1989, doi: 10.1109/23.34427.
29. J. S. Karp and M. E. Daube-Witherspoon, "Depth-of-interaction determination in NaI(Tl) and BGO scintillation crystals using a temperature gradient," *Nucl. Inst. Methods Phys. Res. A*, 1987, doi: 10.1016/0168-9002(87)90124-0.
30. C. L. Melcher *et al.*, "Scintillation properties of LSO:Ce boules," *IEEE Trans. Nucl. Sci.*, 2000, doi: 10.1109/23.856727.
31. A. Saoudi and R. Lecomte, "A novel APD-based detector module for multimodality PET/SPECT/CT scanners," *IEEE Trans. Nucl. Sci.*, 1999, doi: 10.1109/23.775566.
32. E. V.D. Van Loef, P. Dorenbos, C. W. E. Van Eijk, K. Krämer, and H. U. Güdel, "High-energy-resolution scintillator: $Ce^{3+}$ activated $LaBr_3$," *Appl. Phys. Lett.*, 2001, doi: 10.1063/1.1385342.
33. M. A. Saizu and G. Cata-Danil, "Lanthanum bromide scintillation detector for gamma spectrometry applied in internal radioactive contamination measurements," *U.P.B. Sci. Bull., Ser. A*, vol. 73, 2011.
34. M. E. Bedir, B. R. Thomadsen, and B. P. Bednarz, "Development and characterization of a handheld radiation detector for radio-guided surgery," *Radiat. Meas.*, p. 106362, Apr. 2020, doi: 10.1016/j.radmeas.2020.106362.
35. S. Gundacker and A. Heering, "The silicon photomultiplier: Fundamentals and applications of a modern solid-state photon detector," *Phys. Med. Biol.*, vol. 65, no. 17. Institute of Physics Publishing, pp. 17–18, 07-Sep-2020, doi: 10.1088/1361-6560/ab7b2d.
36. "SensL, 'J-Series High PDE and Timing Resolution, TSV Package Date Sheet,' SensL," 2017.
37. "©2004-2016 Saint-Gobain Ceramics & Plastics, 'Brillance®380 Scintillation Material,' Saint Gobain Crystals and Detectors, Newberry, OH, USA," Newberry, OH, 2009.
38. M. E. Bedir, Doctoral Dissertation *"Design, Development and Characterization of a Handheld Radiation Detector for Radioguided Surgery,"* University of Wisconsin-Madison, 2020.
39. D. Demir, A. Un, and Y. Sahin, *Instrumentation Science and Technology Efficiency Determination for NaI (Tl) Detectors in the 23 keV to 1333 keV Energy Range Efficiency Determination for NaI (Tl) Detectors in the 23 keV to 1333 keV Energy Range*, 2008, doi: 10.1080/10739140801944092.

# 7

# *Development of Energy-Integrating Detectors for Large-Area X-Ray Imaging*

**Ho Kyung Kim**

**CONTENTS**



DOI: 10.1201/9781003219446-7

## 7.1 Introduction

In 1895, Wilhelm Conrad Röntgen discovered a new type of rays, which were later named "x-rays." These rays, in combination with photographic films, became the most important diagnostic tool in medicine, allowing doctors to see inside the human body without surgery. Surprisingly, this x-ray film technique has been in use for more than a century and it was the last non-digital diagnostic imaging modality used, until the end of the 20th century. Its slow progress was predominantly due to the difficulty in fabricating digital imaging devices as *large* as radiographic films; the size of a standard radiographic film cassette for a human chest reaches 360 mm × 430 mm.

Photostimulable phosphor (PSP) or storage phosphor technology first received attention in the medical field before the flat-panel detector (FPD) technology emerged in the 1990s. The PSP stores image information created by the absorption of incident x-rays as a latent image; this image is a distribution of trapped charges in metastable traps and can be converted into blue-green or ultraviolet light signals via optical stimulation, typically with a red or near-infrared laser beam. Storage phosphor detector technology normally requires human intervention because the storage phosphor cassette must be transferred to a laser scanning station, which converts the latent image into a digital image. The first commercial system was introduced by Fuji Photo Film Co., Ltd. in the early 1980s [1]. Storage-phosphor-based imaging systems have been given the name computed radiography (CR) to distinguish them from other digital technologies and the term digital radiography (DR) is used to describe digital-imaging equipment designed to capture and process images directly, without the need for film processing or user intervention. This chapter will not discuss CR systems but the interested reader can find a comprehensive review article for these systems elsewhere [2].

During the evolution from analog to DR, as Yaffe and Rowlands [3] addressed, numerous alternate technologies have been tried for large-area imaging, such as the optical coupling between large-area phosphors and small-sized photosensitive arrays (e.g., charge-coupled devices or CCDs), scanning of linear or slot detector arrays, and a mosaic of small-sized detectors. However, FPDs based on active-matrix pixel arrays are currently the most successful. The FPD concept was motivated by the flat-panel display industry, which had high ambitions to fabricate large-area active-matrix liquid crystal displays (AMLCDs) in the mid-1980s. While the first FPD prototype was 64 × 40 pixels in size [4], an FPD with more than ten million pixels and 17″ × 17″ in size, matching that of a standard radiographic film cassette, was available after a decade [5, 6].

While the FPD is realized with several multilayered structures, its configuration can be simplified into two physical building-block layers [7]; an x-ray converter and a signal-readout pixel-array layer.[1] The former detects incident x-ray quanta and converts them into secondary quanta. The latter collects the secondary quanta and stores them as they are, or as other forms of secondary quanta through a further conversion process.

Two technical schemes are used in digital x-ray imaging with FPDs. The first is the *direct*-conversion scheme [8, 9], using photoconductors that permit the conversion of incident x-rays into signal charges, as shown in Figure 7.1(a). The second is the *indirect*-conversion scheme [10], using scintillators that convert incident x-rays into optical quanta, as shown in Figure 7.1(b). The direct-conversion scheme – a signal-readout pixel array – uses a two-dimensional (2D) array of pixel electrodes and storage capacitors to collect and readout latent image charges formed on the photoconductor surface. The indirect-conversion array uses a 2D-photodiode array to collect optical photons emitted from an overlying scintillator and read the

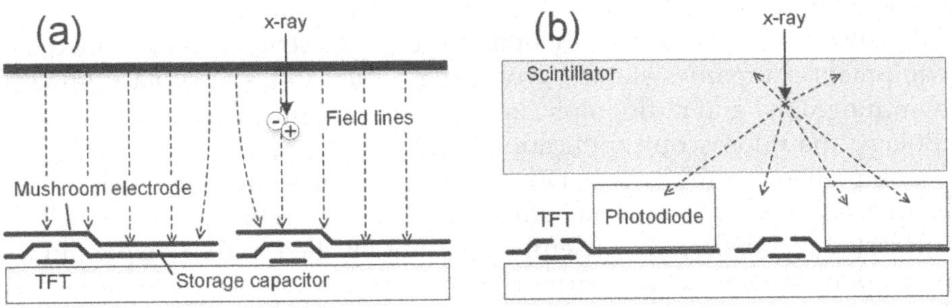


**FIGURE 7.1**
Schematic of x-ray detection principles with FPDs. (a) Direct-conversion detection. The x-rays absorbed by the photoconductor release electron-hole (*e–h*) pairs, which drift to the photoconductor surfaces along the applied field lines. There is minimal lateral diffusion of the charge to reduce the spatial resolution. The charges are collected by the pixel electrodes and distributed onto the storage capacitors. The integrated charges are transferred to the external charge-integrating amplifiers through the data lines by enabling the pixel TFT. (b) Indirect-conversion detection. The scintillator converts incident x-ray quanta into visible light quanta that are absorbed in a thin photodiode layer realized on the readout pixel array, creating *e–h* pairs. The drift motion of these charge carriers due to the external field applied across the photodiode layer discharges the photodiode and these integrated charges are transferred to the charge amplifiers through the TFT.

signal charges converted in a photodiode. Both readout pixel arrays incorporate active switching devices, such as diodes or transistors, in each pixel to temporarily integrate signal charges in storage capacitors or photodiodes during the off-state of switching devices and transfer them to the external readout integrated circuits (ROICs) during the on-state of switching devices. Ultimately, an FPD consists mainly of x-ray converters (i.e., scintillators for indirect-conversion and photoconductors for direct-conversion schemes) and readout pixel arrays. Unlike the direct-conversion configuration, optical coupling between the scintillator and readout pixel array in the indirect-conversion scheme is important because the mismatch of refractive indices between the two components results in a significant loss of optical photons, reducing the x-ray sensitivity.

Large-area FPDs have become well established in the field of medical imaging [11]. Recent medical imaging applications of large-area FPDs include radiography, mammography, fluoroscopy, cardiology, and oncology [12]; this wide spectrum of applications requires a detector with high resolution, wide dynamic range, low noise, fast readout rate, high contrast sensitivity, and scalability. For example, mammography requires the highest resolution and widest dynamic range owing to the requirements of imaging tissue very close to the skin and also at the chest wall; moreover it requires the highest sensitivity as cancerous cells must be contrasted with normal cells. Conversely, the cardiology application requires the fastest readout rate, such as 30 frames per second, and continuous operation during an exam. There is, however, no universal detector to cover all of the applications mentioned. Instead, a large amount of effort has been invested to develop detectors tailored to each relevant application. As a result of extensive research and development efforts, several major medical-equipment companies started producing digital FPDs to replace films in mammography and radiography applications and image intensifiers in cardiology and fluoroscopy applications [6].

Comprehensive reviews of DR detectors are presented in articles, book chapters, and texts, some of which are listed in references [3, 7, 13–19]. Interested readers can find useful topics, not touched in this chapter, in the listed publications and references therein. Major developments in large-area DR detectors have been accomplished in the early 2000s, but efforts to improve their performance are still ongoing. In this chapter,[2] the FPD configuration is decomposed into two components – x-ray converters and readout pixel arrays – and the operating principles, some important design parameters, and developments of the two components are described. The development of FPDs is then briefly discussed and some new advanced concepts of detector development expected in the near future are introduced. Subsequently, the cascaded linear-systems method, a very powerful tool for the design and assessment of x-ray imaging detector systems, is highlighted. Finally, some comments are given regarding the future of DR as an energy-discriminated imaging process.

## 7.2 Flat-Panel Detectors

### 7.2.1 X-Ray Converters

DR detector systems can be simply modeled as multiple cascaded image-forming stages [21, 22]. These stages include the detection of individual x-rays and their conversion into secondary quanta (i.e., optical quanta in indirect-conversion detectors or electrical charges in direct-conversion detectors) within an x-ray converter, the escape of secondary quanta from the converter, the collection of the secondary quanta in pixel sensing or storage elements, and the digitization of the stored signals. However, all detectors extract information relating to the x-ray image in just the first stage, i.e., the detection of x-rays. Subsequent stages may result in a loss of information, even if they provide large amplification. This occurs, for example, during the conversion of optical quanta in the scintillator as the information statistically recorded in an image is still limited by the number of x-rays absorbed within the x-ray converter. However, it remains important to maintain a high level of amplification at each stage to prevent information loss, although no information can be added to the image.

The *quantum sink*, the stage with the fewest quanta (the largest statistical uncertainty), defines the limiting value of the image signal-to-noise ratio (SNR), which is no greater than the square root of the number of quanta at that stage [21, 22]. Information-carrying quanta are lost irrecoverably in the quantum sink and thus amplification stages following a quantum sink cannot improve the SNR. A well-designed system will have a quantum sink located at the very first image-forming stage, i.e., a *primary* quantum sink, so that the information encoded in the image depends only on the number of x-rays detected by the x-ray converter. Such a system is said to be *x-ray quantum limited*. If the quantum sink corresponds to any other stage, i.e., a *secondary* quantum sink, further losses of information-carrying quanta occur in the imaging system due to the suboptimal design and image quality is reduced. For example, lens-coupled detector systems are likely to have a secondary quantum sink at the light collection stage, mainly because of the geometrical inefficiency in the luminous flux collected over the solid angle subtended by the lens assembly [21, 23–26]. X-ray conversion is the first stage in the cascaded imaging chain transferring information to the final user, and hence is the primary factor that determines the overall performance of a detector system.

#### *7.2.1.1 Scintillators*

The most important requirements in a scintillator used as an x-ray converter in the indirect-conversion scheme are the availability of a large area, high light output, and high resolving power. Terbium-doped gadolinium

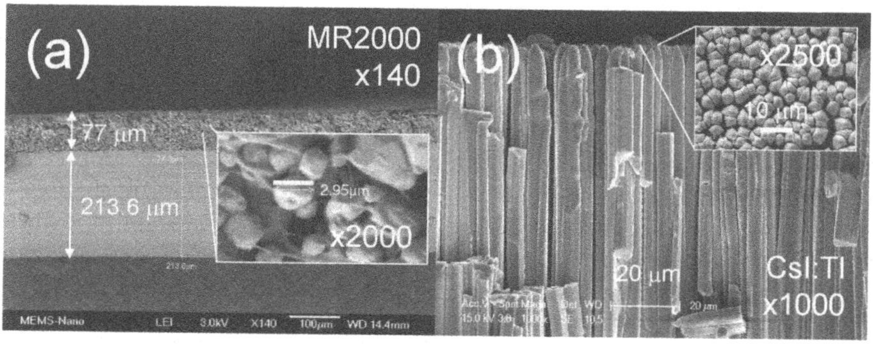


**FIGURE 7.2**
Scanning electron microscope (SEM) pictures showing cut-views of (a) a commercial phosphor screen (MinR-2000™, Carestream Health Inc., Rochester, NY) and (b) a CsI:Tl scintillator. The respective granular and columnar structures of scintillators are clearly compared. In the commercial screen, the thickness of the phosphor layer is measured to 77 μm. More structural information of the screens can be found in reference [27] in detail.

oxysulfide ($Gd_2O_2S$:Tb) and thallium-doped cesium iodide (CsI:Tl) are the best materials used in practical commercial systems to achieve this. Figure 7.2 compares the typical structures of $Gd_2O_2S$:Tb and CsI:Tl scintillators available for DR.

The $Gd_2O_2S$:Tb granular phosphor screen is a popular x-ray converter because its technology is well known, and its size, thickness, and flexibility can be handled easily. Furthermore, it is cost-effective. The main variable that determines the performance of commercially available $Gd_2O_2S$:Tb screens is the coverage or mass thickness. The coverage is the mass of the phosphor-coat per unit area (g $cm^{-2}$). A typical screen is structured as follows [27]: The screen has a thin overcoat, which is optically diffusive. A phosphor layer of the desired thickness is coated on a polyester support. The support contains $TiO_2$ to provide a reflectance of approximately 88% for the 545-nm emission of phosphor. The purpose of this reflector is to enhance screen speed. One type of screen has no reflector to maximize spatial resolution. On the back of the support is an anti-curl layer to ensure the screen is flat. The phosphor layer is made of $Gd_2O_2S$:Tb, with a density of 7.3 g $cm^{-3}$, which is the same as that of the phosphor. The binder is a polyurethane elastomer. Therefore, the effective density of the phosphor layer may be determined by the ratio between the phosphors and organic binders used.

The CsI:Tl scintillator has received much attention for DR applications owing to its multiple advantages. First, it can be readily deposited by thermal evaporation at a low-substrate temperature in the range of 50–250°C [28] and can be directly evaporated onto a readout pixel array, without degrading the properties of the active devices in the array (refer to Figure 7.7(b)). This direct deposition avoids the use of optical coupling agents, such as optical grease or coupling fluid, between the scintillator and the readout pixel array.

Therefore, an FPD could be created at a lower cost without concern for further loss or light spreading through the coupling agents. Second, the CsI:Tl scintillator can be formed in columnar or needle-like structures [29, 30], as shown in Figure 7.2(b). This unique structure restricts the sideways diffusion of optical photons and thus allows high spatial resolution. Third, the optical photon spectrum emitted by CsI:Tl matches well with the absorption response of hydrogenated amorphous silicon (*a*-Si:H) material [31], which is usually used to fabricate a photodiode array. Finally, the CsI:Tl scintillator gives the highest light output of any known scintillator [32].

van Eijk [33] reviewed articles regarding the development of inorganic scintillators in a wide range of medical imaging applications. Nikl [34] reviewed the phosphors and scintillator materials used specifically for x-ray imaging. Their articles and the references they cite will be very helpful to the reader.

#### 7.2.1.2 Photoconductors

The performance of an x-ray photoconductor is mainly limited by x-ray sensitivity, noise, and image lag. For higher x-ray sensitivity, a photoconductor should be made of materials with a high atomic number $Z$ because the x-ray interaction probability (a linear attenuation coefficient, $\mu$ [cm$^{-1}$]) increases with higher values of $Z$. More specifically, it is proportional to $Z^n$, where $n$ is in the range 4–5 [35]. Photoconductors with a higher density and thickness greater than $\mu^{-1}$ can have a larger x-ray interaction probability, and hence a higher x-ray sensitivity. In addition, a photoconductor should have the lowest possible *W*-value. The *W*-value is defined as the average energy required to create a single electron-hole (*e–h*) pair and is proportional to the bandgap energy, $E_G$. From Klein's empirical relation [36], $W$ is approximately 2.8 $E_G$ for semiconductors. For amorphous semiconductors, Que and Rowlands [37] claimed that this should be reduced to approximately 2.2 $E_G$.

With respect to the noise property, the photoconductor should have a negligible dark current [38]. Dark current has two sources: thermal generation and injection through electrical contacts. Blocking contacts on the photoconductor can limit the dark current due to thermal generation only. However, the thermal generation current increases as the bandgap narrows. Therefore, there is a tradeoff between x-ray sensitivity and dark current.

Amorphous materials, which are appropriate for preparation in large areas, suffer from incomplete charge collection and large fluctuations due to the trapping and detrapping of charge carriers by various traps or defects in the bandgap [39–41]; this charge trapping in photoconductors can cause temporal artifacts, such as image lag and ghosting [42–46]. Image lag is the carryover of image charge generated by previous x-ray exposures into subsequent image frames, while ghosting is the change in x-ray sensitivity (i.e., long-term image persistence) as a result of previous exposures [43, 47]. The trapped charges may be detrapped and contribute to the signal in the next readout, producing an image lag. In addition, trapped charges can act as recombination centers for subsequently generated charges, reducing the effective lifetime of mobile charge carriers and, in turn, reducing x-ray sensitivity (ghosting). To avoid or reduce this image lag and ghosting, the photoconductor should generate charge-carrier drift lengths greater than its thickness. The mean drift length $\mu_j\tau_jF$ (also known as the *schubweg*) is defined as the mean distance traversed by a carrier in an electric field before it is trapped [13]. Here, $\mu_j$[3] and $\tau_j$ describe the mobility and lifetime of the charge carrier *j*, respectively, and *F* is the electric field intensity. In addition to this bulk trapping within a photoconductor layer, image lag and ghosting in a photoconductor-based or direct-conversion FPD can arise from other sources, such as charge trapping at the interfaces of the photoconductor layer, charge trapping between pixel electrodes, and incomplete readout of signal charges by electronics [43, 48, 49]. Amorphous silicon, typically used as photodiode materials in indirect-conversion detectors, has similar problems [47]. However, one study reported that the magnitudes of image lag and ghosting are much lower than those of photoconductor-based detectors [50].

Although the x-ray sensitivity of amorphous selenium (*a*-Se) is not higher than that of the discussed scintillators,[4] *a*-Se has proven to be effective as a photoconductor in the direct-conversion scheme, with the distinct advantage that it can be readily prepared as a thick film or layer over large areas via straightforward thermal evaporation in a conventional vacuum coater, without altering its physical properties [13]. It was found that the semiconductors $PbI_2$ and $HgI_2$ achieved close to the theoretical x-ray sensitivity, i.e., a *W*-value of 5 eV, which is nearly a tenfold improvement [51]. Finding materials appropriate for a direct-detection detector is still an important ongoing issue [52, 53].

### 7.2.2 Readout Pixel Arrays

The structure of 2D readout pixel arrays for reading out optical photons emitted from a scintillator, or charge carriers generated in a photoconductor, is similar to a liquid-crystal-skimmed AMLCD, in which each picture element (pixel) is driven by active devices, such as diodes or transistors, arranged in rows and columns to control each pixel. The main structural difference is that the 2D readout pixel array has a photodiode or storage capacitor near the switching device in each pixel element.

In the direct-conversion scheme, the sensing element in each pixel is a simple charge storage capacitor and collection electrode, which can be easily fabricated by dielectric and metal layers. In the indirect-conversion scheme, the sensing element is a thin photodiode, in which the absorbed visible light creates *e*–*h* pairs. This subsection presents the readout pixel array for the indirect-conversion scheme; the details can be found in the references [15, 16, 54]. Additional references for readers particularly interested in the design

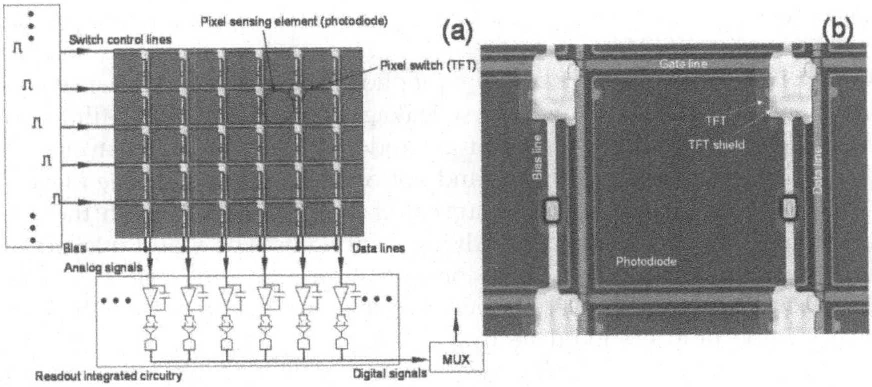


**FIGURE 7.3**
(a) Schematic of the readout pixel array based on *a*-Si:H photodiode/TFT technology. The reverse-biased photodiode integrates the signal charge carriers during a scanning cycle. The integrated charge signal is transferred and converted to a voltage signal through the charge-sensitive preamplifier when the TFT is turned on. The voltage signals are converted into digital signals through analog-to-digital converters and then multiplexed column-by-column. The photograph in the figure is courtesy of Samsung Electronics Co. and Vatech Co., Ltd. (b) A magnified microphotograph of the pixel. The pixel pitch is 143 μm and detailed architectures consisting of the pixel are annotated.

considerations of readout pixel arrays for the direct-conversion scheme are also listed in references [8, 15, 55–59].

Figure 7.3(a) shows a readout pixel array based on an *a*-Si:H photodiode and thin-film transistor (TFT) technology. The photodiode is reverse-biased through a common bias line, and signal charges are accumulated and stored in it during the scanning cycle. When the clock pulse generator of the gate driver emits a gate pulse to each row of the pixel array in sequence, signals from the pixels in the row are readout directly through each data line and then amplified and converted into voltage signals via the array of charge integrating amplifiers. The voltage signals are then converted into digital signals through an analog-to-digital converter (ADC) and multiplexed. Figure 7.3(b) shows a magnified photograph of a pixel of the *a*-Si:H photodiode/TFT array.

The fabrication of readout pixel arrays is mainly based on the *a*-Si:H process because the *a*-Si:H, an alloy of silicon and hydrogen, can be deposited on a large-area glass substrate from the precursor gas state and it exhibits some of the desirable properties of its crystalline counterparts. Amorphous silicon has been known of for a long time, but it only attracted attention as an electronic material once doping to produce *p*-type and *n*-type elements was first demonstrated by Spear and LeComber in 1975 [60]. They demonstrated that *p*- and *n*-type semiconductor materials could be produced by adding dopant gases, such as diborane ($B_2H_6$) or phosphine ($PH_3$), to silane gas ($SiH_4$). Of the various methods of production from the precursor gas state, plasma-enhanced chemical vapor deposition (PECVD) is known to produce the best quality *a*-Si:H layers.

As a pixel photodiode, the *a*-Si:H *p-i-n* photodiode is typical because it has high quantum efficiency, low reverse leakage current, and a thin-film structure. The configuration consists of an undoped or intrinsic (*i*) layer sandwiched between the top *p*-doped and bottom *n*-doped layers. The *i* layer is relatively thick (approximately 1 μm) as it mainly interacts with the optical photons emitted from the overlying scintillator. The *n* and *p* layers are very thin and provide rectifying contacts; ohmic for the majority carriers and blocking for the minority carriers. Optimization of the *p*-layer thickness is very important. It should be thin enough not to significantly attenuate the optical photons and thick enough to prevent electron injection through the *p–i* interface. Metal electrodes should be located on the top and bottom sides to apply a reverse bias to the photodiode. For effective transmission of the scintillation optical photons, a thin metal layer or a transparent conducting layer, such as indium tin oxide, is used for the top of the photodiode. Conventional AMLCD production lines incorporate only the *n*-doping process because the *a*-Si:H TFT does not require a *p*-layer. Modified facilities capable of *p*-doping are required if the *a*-Si:H photodiode/TFT arrays are to be fabricated on a conventional AMLCD line. As an alternative, ion-shower doping, rather than PECVD, has been used to fabricate the *p* layer [61]. This method can avoid contamination of the PECVD chambers due to the *p*-doping process. Other approaches to creating *a*-Si:H photodiodes are the use of Schottky barrier diodes [62] and metal-insulator-semiconductor diodes [63]. These designs do not require *p*-doping capability and are more compatible with the process technology required for thin-film *a*-Si:H TFTs, which only require *n* doping. However, the best quality of photodiodes can be obtained using the PECVD method.

Either a diode or TFT can be used as a switching device. The transistor-based readout scheme of the stored charges in a photodiode was originally suggested by Weckler in 1967 [64] and Street et al. [4] introduced and successfully demonstrated the *a*-Si:H photodiode/TFT sensor array for digital radiation imaging in 1990. The *a*-Si:H TFT is an inverted staggered type with a gate insulator made from amorphous nitride as the first layer, which is deposited onto a substrate. The next layer is a thin intrinsic *a*-Si:H layer as the main current channel. The other layers are *n*-type *a*-Si:H and metal source and drain contacts, dielectric, and passivation layers. The source is connected to a pixel photodiode and the drain is connected to a common data line, followed by the input of the charge-integrating amplifier of the external ROIC. The gate contact located below the gate insulator is connected to a common gate line, which delivers pulse signals to turn on or off the *a*-Si:H TFT from the external gate drivers. Only *n*-channel *a*-Si:H TFTs are practical because of their low hole mobility.

If a diode is used as a switching element, a diode switch is connected back-to-back with a pixel photodiode [65]. The diode switch is normally off and the pixel photodiode is reverse-biased during integration. The signal charges produced during this time are accumulated on the capacitance of the photodiode. During the readout time, the diode switch is forward-biased by applying a gate pulse and the charge stored on the photodiode is then discharged. The integration of the discharging current in the charge-integrating amplifier produces an output voltage pulse. The switching diode can be of the same type as the pixel photodiode for cost savings in fabrication and high device yield. However, the use of diodes as a switching element has some drawbacks, including highly nonlinear characteristics in the forward bias region, and hence severe signal variations from pixel to pixel, as well as the interference of signal currents due to a large feed-through transient at the time of the on-off transition. A double-diode switch can be used instead of a single-diode switch. In this case, the feed-through transient charges can be canceled. Even so, the complex fabrication requirements, such as a large number of interconnections, remain an important drawback. Nevertheless, an FPD operated by diode switches is commercially available [66].

As shown in Figure 7.4, the readout electronics of a large-area FPD are typically composed of a charge-integrating amplifier, correlated double sampling (CDS) circuit, ADC, and multiplexer [55, 67–69]. The typical principles of operation are as follows: The charge-integrating amplifier is reset before reading the pixel signal, bringing the data line to the amplifier reset potential. This pre-sample voltage, $-V_O$, captures the reset noise (i.e., the instantaneous noise sampled on the amplifier feedback capacitor $C_F$ and the parasitic stray capacitance $C_S$ of the readout data line), as well as any offset, and is then stored on the first sample-and-hold capacitor. When the TFT in a pixel is turned on, the signal charge held in the pixel $C_P$ is transferred to $C_F$. When the charge transfer is completed, a second sample voltage, $+V_O$, is stored, which also contains the reset noise and offset. The difference between the two voltage signals is presented to the output through a column multiplexing switch. This CDS cancels the reset noise of $C_F$ and $C_P$, the offset, and any

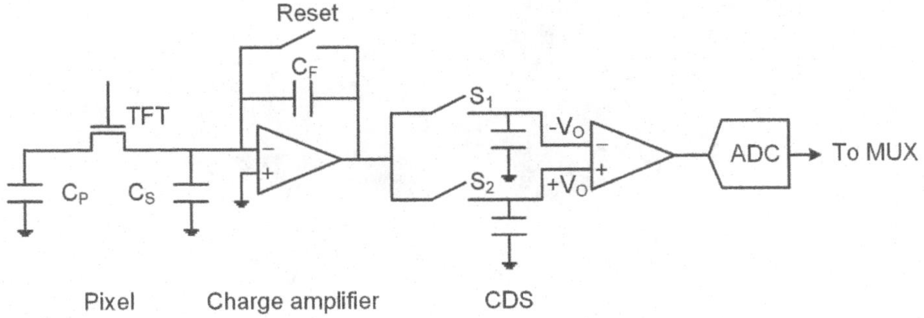


**FIGURE 7.4**
Schematic of the readout electronics typically used in FPDs. The main components are the charge-integrating amplifier, the CDS circuit, the ADC, and the multiplexer.

noise whose predominant contribution lies at frequencies less than the reciprocal of the time interval between the two sampling processes. The output multiplexer delivers signals from all columns onto an output bus before the next row of acquisition can begin. To reduce the demands on the output multiplexer, the sampling and multiplexing operations can be pipelined [70] so that the double samples from the previous row can be multiplexed out while the charge from a new row is being integrated.

### 7.2.3 Developments of Flat-Panel Detectors

GE Healthcare, a leading medical equipment company, pioneered the development of indirect-conversion FPDs with a CsI:Tl scintillator in mammography (1999) [71], radiography (1999) [72], and cardiology (2000) [73]. The structure of the GE FPD is shown in Figure 7.5. GE Healthcare asserts that indirect-conversion architecture allows independent optimization of scintillator and photodetection, which is not possible for *a*-Se technology. Furthermore, they also assert that existing LCD fabrication technologies can be easily adapted.

The *a*-Se technology developed by Hologic, a leading mammography manufacturer, is based on the direct-conversion mechanism [74, 75]. Systems based on the *a*-Se platform are sometimes claimed to have better imaging characteristics as the thickness of the *a*-Se layer may be increased without sacrificing spatial resolution. Scintillator-based systems should have

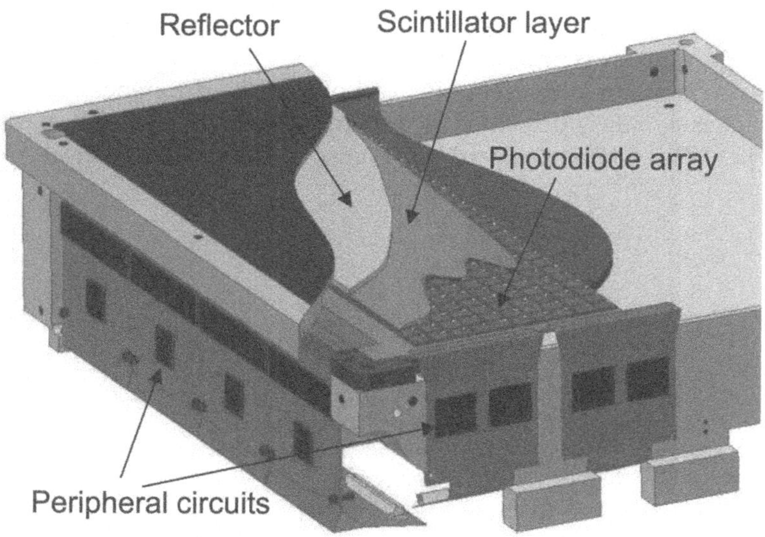


**FIGURE 7.5**
Cutout diagram of GE FPD (image courtesy of GE Global Research). The peripheral circuits are normally tape-carrier packing-bonded to the panel to reduce the volume of the system and prevent the direct irradiation of x-rays onto the electronics.

a scintillator thick enough to stop x-rays and thin enough to reduce the degree of light spreading in that layer, which affects the spatial resolution. Conversely, the detection efficiency of a system based on *a*-Se decreases as the *a*-Se layer becomes thicker as the charge carriers may be recombined (or absorbed) by the *a*-Se layer itself and thus the number of charges arriving at the charge-collection electrodes decreases. The *a*-Se-based technology also has cost benefits since the detectors can be manufactured fairly easily by depositing *a*-Se onto commercially available LCD panels; hence, a full fabrication facility is not required.

Traditionally, spatial resolution is measured using the modulation-transfer function (MTF); this does not consider the signal-to-noise characteristics of a system, which might dominate the image quality in applications with low or medium contrast. Detective quantum efficiency (DQE) is a more useful metric because it takes both noise and resolution into account to determine if a feature is detectable [76, 77]. The spatial-frequency-dependent DQE, DQE($u$) where $u$ denotes the spatial frequency,[5] is dependent on $MTF^2(u)$ and reciprocally dependent on the noise-power spectrum, NSP($u$). Figure 7.6

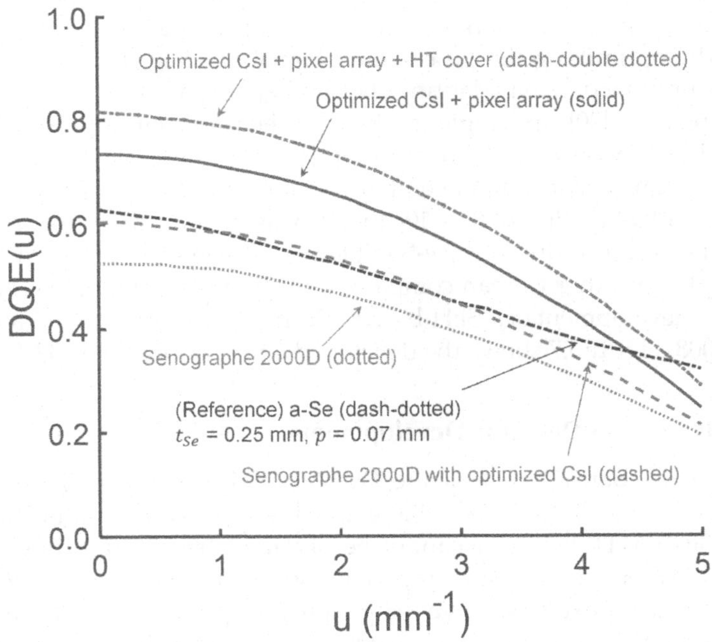


**FIGURE 7.6**
DQE performance at an exposure of 8.5 mR (for a spectrum from Mo 28 kVp + 25 μm Mo/4.2 cm Lucite) for the baseline Senographe 2000D detector (dotted) compared to those of a detector with an optimized scintillator layer but the same *a*-Si:H readout pixel array (dashed); a detector with both optimized scintillator and *a*-Si:H readout pixel array (solid); and the projected results for a detector with an optimized scintillator and *a*-Si:H readout pixel array with a high-transmission cover (dash-double dotted). Also shown are the DQE results from the *a*-Se FPD (250 μm *a*-Se and 70 μm pitch) described in reference [74] (dash-dotted).

shows the high-dose DQE performance of the *a*-Si:H-based GE Senographe 2000D detector system used in mammography and the various modifications. Also shown is the performance of an *a*-Se-based FPD [74]. Based on the study [74], the upper limit to the high-exposure DQE for an *a*-Se-based FPD is estimated to be 70%. An *a*-Si:H-based independent architecture system does not suffer from problems related to charge collection efficiency and is thus able to deliver a high-dose DQE of more than 80% [78]. At low exposures, an *a*-Se-based platform suffers from its fundamental inefficiency in converting x-rays into detectable signals [78]. More detailed comparisons of commercial mammography systems based on FPD technologies can be found in references [79–84]. In addition, comparisons of commercial FPDs for DR have also been reported in references [85–87].

Similar *a*-Si:H-based platforms were also developed independently by other medical equipment companies. In 1997, Siemens Medical Solutions and Philips Medical Systems, two of the leading medical equipment companies, formed a joint venture company, Trixell, with Thales Electron Device. Later, the parent companies of Trixell acquired the majority share in dpiX, further securing a consistent supply of amorphous silicon plates. In 2001, Trixell started providing the *a*-Si:H-based detector Pixium 4600 to Siemens and Philips. dpiX has mainly focused on improving the *a*-Si:H platform detector by optimizing the process and layout [88, 89]. On the other hand, Trixell mainly focuses on adapting the *a*-Si:H platform detector to radiography, fluoroscopy [90], and angiography [91]. The latest developments of FPDs by Trixell can be found in the reference [92].

DRTECH Corp. is a Korean company that develops *a*-Se FPDs. The early application of *a*-Se FPDs of 8″ × 10″ in size was in veterinary practice. Later, DRTECH Corp. introduced the *a*-Se FPD with a size of 17″ × 17″ for general radiography. Another Korean company, Samsung Electronics, Co., has completed the development of *a*-Si:H-based FPD in collaboration with Vatech Co., Ltd. in 2008. Figure 7.7 shows the developed *a*-Si:H photodiode/TFT array.

### 7.2.4 Advances in Detector Developments

Regarding the performance of the DQE($u$), Moy [93] stated that the ideal MTF should be as large as possible below the Nyquist limit and then drop rapidly. Such MTF behavior cannot be obtained from conventional detectors; the MTF of a scintillator drops with spatial frequency $u$, while that of a photoconductor extends above the Nyquist frequency. An ideal MTF may be achieved if the scintillator is structured to match the underlying pixel size of the readout photosensitive elements, as shown in Figure 7.8(a). With such a structured scintillator, very little signal at spatial frequencies above the Nyquist frequency $u_c$ and replication of the spectra would cause little signal distortion. A significant effort has been invested in this new concept of imaging detectors [94–101]. In addition to the pixelization of bulk crystalline scintillator by laser dicing, the pixel-structured or pixelated scintillator

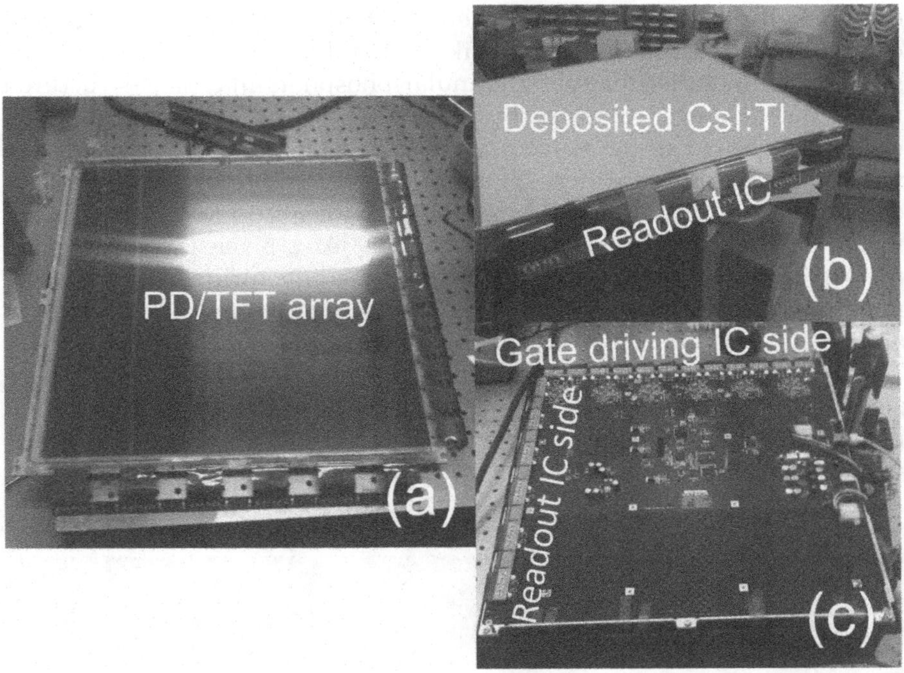


**FIGURE 7.7**
Photographs showing an unpacked FPD (Samsung Electronics Co.): (a) a top view of the *a*-Si:H photodiode/TFT pixel-array panel, (b) an unpacked view of the panel on which a CsI:Tl layer is directly deposited, and (c) a back-side view of the panel. The panel size is 46 cm × 46 cm and the active image format is 3072 × 3072 pixels. Tape-carrier packages containing a gate driver and readout circuits are shown.

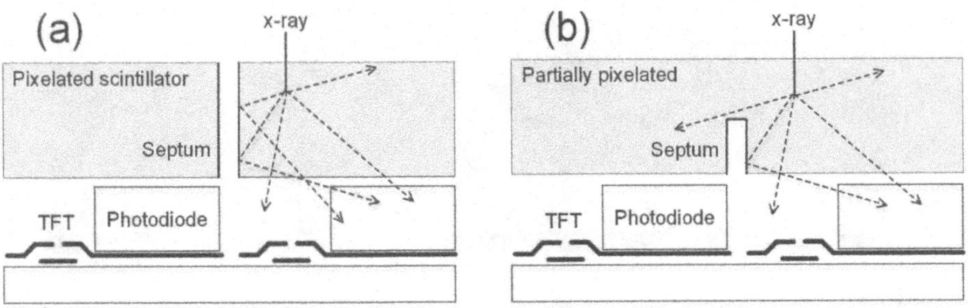


**FIGURE 7.8**
Designs of pixel-structured scintillator-based FPDs. (a) Sketch describing a pixel-structured scintillator in conjunction with the conventional readout pixel array. (b) Conceptual design to reduce the fill-factor effect possibly observed in the pixel-structured scintillator design, which reduces the DQE in the low-frequency band.

can be realized via several other methods. One is to grow the scintillator on patterned substrates in a pixel-like format [100]. Melting powder-type scintillation materials or filling granular phosphors into pixel-structured molds has also been reported by several groups [94–99, 101]. The pixel-structured mold can be realized by deep reactive ion etching (DRIE) or wet etching on silicon wafers [94, 95, 97, 99, 101, 102] or by using a micro-electromechanical system (MEMS) process, with silicon wafers or polymer materials, for example, SU-8 [96, 98, 103, 104]. As a pixel-structured mold, the barrier rib technology, which is used for manufacturing plasma display panels, has recently been introduced [105]. Figure 7.9 shows the laboratory-scale realization of pixel-structured scintillators and a flexible mold.

Kim et al. [106] reported that the pixelated design is appropriate for high-sensitivity and high-resolution imaging systems, and imagers with a small pixel pitch, such as those employed in mammography or intraoral imaging, using theoretical and numerical approaches. However, the most significant

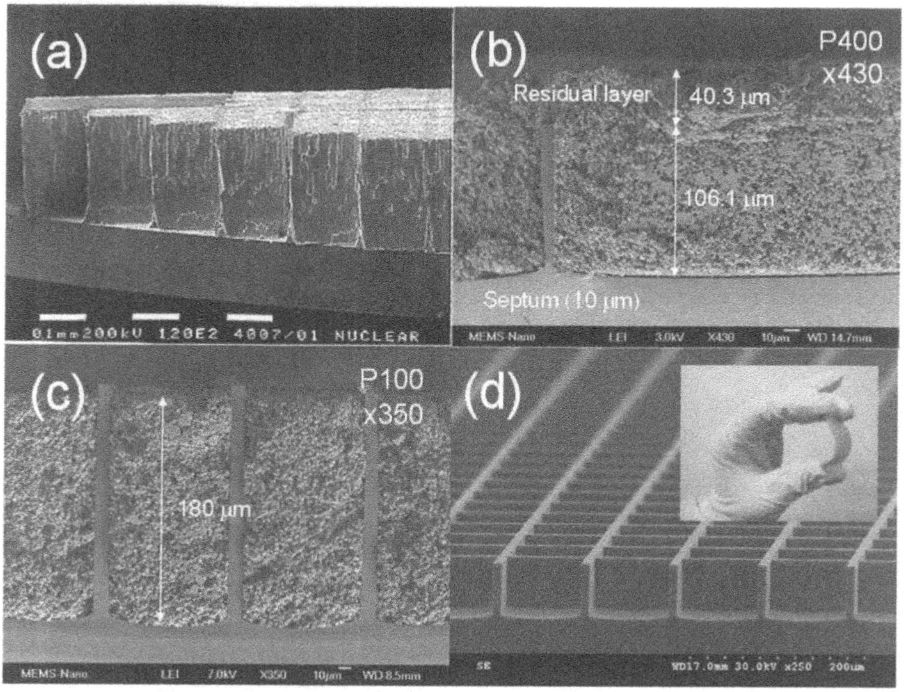


**FIGURE 7.9**
SEM pictures showing the designs of pixel-structured scintillators. (a) Pixel-structured polycrystalline CsI layer manufactured by thermal evaporation onto the pixel-patterned substrate (image courtesy of G. Cho et al., KAIST, Korea). (b) Pixel-structured $Gd_2O_2S$:Tb layer manufactured by sedimentation within the DRIE-processed silicon mold, with a pixel pitch of 400 μm. (c) The same as (b) but a smaller pixel design of 100 μm. (d) Flexible pixel mold manufactured by the MEMS process.

disadvantage of the pixelated scintillator is the reduction of the DQE in the low-frequency band because of the finite fill factor. As a potential solution to mitigate this problem, Kim et al. [106] proposed the partially pixelated scintillator configuration shown in Figure 7.8(b). The concept of a pixel-structured scintillator is attractive to digital x-ray imaging because of the relatively high atomic number and density of the scintillation material and the possibility of designing a band-limited MTF [107]. Research is active in this area but several physical hurdles, for example, significant variations in the optical gain in pixel-structured scintillators [101, 108], still need to be overcome.

One major issue in the development of DR detectors is low-dose imaging; this includes dynamic imaging, such as cardiac imaging and fluoroscopy. Low-dose imaging could be partially achieved by amplifying signals before the signals are delivered to the signal processing electronics. The development of ultra-low-noise electronics is another necessity to be combined. Such a system would be x-ray quantum-noise limited; hence, the noise could not be reduced further and only random x-ray quantum noise existed. Antonuk et al. [109] described strategies to improve the signal and noise performance of FPDs. One approach is to reduce the additive electronic noise through an improved amplifier, pixel, and array design [67, 69, 110, 111]. An alternative approach is to increase the system gain by various means, for example, the use of higher gain x-ray converters for direct-conversion detectors, such as $PbI_2$, PbO, $HgI_2$, and CdZnTe [112–115], and the use of continuous photodiodes for indirect-conversion detectors [116]. Another possible approach to increase the x-ray sensitivity, regardless of the type of detector operation, is to incorporate an amplifier in each pixel [117–119]. Matsuura et al. [117] theoretically investigated the signal and noise characteristics of an FPD with amplified pixels by incorporating pixel-level amplifiers and thus the feasibility of this type of FPD for fluoroscopy. They reported that their design of an amplified pixel detector array could reduce noise more than conventional FPDs but not as low as to reach the ultimate quantum noise limit for fluoroscopy. This concept has been realized by several groups. Karim et al. [118] proposed active pixel architectures based on an amorphous silicon process. Lately, Lu et al. [119] successfully demonstrated large-area compatible FPDs that contain amplifiers at the pixel level based on excimer laser crystallized poly-Si TFTs. Readers can find references [120–124], which discuss potential performance improvements by advanced pixel designs and the use of active pixel architectures.

A new concept of indirect-conversion FPD with avalanche gain and field emitter array (FEA) readout has been proposed for low-dose and high-resolution x-ray imaging [125–129]. It is fabricated by optically coupling a CsI:Tl scintillator to an *a*-Se *avalanche* photoconductor. The optical photons emitted from the scintillator generate *e–h* pairs near the top of the *a*-Se layer and they experience avalanche multiplication under the high electric field intensity applied within the *a*-Se layer. The amplified charge image is then readout, with the electron beams emitted from a 2D FEA, which can have pixel sizes down to 50 μm. Therefore, this detector is referred to as a scintillator avalanche photoconductor with high-resolution emitter readout (SAPHIRE). The use of the conventional TFT array as a readout device has also been considered by this group [125, 126, 130–133]. A similar concept was proposed much earlier [128], but the avalanche gain within the *a*-Se layer was not considered.

The fabrication technologies of readout pixel arrays have advanced considerably since the introduction of organic semiconductors and have made impressive improvements to the performance of optoelectronic applications [134, 135]. Street and his colleagues developed a novel jet-printing approach, or digital lithography, to fabricate TFT and active matrix arrays for x-ray imaging detectors [136–140]. They have shown that the use of jet-printing techniques to manufacture readout pixel arrays provides significant advantages over conventional processes [139]. First, production costs could conceivably be reduced. Second, compared to the heavy and fragile glass substrates used in conventional arrays, the printed arrays on plastic substrates would be lighter and more robust. Third, printing techniques can enable the fabrication of significantly larger monolithic arrays than is presently possible for conventional designs. Fourth, flexible arrays would provide an entirely new degree of freedom from existing arrays (e.g., curved/conformal imaging detectors).

## 7.3 Cascaded Linear-Systems Model

Technical excellence in medical imaging is critical to high-quality medical care due to the health risks associated with exposure to radiation and those due to inconclusive or misleading medical diagnoses. In radiology, image-quality excellence is achieved by ensuring a balance between system performance and patient radiation dose. X-ray systems must be designed to ensure that the maximum image quality is obtained for the lowest consistent dose. This section describes some of the concepts and methods used to quantify, understand, measure, and predict the performance of x-ray detectors and imaging systems. The terms, MTF, NPS, and DQE, were introduced in Sections 7.2.3 and 7.2.4, including some detector developments to improve those metrics. This section brifely summarizes their definitions and the theoretical analysis of the NPS and DQE of a detector, using the cascaded linear-systems theory.

### 7.3.1 Detective Quantum Efficiency

The use of Fourier-transform-based metrics to describe spatial resolution in terms of the MTF was introduced to the medical imaging community by Rossmann and coworkers [141–143] and is now widely used. The one-dimensional (1D) MTF is defined as the Fourier transform of the line-spread function (LSF) [144]:

$$\mathrm{MTF}(u) = \left|\mathcal{F}\{\mathrm{lsf}(x)\}\right| \tag{7.1}$$

where $u$ is the spatial-frequency variable and $\mathrm{lsf}(x)$ is the LSF expressed as a function of position $x$, normalized to unity area.

Image noise can be described in terms of the auto-covariance function $\mathrm{K}(x)$ in the spatial domain [145], or the image Wiener NPS in the spatial-frequency domain [144, 146, 147]. The NPS is the spectral decomposition of the noise variance, describing noise in terms of its frequency components:

$$\sigma_d^2 = \int_{-\infty}^{\infty} \mathrm{NPS}_d(u)\mathrm{d}u \tag{7.2}$$

where $d$ is the x-ray detector signal and $\sigma_d^2$ is the statistical variance in $d$. The NPS is related to the auto-covariance function by [145]:

$$\mathrm{NPS}(u) = \mathcal{F}\{\mathrm{K}(x)\}. \tag{7.3}$$

A limitation of the NPS is that its values depend on scaling factors that may be applied by the imaging system and differ from one system to another. This problem was solved by Shaw, who described the image SNR in terms of a noise-equivalent number of quanta (NEQ) [148–150]. The NEQ is independent of scaling factors and describes how many Poisson-distributed quanta (per unit area) would give the same SNR with an ideal imaging system; it thus describes the number of quanta an image is *worth*. This leads directly to what is referred to as the DQE of an imaging system:

$$\mathrm{DQE}(u) = \frac{\mathrm{NEQ}(u)}{\overline{q}_0} = \frac{\overline{q}_0 G^2 \mathrm{MTF}^2(u)}{\mathrm{NPS}_d(u)} \tag{7.4}$$

$$= \frac{\mathrm{MTF}^2(u)}{\overline{q}_0\left[\mathrm{NPS}_d(u)/\overline{d}^2\right]}, \tag{7.5}$$

where $\overline{q}_0$ is the mean number of x-ray quanta incident per unit area on the detector, $G = \overline{d}/\overline{q}_0$ is the system large-area gain factor, and $\overline{d}$ is the mean detector output value [151]. The MTF and DQE are generally considered to be the most important measures of performance for x-ray imaging systems.

The DQE is sometimes also defined as the transfer of the squared SNR through the imaging system:

$$\mathrm{DQE}(u) = \frac{\mathrm{SNR}_{\mathrm{out}}^2(u)}{\mathrm{SNR}_{\mathrm{in}}^2(u)}. \tag{7.6}$$

However, this is only correct for x-ray imaging where the incident quanta are Poisson distributed (and therefore $\mathrm{SNR}_{\mathrm{in}}^2(u) = \bar{q}_0$) [152] and only when both signal and noise are defined as used in this chapter. In general, either Equation (7.4) or (7.5) should be used as the definition of the DQE.

All Fourier methods assume a linear and shift-invariant imaging system and wide-sense stationary or wide-sense cyclostationary (WSCS) random noise processes [145, 151]. Linearity means that the output of the detector value must scale linearly with the input signal. Image pixel values satisfy this requirement only for linear detectors using dark-subtracted *raw* images. Nonlinear systems (or systems using nonlinear image processing) must be *linearized* before these methods can be used. Otherwise, the DQE results will be corrupted [153, 154]. Shift invariance indicates that the image noise and resolution characteristics are equal for all parts of an image. This assumption may fail near the edges of the images.

Equation (7.5) shows that the DQE of any detector can be determined if the system MTF and the image NPS with associated mean image pixel value and incident number of quanta per unit area can be determined. The cascaded linear-systems theory is powerful because it can be used to determine all these quantities based on system design parameters.

Early use of linear-systems theory used simplistic ideas of noise transfer that ignored the statistical properties of secondary image quanta [144, 155–159]. This changed when Rabbani, Shaw, and van Metter described noise transfer through quantum gain and quantum scattering processes [160, 161]. A comprehensive review of the cascaded linear-systems theory has been published [151, 152]. In this section, we briefly summarize the theory.

### 7.3.2 Elementary Processes in Cascaded Models

Cascaded models of x-ray imaging systems describe how quantum-based images are propagated through a system by cascading simple *elementary* processes. A quantum image is a spatial distribution of quanta, generally in two dimensions. Quantum has negligible size and a quantum image is represented as a random spatial distribution of Dirac impulse functions [151, 152, 162]:

$$\tilde{q}(\mathbf{r}) = \sum_{i=1}^{N} \delta(\mathbf{r} - \tilde{\mathbf{r}}_i) \tag{7.7}$$

where $N$ is the total number of quanta in the image and $\tilde{\mathbf{r}}_i$ is a random vector describing the spatial position of the $i$th quantum (the overhead tilde is used to emphasize a random variable). An imaging system is represented as a series of cascaded processes connecting an input x-ray image $\tilde{q}_0(\mathbf{r})$ to an output image $\tilde{d}(\mathbf{r})$.

#### 7.3.2.1 Quantum Gain

Quantum gain [151, 152, 160, 163] is a random point process in which each input quantum is replaced by $\tilde{g}$ overlapping output quanta, such as replacing each x-ray quanta with $\tilde{g}$ optical quanta where $\tilde{g}$ is an integer random variable with a mean value $\overline{g}$ and standard deviation $\sigma_g$. The output is another quantum image. The values of $\overline{q}_{out}$, $T_{out}(u)$ (the characteristic function), and $NPS_{out}(u)$ are summarized in Table 7.1. The output NPS consists of two terms. The first, $\overline{g}^2 NPS_{in}(u)$, describes the transfer of noise by the mean gain $\overline{g}^2$. The second describes the additional uncorrelated noise (independent of frequency) resulting from random variations in $\tilde{g}$.

#### 7.3.2.2 Quantum Selection

Quantum selection with a probability α is a special case of quantum gain, in which each quantum incident on this selection stage is either transferred or not. An example is the quantum efficiency of an x-ray converter (e.g., radiographic phosphor screen or photoconductor).

#### 7.3.2.3 Quantum Scatter

Quantum scatter is a point process in which each input quantum is randomly relocated to a new position in the image. If the relocation point-spread function (PSF) is psf($x$), the scatter characteristic function is $T(u) = \mathcal{F}\{psf(x)\}$. While scatter will normally cause blurring of the output image by T($u$), scatter differs from the convolution used to describe the blurring of a linear filter in its noise-transfer properties.

#### 7.3.2.4 Spatial Integration of Quanta

Digital detectors operate by producing an electrical signal proportional to the number of accumulated image quanta (such as electronic charge) in individual detector elements. This *binning* process is a spatial integration of points that represent interacting quanta. If all quanta incident on an element of width $a_x$ are detected, the number of quanta interacting in the $n$th element of a detector element at $x$ is:

$$\tilde{d}_{out}(x) = \int_{x-a_x/2}^{x+a_x/2} \tilde{q}_{in}(x')dx' \tag{7.8}$$

$$= \tilde{q}_{in}(x) * \Pi\left(\frac{x}{a_x}\right) \tag{7.9}$$

**TABLE 7.1**

Expressions for the Mean Output Signal ($\bar{q}_{out}$ for Quantum Images and $\bar{d}_{out}$ for Detector Signals), $MTF_{out}$, and $NPS_{out}$ for Simple Elementary Processes in the Cascaded Linear-Systems Theory. Expressions Are Shown in the 1D Form and the Extensions to 2D Form Are Straightforward

| Process | | Parameter | | Process Output | | |
|---|---|---|---|---|---|---|
| | | (Gain) | (Variance) | Mean Signal | MTF | NPS |
| Stochastic | Quantum gain | $\bar{g}$ | $\sigma_g^2 = \bar{g}^2(1/I_g - 1)$ | $\bar{q}_{out} = \bar{g}\bar{q}_{in}$ | $T_{in}(u)$ | $\bar{g}^2 W_{in}(u) + \sigma_g^2 \bar{q}_{in}$ |
| | Quantum selection | $\alpha$ | $\sigma_\alpha^2 = \alpha - \alpha^2$ | $\bar{q}_{out} = \alpha\bar{q}_{in}$ | $T_{in}(u)$ | $\alpha^2 \left[ W_{in}(u) - \bar{q}_{in} \right] + \alpha\bar{q}_{in}$ |
| | Quantum scatter | Characteristic function $T_c(u)$ | | $\bar{q}_{out} = \bar{q}_{in}$ | $\lvert T_c(u) \rvert T_{in}(u)$ | $\left[ W_{in}(u) - \bar{q}_{in} \right] \lvert T(u) \rvert^2 + \bar{q}_{in}$ |
| Deterministic | Quantum integration | Aperture $a_x$ | | $\bar{d}_{out} = a_x \bar{q}_{in}$ | $\lvert \mathrm{sinc}(a_x u) \rvert T_{in}(u)$ | $a_x^2 \mathrm{sinc}^2(a_x u) W_{in}(u)$ |
| | Linear filter | Characteristic function $T_c(u)$ | | $\bar{d}_{out} = T_c(0)\bar{q}_{in}$ | $\left[ \lvert T_c(u) \rvert / T_c(0) \right] T_{in}(u)$ | $\lvert T(u) \rvert^2 W_{in}(u)$ |
| | Sampling | Sampling pitch $p_x$ | | $\bar{d}_{out} = \bar{d}_{in}$ | | $W_{in}(u) + \sum_{i=1}^{\infty} W_{in}(u \pm i / p_x)$ |

where $\Pi(x / a_x)$ is a rectangular function with a value of 1 for $-a_x/2 < x < a_x/2$ and 0 elsewhere and the * symbol denotes the convolution operation.

#### *7.3.2.5 Sampling*

Several authors [151, 164] have presented analog image sampling to produce a discrete digital image as an elementary process in cascaded systems, through multiplication by an infinite train of uniformly-spaced $\delta$ functions at intervals of $x_0$. Thus, in 1D, the output is:

$$\tilde{d}_{\text{in}}^{\dagger}(x) = \tilde{d}_{\text{out}}(x) \sum_{n_x=-\infty}^{\infty} \delta(x - n_x x_0) = \sum_{n_x=-\infty}^{\infty} \tilde{d}_n \, \delta(x - n_x x_0) \tag{7.10}$$

where the superscript † is used to indicate a discretely sampled function represented as an infinite train of scaled and uniformly-spaced δ functions. When $\tilde{d}_{\text{in}}(x)$ is a detector presampling signal, the digital image consists of a set of random values $\tilde{d}_n$.

This sampling process is a linear operation, although it results in a shift-variant output. Fourier-based metrics are still applicable because the output, even in the presence of noise aliasing, is a WSCS random process.

### 7.3.3 Simple Systems

#### *7.3.3.1 X-Ray Converter*

Simple systems are modeled by cascading elementary processes where the output of one process becomes a virtual input to the next. Let us first consider a cascaded model representing a scintillator as a combination of: quantum selection to select x-ray quanta that interact in the scintillator with probability $\alpha$ (i.e., quantum efficiency); quantum gain representing conversion to light quanta with a mean gain $\bar{m}$ and standard deviation $\sigma_m$; and quantum scatter, representing optical scatter in the scintillator with optical MTF $T_c(u)$. The resulting average number of optical quanta is $\bar{q}_0 \alpha \bar{m}$, the MTF is $T_c(u)$, and

$$\text{NPS}(u) = \bar{q}_0 \alpha (\bar{m}^2 - \bar{m} + \sigma_m^2) T_c^2(u) + \bar{q}_0 \alpha \bar{m}. \tag{7.11}$$

Combining these gives the DQE of the scintillator:

$$\text{DQE}(u) = \frac{\alpha}{1 + \dfrac{\sigma_m^2 - \bar{m}}{\bar{m}^2} + \dfrac{1}{\bar{m} T_c^2(u)}}. \tag{7.12}$$

Some observations are: (1) the DQE scales with the quantum efficiency $\alpha$; (2) a large optical gain $\bar{m}$ is required to ensure that the DQE is close to $\alpha$;

(3) if $\bar{m}$ is not sufficiently large, a secondary *quantum sink* exists that degrades the DQE due to an inadequate number of optical quanta. The optical gain must be particularly large to ensure no secondary quantum sink at high spatial frequencies, where $T_c^2(u)$ may be small.

Noting the Swank noise factor $I = \bar{m}^2 / \overline{m^2}$ [165], Equation (7.12) can also be expressed as:

$$\mathrm{DQE}(u) = \frac{1}{\frac{1}{\alpha}\left(\frac{1}{I} - \frac{1}{\bar{m}}\right) + \frac{1}{\alpha\bar{m}\mathrm{T}_c^2(u)}}. \tag{7.13}$$

We note that the upper limit of Equation (7.13) is $\alpha I$. In other words, the DQE of a scintillator may be determined by the detection efficiency of x-ray quanta in the scintillator, $\alpha$, times the Swank noise factor $I$. This simple relationship is widely used as an approximate DQE value of FPDs at zero spatial frequency, $\mathrm{DQE}(u)|_{u=0} \approx \alpha I$. $\alpha$ is an ideal DQE of a scintillator (and a photoconductor) and $I$ is a DQE-degradation factor.

### *7.3.3.2 Readout Pixel Array*

A second simple cascaded model considers an ideal 1D pixel array with unity quantum efficiencies for the conversions from x-ray to light quanta ($\alpha = 1$) and light to electronic quanta ($\eta_e = 1$). The pixel width and aperture are $p_x$ and $a_x$, respectively, and the number of electronic quanta integrated by an aperture is related to the output signal by a deterministic scaling factor $k$. For a uniform distribution of incident x-ray quanta $\bar{q}_0(\mathbf{r})$ [mm$^{-2}$], the resulting output signal is $ka_{xy}\bar{q}_0$, where $a_{xy} = a_x a_y$, and the MTF is:

$$\mathrm{MTF}(u,v)\big|_{v=0} = \mathrm{sinc}(a_x u)\mathrm{sinc}(a_y v)\big|_{v=0} = \mathrm{sinc}(a_x u), \tag{7.14}$$

and

$$\mathrm{NPS}(u,v)\big|_{v=0} = k^2 a_{xy}^2 \bar{q}_0 \left[\sum_{n_x=0}^{\infty}\sum_{n_y=0}^{\infty} \mathrm{sinc}^2\left(a_x\left(u \pm \frac{n_x}{p_x}\right)\right)\mathrm{sinc}^2\left(a_y\left(v \pm \frac{n_y}{p_y}\right)\right)\right]\Bigg|_{v=0}. \tag{7.15}$$

It is shown by Zhao and Rowlands [164] that the sum of $\mathrm{sinc}^2(\gamma_x p_x u)$ and its aliases at harmonics of $u_s = 1/p_x$ is always a constant equal to $1/\gamma_x = p_x / a_x (\geq 1)$, for all frequencies up to the sampling cutoff frequency $u_c = u_s / 2$. From the separability, we have:

$$\mathrm{NPS}(u) = k^2 a_{xy}^2 \bar{q}_0 \left[\frac{p_x}{a_x}\frac{p_y}{a_y}\right] = \frac{k^2 a_{xy}^2 \bar{q}_0}{\gamma_{xy}}. \tag{7.16}$$

Consequently, the DQE is given by

$$\mathrm{DQE}(u) = \gamma_{xy}\mathrm{sinc}^2(a_x u). \tag{7.17}$$

This result demonstrates noise aliasing. It is readily expected that $\mathrm{DQE}(u) = \mathrm{sinc}^2(a_x u)$ for a pixel design with a unity fill factor.

### 7.3.4 Complex Systems

Combining the two simple models described earlier gives a more complete and appropriate model for the analysis of typical FPD structures. Referring to Figure 7.10, propagation of the mean and NPS through a cascade of elementary processes in an FPD, including the addition of electronic noise, is developed in Appendix A and the resulting DQE formalism is:

$$\mathrm{DQE}(u) = \frac{\mathrm{T}_c^2(u)\mathrm{sinc}^2(a_x u)}{\dfrac{1}{\gamma_{xy}\bar{g}} + \dfrac{1}{\alpha}\left(\dfrac{1}{I} - \dfrac{1}{\bar{m}}\right)\left[\mathrm{T}_c^2(u)\mathrm{sinc}^2(a_x u) + \mathrm{T}_{\mathrm{alias}}^2(u,v)\Big|_{v=0}\right] + \dfrac{\bar{q}_0 p_{xy}\sigma_{\mathrm{add}}^2}{\bar{d}^2}}. \tag{7.18}$$

The spreading of secondary quanta is dependent upon depth in an x-ray converter – where x-ray interaction occurs – which is known as the Lubberts' effect [166]. Furthermore, all x-ray and secondary quanta interactions are depth-dependent. The parameters shown in Equation (7.18) are assumed to be the averaged values considering the interaction depths.

Aggressive assumptions may provide us with an insight into some fundamental DQE differences between direct- and indirect-conversion detectors. First, consider an indirect-conversion detector employing a low-resolution scintillator so that $\mathrm{T}_{\mathrm{pre}}(u) \approx \mathrm{T}_c(u)$, where $\mathrm{T}_c(u)$ is band-limited. In addition, $\bar{m} \gg 1$ to ignore the additive noise. Then, Equation (7.18) can be simplified to:

$$\mathrm{DQE}(u) \approx \alpha I, \tag{7.19}$$

which corresponds to the upper limit of an x-ray converter DQE [53], as shown in Section 7.3.3.1.

Now, let us apply an opposite assumption to a direct-conversion detector with a high-resolution capability such that $\mathrm{T}_{\mathrm{pre}}(u) \approx \mathrm{sinc}(a_x u)$. Then, the DQE is approximated as:

$$\mathrm{DQE}(u) \approx \gamma_{xy}\alpha I\mathrm{sinc}^2(a_x u). \tag{7.20}$$

In this case, the pixel array property, as shown in Equation (7.17), is explicitly revealed in combination with a photoconductor DQE performance; $\gamma_{xy}$ acts as an additional DQE-degradation factor as a result of noise aliasing due to the infinite bandwidth of $\mathrm{sinc}(a_x u)$. When the spreading of secondary quanta in a photoconductor is negligible, the spatial resolution is determined only

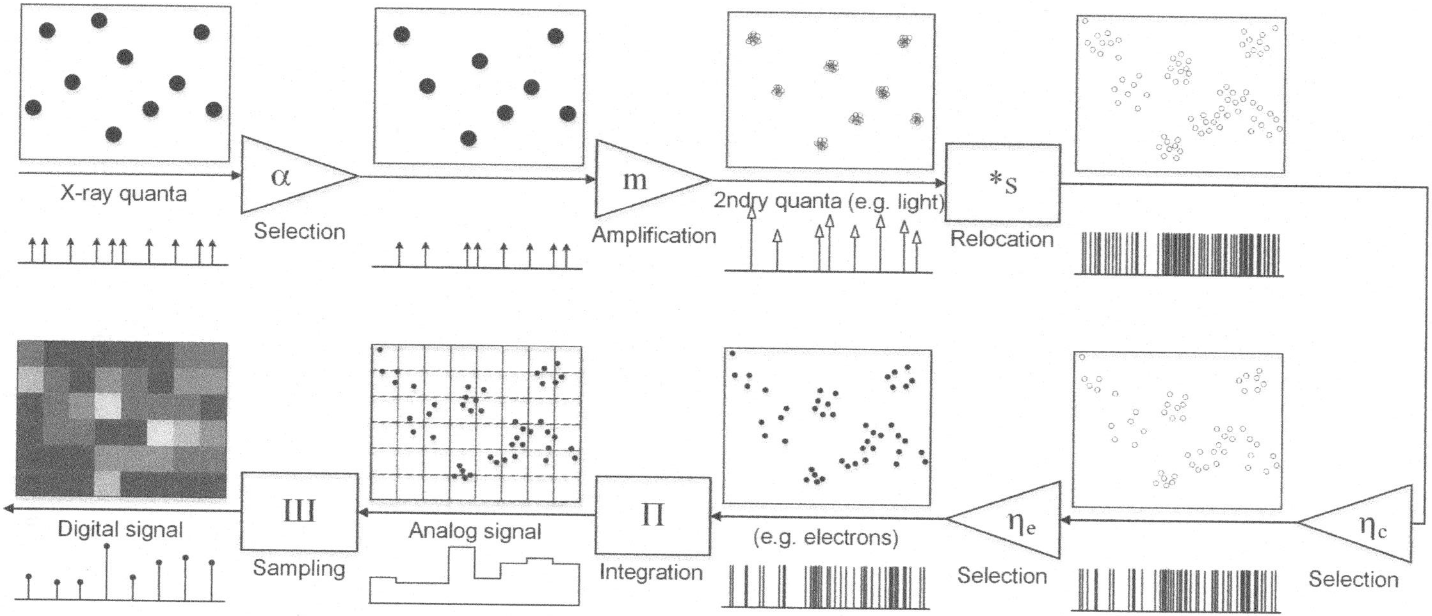


**FIGURE 7.10**
Schematic illustration of a cascaded model describing the propagation of information-carrying quanta in a detector system. The model shows the detection of x-ray quanta (α), the conversion of the detected x-ray quanta into secondary quanta (*m*) (e.g., the optical quanta in a scintillator or the electronic quanta in a photoconductor), the random relocation of the secondary quanta (*s), and their collection or escape ($\eta_c$) in an x-ray converter. Then, the model shows the detection of secondary quanta ($\eta_e$) in the readout pixel array, the spatial integration of the detected quanta over an aperture (Π) and the digital sampling process (III). An additional stage describing the addition of electronic noise ($\sigma_{add}$) can follow the digital output stage.

by the pixel aperture size. However, a noise-power penalty, with a magnitude corresponding to the reciprocal of the pixel fill factor, is reflected in the DQE performance. In the design of direct-conversion detectors, the geometric fill factor is less than unity but the effective fill factor viewed by secondary quanta is generally unity because charge carriers drift along the electric field lines, including their lateral diffusion. Therefore, the reduction of the DQE due to $\gamma_{xy}$ may not be expected in the measurement.

Although Equation (7.18) results from several assumptions and approximations, it can suggest the correct direction for improved detector designs. This can be clearly observed by investigating the additive NPS term normalized by the squared detector output signal in Equation (7.18):

$$\frac{W_{\text{add}}(u)}{\bar{d}^2} = \frac{p_{xy}(\sigma_{\text{add}} / k)^2}{(\bar{q}_0 a_{xy})^2 (\alpha \bar{m} \eta)^2}, \tag{7.21}$$

where $\eta = \eta_c \eta_e$, or total coupling efficiency. Uncorrelated white spectral additive noise, as shown in Equation (7.21), is lethal to the DQE performance, particularly at high spatial frequencies where the number of secondary quanta decreases. The smallest possible $W_{\text{add}}(u) / \bar{d}^2$ may be crucial for detector design. Negligible $W_{\text{add}}(u) / \bar{d}^2$ implies the quantum-noise-limited operation of detectors and the DQE is independent of the dose used for imaging. Let us briefly discuss how to make $W_{\text{add}}(u) / \bar{d}^2$ negligible.

- Increase the number of x-ray quanta incident to the pixel aperture, $\bar{q}_0 a_{xy}$. However, this approach is strictly prohibited when considering the patient dose. A recent trend in radiology is the pursuit of imaging with lower levels of $\bar{q}_0 a_{xy}$, which makes detector design challenging.
- Reduce the additive electronic noise $\sigma_{\text{add}} / k$ (in units of the number of electrons per pixel). The additive electronic noise contains various components, such as pixel, data line thermal, externally coupled, preamplifier, and ADC digitization noise [167]. Maolinbay et al. [168] demonstrated that pixel and digitization noise increased with increasing pixel pitch due to the corresponding increase in pixel charge capacity. In contrast, the data line and preamplifier noise decreased with increasing pixel pitch owing to the reduction in data line capacitance as the number of pixels along the data line decreased for a given detector panel size; this competing effect of various noise contributions requires an optimal pixel pitch. Moreover, the coupling capacitance between metal lines, such as between TFT gate and source/drain and between bias lines and gate/data lines, can be a significant contributor to the total data line capacitance seen by the external ROIC. Therefore, feed-through signal charge through TFTs, variations in gate voltage, fluctuations in the bias voltage, and changes in TFT threshold voltages all appear as noise. Efforts

to reduce the overlapping area between metal lines are usually performed, as shown in Figure 7.3(b). In fact, the development history of large-area DR detectors has always been accompanied by electronic noise reduction and this companionship will continue.

- Improve the conversion gain $\bar{g} = \alpha\bar{m}\eta$. We may separately treat $\alpha\bar{m}$ and $\eta$ as the source (generation) and sink (loss), respectively, of secondary quanta. It is fundamental to find and design an x-ray converter with a larger $\alpha\bar{m}$, as discussed previously, but it is also important to preserve the secondary quanta produced and deliver them to the readout pixel aperture without loss. Therefore, careful attention must be given to the coupling between the x-ray converter and readout pixel array. $\alpha$ and $\bar{m}$ are mainly determined by x-ray converter materials, including the doping processes. $\eta_c$ is largely dependent upon the converter internal structures and impurities contained within it, and its related noise is not negligible, even for the DQE of direct-conversion detectors [169]. In indirect-conversion detector designs, $\eta_e$ is mainly related to the performance of photodiodes considering spectral information of light quanta.

In most cases, the detector design parameters are interrelated. For example, let us consider the Swank factor $I$ in a scintillator. Even if the incident x-ray quanta are monoenergetic, the Swank factor cannot be unity because of the loss of energy carried by Compton scattered and fluorescent photons, resulting in a distribution of absorbed energies [170].[6] The light collection efficiency is known to increase linearly with increasing depth [171–174] and this depth-dependent light collection property further distorts the optical pulse-height distribution in the photodiode (if it is possible to measure it), hence further reducing the Swank factor. The slope of the linear depth dependency of light collection can be alleviated by introducing reflectance at the scintillator backing [172] but the degradation in spatial resolution due to larger light spreading should be taken as a collateral consequence.

The effects of the detector design parameters on the DQE are demonstrated using Equation (7.18) for a representative RQA-5 spectrum [175]. The reference parameter values used for the calculations are summarized in Table 7.2. The pixel pitch $p_x$ was assumed to be 0.15 mm. While the MTF of the *a*-Se FPD was simply modeled using the sinc function, that of CsI:Tl FPD was based on the measured data for an FPD with a pixel pitch of 0.143 mm, manufactured by Toshiba. As shown in Figure 7.11(a), the measured MTF was analyzed via the least-squares regression using the Lorentzian function, $\left(1+(2\pi/\sigma_L)^2u^2\right)^{-1}$, which corresponds to an exponential function, $(\sigma_L/2)e^{-\sigma_L|x|}$, in the space domain. To accommodate the different pixel pitches, the blurring parameter $b = p_x/\sigma_L$ was defined and $b = 0.9$ was determined to be reasonable for $p_x = 0.15$ mm. To consider higher and lower resolution characteristics, $b = 0.6$ and 1.2 were applied to the Lorentzian function, respectively, and the corresponding LSFs are shown in Figure 7.11(b).

**TABLE 7.2**

Detector System Parameters and Corresponding Values Used for the Cascaded-Systems Analysis of Indirect- and Direct-Conversion Detectors

| Parameters | | Value | |
|---|---|---|---|
| **Symbol** | **Description** | **CsI:Tl** | ***a*-Se** |
| $\bar{q}_0 / X$ | Fluence per unit exposure [$mm^{-2}$ $mR^{-1}$] | $2.71 \times 10^{3}$ [a] | |
| $X$ | Exposure [mR] | 0.10[b] | |
| $p_x$[c] | Pixel pitch [mm] | 0.15[b] | |
| $\gamma_{xy}$ | (Effective) Pixel fill factor | 0.70[b] | 1.00[b] |
| $\alpha$ | Quantum efficiency | 0.85[d] | 0.53[d] |
| $\bar{m}$ | Secondary quantum gain | $2.71 \times 10^{3}$ [e] | $0.93 \times 10^{3}$ [e] |
| $I$ | Swank noise factor | 0.85[f] | 0.95[f] |
| $\eta_c$[g] | Collection efficiency | 0.60[b] | 0.90[b] |
| $\eta_e$[g] | Conversion efficiency | 0.60[b] | 0.90[b] |
| $\sigma_{add}$ | Additive electronic noise [$e^-$] | $3.00 \times 10^{3}$ [b] | |

[a] Value for the RQA-5 spectrum.
[b] Assumed value.
[c] Assumed as a squared pixel geometry, hence $p_x = p_y$.
[d] Calculated value (also confirmed by Monte Carlo simulation).
[e] Calculated value.
[f] Estimated value from Monte Carlo simulation (including the optical transport for CsI).
[g] Coupling efficiency $\eta = \eta_c \eta_e$.

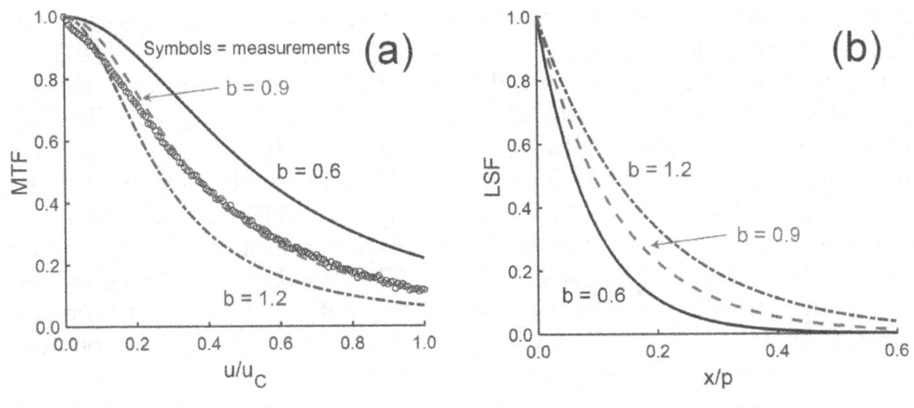


**FIGURE 7.11**
(a) Lorentzian fit analysis of the measured MTF obtained for a CsI:Tl-based FPD. While the FPD has the pixel pitch of 0.143 mm, the Lorentzian function is based on a pixel pitch of 0.15 mm. The blurring parameters $b = 0.6$ and 1.2 correspond to high- and low-resolution characteristics. The abscissa is normalized by the Nyquist frequency. (b) LSFs corresponding to the Lorentzian functions with different blurring parameters. The abscissa is normalized by the pixel pitch of 0.15 mm.

The calculation results are graphically tabulated in Figure 7.12. As plotted in Figure 7.12(a), the spatial frequency-dependent DQE, DQE($u$), decreases gradually with increasing $u$ and the MTF performance with respect to $b$ is well reflected in the DQE($u$) of CsI:Tl FPD. While the approximate DQE formula for a direct-conversion detector, Equation (7.20), effectively describes the DQE($u$) of $a$-Se FPD, that for an indirect-conversion detector, Equation (7.19), estimates the level of the DQE of CsI:Tl FPD around the zero frequency. Figure 7.12(b) shows the DQE value at the zero-frequency and Nyquist limit; i.e., DQE(0) and DQE($u_c$), respectively, at each stage number, as described in detail in Appendix A. The DQE(0) never surpasses that of the previous stage. On the other hand, DQE($u_c$) abruptly drops at the sampling stage. Noise aliasing is detrimental to the DQE performance of the hypothetical CsI:Tl FPD because of non-negligible MTF values beyond the Nyquist limit, as expected from Figure 7.11(a). The plot of the DQE as a function of stage number shows that the CsI:Tl FPD is more sensitive to additive noise than the $a$-Se FPD. The effect of x-ray exposure on the DQE(0) is shown in Figure 7.12(c), where the DQE(0) is normalized by the ideal DQE(0) = $\alpha I$. For given reference parameters, as listed in Table 7.2, the quantum-noise-limited exposure levels of CsI:Tl and $a$-Se FPDs are ~$7 \times 10^{-4}$ and ~$1 \times 10^{-3}$ mR, respectively.[7] The effects of the detector design parameters on the DQE(0) are shown in Figure 7.12(d)–(f). The cascaded model demonstrates that the DQE(0) of CsI:Tl FPD is relatively less sensitive to the coupling efficiency $\eta$ and pixel fill factor $\gamma_{xy}$ compared to that of $a$-Se FPD and both detectors are very sensitive to the additive electronic noise greater than ~$10^3$ $e^-$.

Cascaded-systems analysis, using cascaded linear-systems theory, was widely used in the late 1990s and early 2000s when the development of FPDs was at its height [109, 164, 176–179]. It has been successfully used to describe several systems of complex design and using different technologies: CR systems [180]; video-based electronic portal imaging devices (EPIDs) [181]; indirect- and direct-conversion EPIDs [182, 183]; slot-scanning photodiode/CCD and CdZnTe hybrid digital mammography systems [184, 185]; fiber optics-coupled mammography systems [186]; video-based fluoroscopy systems [187]; solid-state fluoroscopic system [188]; micro-angiography systems [189, 190]; lens-coupled chest radiography systems [191]; contrast enhanced mammography [192]; FPD-based breast tomosynthesis [193]; FPD-based dual-energy (DE) imaging systems [194]; FPD-based cone-beam computed tomography (CBCT) systems [195–197]; solid-state avalanche detector systems [130, 133, 198, 199]; organic device-based FPDs [135]; and photon-counting detector (PCD) systems [200].

Many of these require parallel cascades of simple processes, which can be achieved with the use of a cross-spectral density term [163, 201, 202]. For example, the reabsorption of characteristic x-ray emissions in CsI:Tl can be modeled using parallel cascades and can have a substantial effect on image noise, and hence the DQE [203–205]. Akbarpour et al. [206] further extended this approach to the spatio-temporal domain for the description of fluoroscopic systems. Figure 7.13

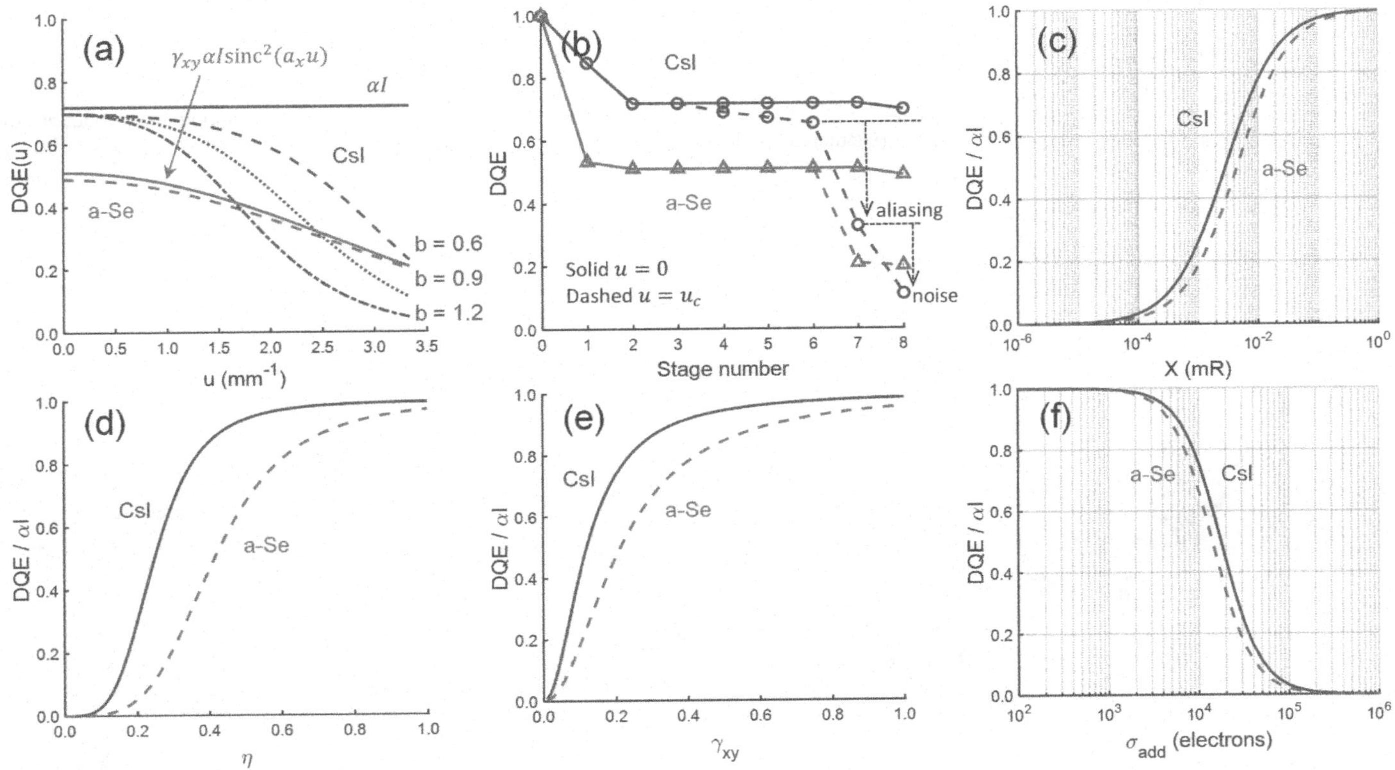


**FIGURE 7.12**
Summary of cascaded-systems analysis for hypothetical CsI:Tl and *a*-Se FPDs, with design parameters listed in Table 7.2. Calculated DQE($u$) results are compared in (a). For comparisons, the calculation results using approximate DQE formulas are also plotted. In (b), DQE values at two extreme spatial frequencies, i.e., the zero and Nyquist frequencies, as a function of stage number, are compared. (c) through (f), the DQE(0) normalized by the ideal DQE(0) = $\alpha I$ is plotted for various parameters, such as x-ray exposure, coupling efficiency, pixel fill factor, and additive electronic noise, respectively.

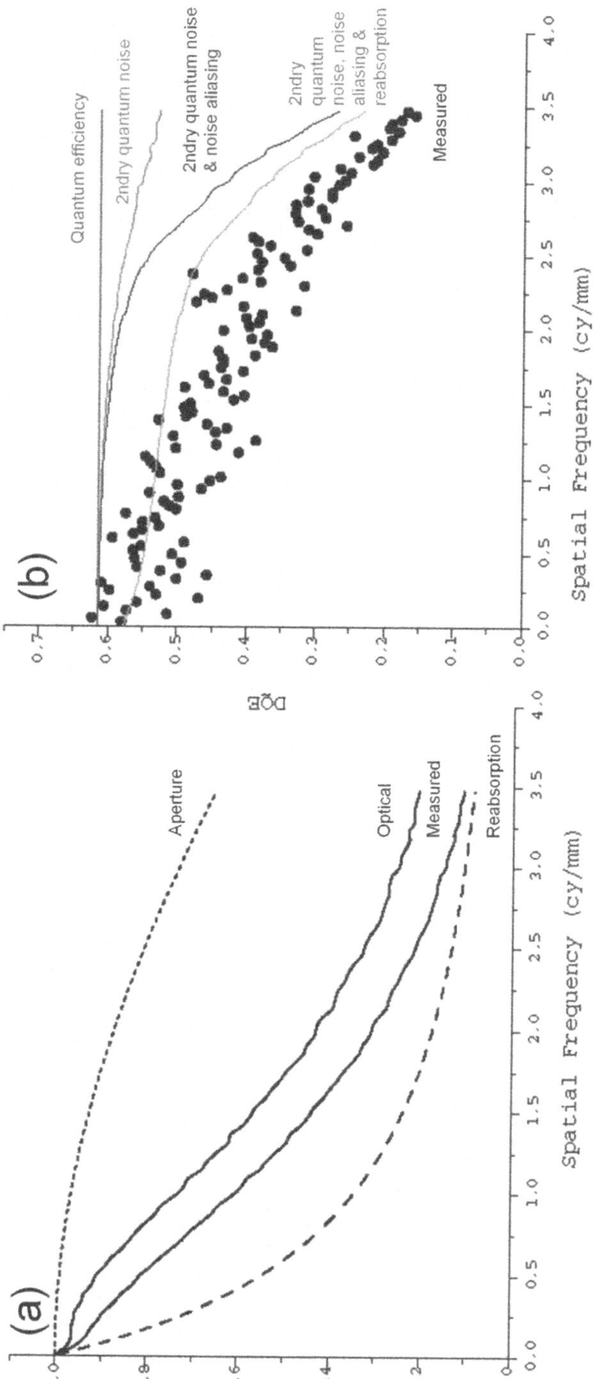


**FIGURE 7.13**
(a) Comparison of measured, theoretical reabsorption, estimated optical, and aperture (sinc) MTFs. (b) Comparison of measured DQE (117 kV, 4 cm Al) with the theoretical model.

shows both the calculated and measured DQE values of an indirect-conversion FPD employing CsI:Tl [207]. The cascaded model was used to determine the effect of various processes on the DQE and observe the following:

1. A mild secondary quantum sink exists at high spatial frequencies, degrading the DQE by approximately 10% at the sampling cutoff frequency. This loss can be recovered only by increasing the light output from the CsI:Tl or by improving the light collection efficiency of the FPD.
2. Noise aliasing starts to degrade the DQE at 2.5 cycles/mm and decreases the DQE by approximately 50% at the cutoff frequency. This could be improved by decreasing the detector element size but this may be impractical.
3. Reabsorption of characteristic x-ray emissions degrades the DQE by approximately 10% over most spatial frequencies. It may not be possible to improve this.

Measured DQE values remain below the cascaded-model prediction by up to 15% at frequencies greater than approximately 1 cycle/mm. While the reason for this discrepancy is not yet known, it has been shown that x-rays transmitted through the scintillator may interact directly in the readout pixel array and may account for this discrepancy [208, 209].

## 7.4 Limitations

The detector *projection* signal for an x-ray beam attenuation through a ray $s$ directed to a pixel positioned at $\mathbf{x}$ in the detector can be approximately given by:

$$\bar{d}(\mathbf{x}) \approx (1+\mathrm{SPR})kp_{xy}\int_0^{\infty}\bar{q}_0(E)e^{-\int_{s(\mathbf{X}')}\mu(E,\mathbf{X}')ds}\alpha(E)w(E)dE, \tag{7.22}$$

where $\mathbf{x}'$ describes the object space and $w(E)$ represents an energy-dependent detector response. SPR stands for the scatter-to-primary ratio. The excellence of x-ray radiography may be in the precise extraction of $\mu(\mathbf{x}')$ from projection images. However, it is impossible to extract $\mu(\mathbf{x}')$ directly as it is projected into *space-* and *energy-averaged* forms, as shown in Equation (7.22); hence, the conspicuity of lesions that describes the extent to which a stimulus is detectable, characterizable, or otherwise *well seen* is reduced, relative to confounding factors in the image. Some limitations of the projection theory can be summarized as follows:

1. The projection of a three-dimensional (3D) human anatomy on a 2D-image results in the superimposition of normal tissue that generates structural noise and obscures the abnormalities to be detected. In some frequent cases, superimposed normal tissue can be misread as abnormalities (i.e., lesions). It is known that the anatomical background itself is the dominant factor in conspicuity by an order of magnitude compared to the image quantum noise [210].
2. While the linear attenuation coefficient $\mu$ is energy- and material-dependent, the energy-averaging projection can yield the same $\mu$ for different materials in the image [211].

Detection and classification of pulmonary nodules are challenging tasks in chest radiography. While the inherent contrast of a 3-mm nodule at 120 kVp is approximately 1.5% using an SPR of 2 [i.e., $C = (1 - e^{-\Delta\mu \times 0.3\text{cm}}) / (1 + \text{SPR})$], for example, this contrast is rarely visible in practice on chest radiographs; this is because the overlying rib and vasculature structures produce a background that masks the presence of small nodules. Conversely, the confusing background of overlying anatomic structures fundamentally limits the visibility of pulmonary nodules via chest radiography, as claimed by Samei et al. [212].

These issues may be mitigated by advanced x-ray imaging techniques capable of *depth* and *energy* discrimination, such as CT, and DE or multi-energy imaging, respectively. These advanced x-ray imaging techniques require a linear approximation of Equation (7.22) for their reconstruction. Otherwise, more elaborate numerical approaches, which usually require a large computational effort, are required.

### 7.4.1 Depth Discrimination

CT can provide high-conspicuity images by isolating specific planes of anatomic structures from the overlying background, and CT has become the gold standard for detecting pulmonary nodules. In principle, CT acquires many *forward* projections around the patient and then projects them *backward* to the patient space; hence, medical CT requires a specialized gantry system, including a high-power x-ray source and a fast detector system. It is well-known that CT causes a large patient dose. Therefore, the CT system cannot substitute for or replace the role of a simple radiographic system.

With a small number of projections obtained in limited angular ranges, tomographic image reconstruction is possible. This limited-angle tomography is regarded as a new genre in medical imaging techniques, popularly known as *digital tomosynthesis*. Tomosynthesis is a technique that produces section or slice images parallel to the pivot axis or axis of rotation from a series of projection images, acquired as the x-ray source moves over a prescribed path. Therefore, it can provide 3D information at lower x-ray irradiation and potentially lower cost than conventional CT in certain imaging situations. Although the theoretical framework describing tomosynthesis was established in the 1930s [213], it has received renewed interest since large-area FPDs became commercially available in the late 1990s.

The tomosynthesis clearly shows lesions that are not explored by conventional radiographs [214]. Although a narrow angular scan range can result in poor depth resolution and cause out-of-plane blur artifacts, it gives rise to a better spatial resolution in planes than the conventional CT scan. For example, Flynn et al. [215] reported that the spatial resolution of digital tomosynthesis was about three times better than that of conventional volumetric CT and further confirmed their claim by investigating the knee joint tomographic images of a human cadaver reconstructed from both modalities. Interested readers can find comprehensive review articles on digital tomosynthesis [216, 217] and references therein. Tomosynthesis is most successful for breast imaging and readers are recommended to refer to a series of review articles [218, 219].

### 7.4.2 Energy Discrimination

Another method for reducing background clutter in some imaging tasks is to use an energy-discriminating technique or DE imaging. The DE method enhances material content (e.g., bone or soft tissue) within a 2D radiograph by combining two images obtained at different x-ray energies [211]. The enhanced soft-tissue signal with less bone-signal background in diagnostic radiography has increased the sensitivity in the detection of lesions [220–222]. In contrast, bone-signal enhancement has been used to improve specificity in characterizing benign lesions [223]. Receiver-operating characteristic studies have shown that the DE method can improve the conspicuity of lesions in particular examinations, such as detecting and characterizing small lung lesions, compared to conventional DR for the same patient dose [224].

The first literature describing DE imaging was presented in 1953 when Jacobson investigated the practical possibilities for *dichromatic* absorption radiography, briefly *dichromography* [225]. However, DE chest imaging was first practically applied when CR technology became available in the early 1980s [1]. On the basis of promising findings by several investigators, a commercial version of the DE technique based on CR plates was released in the 1990s. However, DE images obtained using CR plates with certain technical factors suffered from poor noise characteristics. Again, the advent of large-area FPDs in the late 1990s made available chest radiography with higher DQE and improved noise properties. A commercial version of a DE product based on FPDs is available [223, 226, 227].

Commercial DE imaging systems employing FPDs currently use a *dual-shot* or dual-exposure approach that acquires low- and high-energy projections in successive x-ray exposures by rapidly switching the kilovoltage (kVp) applied to the x-ray tube. However, the time interval between exposures can result in "motion artifacts," which must be addressed [228]. To mitigate the motion-artifact issue, DE imaging systems may incorporate an additional cardiac-gating system designed to deliver both low- and high-energy exposures at the same phase of the cardiac cycle (e.g., diastole). The gating system may measure an electrocardiogram [229] or a plethysmogram using a fingertip pulse oximeter [230]. In addition to hardware-based registration, software-based image registration will further enhance the alignment between the low- and high-energy images. For non-gated acquisitions, motion artifacts can restrict the successful application of dual-shot methods to relatively stationary and cooperative patients.

An alternative method is to use a *single-shot* approach to DE imaging by acquiring two images simultaneously, such as by stacking PSPs in a sandwich configuration [231, 232]. The front layer absorbs (primarily) low-energy x-ray photons, while the rear layer absorbs (primarily) high-energy photons. Early investigators examined the use of different detector systems. Speller et al. [233] used film-screen pairs but found their dynamic range and speed insufficient for practical use. Brooks and Di Chiro [234] and Fenster [235] investigated a xenon split-detector design for CT. Barnes et al. [236] demonstrated single-shot images using a pair of scintillator-coupled linear photodiode arrays that required a time-consuming scanning procedure for area images. Later, Allec et al. [237] reported on the feasibility of using amorphous-selenium layers. Their demonstration single-pixel detector showed good agreement with a theoretical model of signal and noise and suggested practical feasibility.

While single-shot methods are more tolerant of patient motion and less susceptible to motion artifacts, they generally suffer from reduced contrast-to-noise ratio compared to dual-shot methods for the same total patient dose due to poor spectral separation [238, 239]. Spectral separation can be improved using layers of differing atomic numbers, but this is not always practical. Ergun et al. [240] showed that image quality could be improved using PSPs by increasing the energy separation between the front and rear layers and using scatter and beam-hardening corrections. Alvarez [241] described a hybrid single-dual-shot method using fast kVp switching with a novel PSP-based sandwich detector in which the sensitivity of the front PSP was modulated by a custom electro-optical system that erased the high-kVp signal to achieve greater spectral separation.

Kim and his colleagues have developed sandwich-like multilayered detectors for single-shot DE radiography and tomography by stacking two indirect-conversion detectors (the $Gd_2O_2S$:Tb phosphor screen plus photodiode array format) [242–244]. For the same photodiode array, the front detector layer uses a thin phosphor while the rear detector layer employs a thicker phosphor. Such detector configuration allows us to obtain two different energy-like images as well as two different spatial-resolution images at a single exposure [245, 246]. The sandwich detector suffered from significant quantum noise in the rear detector layer [247] and an updated version of a

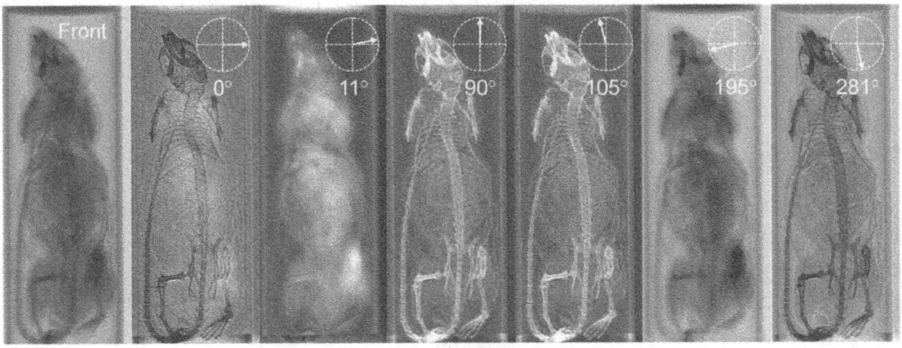


**FIGURE 7.14**
Vector images of a postmortem mouse using a laboratory-scale sandwich detector. The left-most image corresponds to conventional radiography as it was obtained from the front detector layer of the sandwich detector. Angles of 0° and 90° correspond to the basis materials of PMMA and Al, respectively. We note that the image is inverted for a 180° angulation.

laboratory-scale sandwich detector was developed [248], which employed a columnar structured CsI:Tl scintillator with a higher quantum efficiency for the rear detector. Compared to the field of view (FOV) of previous designs (~50 × 20 $mm^2$), the FOV was increased up to ~140 × 130 $mm^2$. Example mouse images obtained using the sandwich detector are shown in Figure 7.14. The DE images were reconstructed using the basis material decomposition approach [249]. Polymethyl methacrylate (PMMA) and aluminum (Al) were chosen as the basis materials. The combination of the phosphor and photodiode array layers for the indirect-conversion FPD configuration has an advantage regarding design flexibility and performance optimization as the two phosphor and photodiode parts are physically separate and the corresponding x-ray and light conversions are independent of each other. A theoretical framework describing the signal and noise generation in the sandwich detector has been introduced [244] and this approach will be useful for designing a detector in terms of various detector design and operation parameters (e.g., the intermediate filter material/thickness and the applied tube voltage or kVp).

A three-layered detector design for DE imaging is available [250].[8] Each detector layer is based on the *a*-Si:H photodiode/TFT pixel array technology. Conventional radiography images are reconstructed by summing all of the layers' images while tissue-subtracted images are generated using logarithmic subtraction. Such a design does not require the use of an intermediate metal filter layer; hence, it may enable the potential for high dose efficiency while maintaining a good spectral separation. The author has found the latest article that describes the development of a prototype large-area (17-inch) flat-panel sandwich detector and characterizes the performance of single-shot DE radiography/fluoroscopy and DE-CBCT [251]. The prototype consists of a 0.2-mm thick CsI:Tl-coupled *a*-Si:H photodiode/TFT panel with a pitch of 0.15 mm for the front detector layer and a 0.55-mm thick CsI:Tl-coupled same panel for the rear layer.

### 7.4.3 Energy Weights

The DQE of conventional detectors includes the detector contrast modulation $T_c(u)$ in the spatial-frequency domain, as described in Equation (7.18), but it ignores the detector contrast modulation in the *energy* domain. While the x-ray beam used for imaging is usually a distribution of the number of x-ray quanta as a function of energy $q_0(E)$ [mm$^{-2}$ keV$^{-1}$], conventional *energy-integrating* detectors (EIDs) do not portray incoming spectral information to output images.

Let us consider a simple large-area SNR measurement task of a target material with a thickness $x$ embedded in background material. The SNR may be given by the signal difference between the background and target regions divided by the corresponding noise, or $(\bar{d}_b - \bar{d}_t)/(\sigma_b^2 + \sigma_t^2)^{1/2}$. Assuming that the measurements are Poisson-distributed, we have:

$$\mathrm{SNR}_d^2 = A \frac{\left[ \int_0^\infty \bar{q}_b(E)\left(1 - e^{-\Delta\mu(E)x}\right)\alpha(E)\langle w(E)\rangle \mathrm{d}E \right]^2}{\int_0^\infty \bar{q}_b(E)\left(1 + e^{-\Delta\mu(E)x}\right)\alpha(E)\langle w^2(E)\rangle \mathrm{d}E} \tag{7.23}$$

$$\leq A \int_0^\infty \bar{q}_b(E) \frac{\left(1 - e^{-\Delta\mu(E)x}\right)^2}{1 + e^{-\Delta\mu(E)x}} \alpha(E) \frac{\langle w(E)\rangle^2}{\langle w^2(E)\rangle} \mathrm{d}E, \tag{7.24}$$

where $\Delta\mu$ is the difference between the linear attenuation coefficients of the target and background materials, and

$$\langle w(E)\rangle = \int_0^\infty w(\varepsilon)\mathcal{R}(\varepsilon, E)\mathrm{d}\varepsilon. \tag{7.25}$$

$\bar{q}_b(E)$ represents the attenuated x-ray beam through the background and $A$ denotes the imaging area. The response function $\mathcal{R}(\varepsilon, E)$ gives the probability density of depositing energy $\varepsilon$ given an interacting quantum with energy $E$. Equation (7.24) can be obtained using the Schwartz inequality. Since $\langle w(E)\rangle = \bar{\varepsilon}(E)$ for EIDs, we have:

$$\mathrm{SNR}_{\mathrm{EI}}^2 \leq A \int_0^\infty \bar{q}_b(E) \frac{\left(1 - e^{-\Delta\mu(E)x}\right)^2}{1 + e^{-\Delta\mu(E)x}} \alpha(E) \frac{\bar{\varepsilon}^2(E)}{\overline{\varepsilon^2}(E)} \mathrm{d}E. \tag{7.26}$$

It is noted that the term $\overline{\varepsilon}^2(E)/\overline{\varepsilon^2}(E)$ is the Swank factor [252]. Since the DQE can be defined as the ratio of ideal and actual NPSs, the task-dependent DQE can be similarly defined as [253]:

$$\mathrm{DQE}_{\mathrm{task}} = \frac{\mathrm{SNR}_{\mathrm{EI}}^2}{\mathrm{SNR}_{\mathrm{ideal}}^2}, \tag{7.27}$$

where

$$\mathrm{SNR}_{\mathrm{ideal}}^2 = A\int_0^{\infty} \overline{q}_b(E)\frac{\left(1-e^{-\Delta\mu(E)x}\right)^2}{1+e^{-\Delta\mu(E)x}}\,\mathrm{d}E \tag{7.28}$$

that is achieved by an *ideal* detector, where both the quantum efficiency and Swank noise factor are equal to unity.

In the case of *photon-counting* detectors or PCDs, the weighting factor might be an energy-independent constant, and Tapiovvarra and Wagner [253] demonstrated that the EIDs were not as efficient as the PCDs in terms of task-dependent DQE. For various mammographic imaging tasks, Cahn et al. [254] claimed that the DQE of PCDs was larger than that of the EIDs by ~10%.

The optimal detector response that gives rise to the equality in Equation (7.24) is [253]:

$$\overline{w}^*(E) = c\frac{1-e^{-\Delta\mu(E)x}}{1+e^{-\Delta\mu(E)x}}, \tag{7.29}$$

where $c$ is an arbitrary constant and, therefore, a low-contrast limit yields $\overline{w}^*(E) \propto \Delta u(E)$. Recognizing this, Cahn et al. [254] further demonstrated that when PCDs were operated with energy-dependent weighting proportional to $E^{-3}$, the DQE improvement could be around 30% in mammography, compared to the EID.

The development of PCDs with energy-resolving capability is of great interest in x-ray radiology [252, 255, 256]. Expected benefits include improved contrast of specific materials (e.g., iodine) with the development of energy-weighting algorithms [253, 257, 258] and improved SNR performance using energy thresholds to reject electronic noise [254, 259, 260]. In fact, the photon-counting approach has been used in gamma-ray or nuclear medicine imaging modalities, such as scintigraphy, single-photon emission computed tomography, and positron emission tomography since the 1970s but, until recent developments, higher count rates and stricter spatial-resolution requirements have restricted their use in x-ray imaging.

An important milestone in the development of PCDs for high-resolution x-ray imaging is the development of a pixel readout chip, referred to as Medipix2, having a pitch of 55 μm in a format of 256 × 256 pixels for reading out charge signals from the overlying x-ray converter [261]. The x-ray converter could be gaseous [261] or semiconductor detectors [262]. Each pixel contains 504 transistors, constituting a charge preamplifier, discriminators, shift registers, digital-to-analog converters, and digital logic. Readers can find a review article [263] covering the early developments in the area of semiconductor PCDs, including the development of other versions of photon-counting pixel readout chips. It is typical to use a bump-bonding technique for the electronic coupling between semiconductor conversion layers and photon-counting readout chips, which limits the detector size. As witnessed in the DR detector development history, photon-counting imaging systems employ scanning of linear or slot detector arrays and a mosaic of small-sized detectors for large-area x-ray imaging. Problems remain to be solved for DR, such as large-area coverage.

PCDs consist of a matrix of sensor elements with pulse-height data obtained from each. Incorrect determination of photon energy results when x-ray energy is deposited into more than one element or more than one readout interval [264, 265]. For example, charge summing has been used to address charge diffusion and signal sharing (Medipix3) [266] and signal processing methods to address pulse pileup [267]. The random nature of x-ray interactions and energy deposition causes additional pulse-height errors that remain a problem for the reliable use of PCDs. Escape or reabsorption of fluorescent photons after a photoelectric interaction, or Compton scatter, causes errors in measured photon energy and false counts [252, 268]. With conventional EIDs, this contributes to the Swank noise that, when multiplied by the quantum efficiency, gives the DQE of the detector. If not corrected in PCDs, the degraded x-ray interaction-induced SNR is an upper limit to the SNR that can be achieved [53, 205]. Recently, research to understand the signal and noise characteristics, or DQE, of PCDs in the coupled space-energy domain is active for a better design and operation of PCDs [264, 269–275].

## 7.5 Concluding Remarks

The highest DQE performance of a detector system that can be achieved is the x-ray quantum detection efficiency determined by the first interaction in an x-ray converter. The ideally achievable DQE is then gradually degraded by fluctuations in subsequent processes conducted by the x-ray converter and readout pixel array, even if the process is signal amplification; this entire DQE degradation due to several complicated factors may be simply categorized as the Swank factor [276]. For example, imagine a detector having DQE = $\alpha I$, which is not appropriate for a certain imaging task requiring an image SNR of $\sqrt{\bar{q}_0 \alpha}$. To accomplish the imaging task successfully, the detector should be redesigned by employing a new x-ray converter, $\alpha_{\text{new}} = \text{DQE} / I$. Otherwise, the detector should be designed to have

$I=1$. Both approaches are challenging. Efforts are still being made to enhance the designs of readout pixel arrays, photoconductors, and scintillators to produce images to detect lower-contrast and smaller lesions using the lowest possible patient dose.

In the history of digital radiographic detector developments, the large-area imaging capability of a detector was a barrier, particularly for clinical applications, and amorphous silicon technology was a breakthrough. It took about one-hundred years to carve x-ray images onto a 17-inch single electronic panel instead of a photographic film. However, the scenes viewed by the amorphous silicon-based detectors operated in the energy-integrating mode are black and white, and x-ray beam spectral information has not been properly utilized. Photon-counting detectors in cooperation with multi-energy bins explore true colored imaging but the same barriers prevent their realization in large areas as with the EIDs.

## Acknowledgment

I would like to thank Prof. Gyuseong Cho (KAIST, Daejeon, Korea) for sharing his experience in the development of flat-panel detectors, Dr. Zhye Yin (GE Global Research Center, Niskayuna, NY, USA) for her advice in view of the industry, Dr. Ian Cunningham (Robarts Research Institute, London, ON, Canada) for sharing his knowledge on the cascaded linear-systems theory. My gratitude extends to former students, Dr. Dong Woon Kim and Dr. Junwoo Kim, for their help in preparing some figures and for their useful discussions. Lastly, I would like to thank Dr. Jinwoo Kim who assisted in editing the manuscript and gave me valuable words of advice. This work was supported by the National Research Foundation of Korea (NRF) grants funded by the Korea government (MSIP) (No. 2017M2A2A6A01019930 and No. 2021R1A2C1010161).

## Appendix A: Signal and Noise Transfer through a Cascade of Simple Elementary Processes in Flat-Panel Detectors

For a cascaded model of an x-ray imaging detector in a combination of several elementary stages, each of which describes a physical transfer mechanism of information-carrying quanta, as depicted in Figure 7.10, the linear-systems theory of signal and noise transfer, including the addition of electronic noise at the digital output, begins with a Poisson-distributed incident fluence $q_0$ with mean $\overline{q}_0$ [mm$^{-2}$] and NPS $W_0(\mathbf{u}) = \overline{q}_0$ [mm$^{-2}$], where $\mathbf{u} = (u, v)$ [162].

### Stage 1: X-Ray Detection

The first stage describes the binomial detection of x-ray quanta in a converter with mean $\alpha$ and variance $\sigma_\alpha^2 = \alpha - \alpha^2$. The quantum selection process shown in Table 7.1 gives rise to:

$$\bar{q}_1 = \bar{q}_0 \alpha \tag{7.30}$$

in [mm$^{-2}$] and

$$W_1(\mathbf{u}) = \bar{q}_0 \alpha \tag{7.31}$$

in [mm$^{-2}$].

### Stage 2: Conversion to Secondary Quanta

In this stage, the detected x-ray quantum is converted to $\bar{m}$, the number of secondary quanta (e.g., charge carriers in a photoconductor or light quanta in a scintillator) per detection on average with $\sigma_m^2 = \overline{m^2} - \bar{m}^2$. The quantum gain process shown in Table 7.1 gives rise to:

$$\bar{q}_2 = \bar{q}_0 \alpha \bar{m} \tag{7.32}$$

in [mm$^{-2}$] and

$$W_2(\mathbf{u}) = \bar{q}_0 \alpha \overline{m^2} = \bar{q}_0 \alpha \frac{\bar{m}^2}{I} \tag{7.33}$$

in [mm$^{-2}$].

### Stage 3: Collection of Secondary Quanta

This stage describes the survival of secondary quanta (or their escape) to the converter bottom surface during their transport. The assumption of binomial quantum selection with a mean $\eta_c$ results in:

$$\bar{q}_3 = \bar{q}_0 \alpha \bar{m} \eta_c \tag{7.34}$$

in [mm$^{-2}$] and

$$W_3(\mathbf{u}) = \bar{q}_0 \alpha^2 \bar{m}^2 \eta_c^2 \left[ \frac{1}{\alpha \bar{m} \eta_c} + \frac{1}{\alpha}\left( \frac{1}{I} - \frac{1}{\bar{m}} \right) \right] \tag{7.35}$$

in [mm$^{-2}$].

**Stage 4: Random Relocation of Secondary Quanta**

This stage assumes that all the secondary quanta reaching the converter bottom surface are relocated independently and with the same probability described by the PSF, which corresponds to the converter characteristic function $T_c(\mathbf{u})$ in the Fourier domain. Application of the quantum scatter process shown in Table 7.1 results in:

$$\bar{q}_4 = \bar{q}_0 \alpha \bar{m} \eta_c \tag{7.36}$$

in [mm$^{-2}$] and

$$W_4(\mathbf{u}) = \bar{q}_0 \alpha^2 \bar{m}^2 \eta_c^2 \left[ \frac{1}{\alpha \bar{m} \eta_c} + \frac{1}{\alpha} \left( \frac{1}{I} - \frac{1}{\bar{m}} \right) T_c^2(\mathbf{u}) \right] \tag{7.37}$$

in [mm$^{-2}$].

**Stage 5: Selection of Secondary Quanta**

This stage describes the selection of secondary quanta that interact with a detector element (or pixel). The photodiode quantum efficiency, which describes the conversion of light quanta into charge carriers, corresponds to this process. The binomial quantum selection with mean $\eta_e$ for this stage gives rise to:

$$\bar{q}_5 = \bar{q}_0 \alpha \bar{m} \eta_c \eta_e = \bar{q}_0 \bar{g} \tag{7.38}$$

in [mm$^{-2}$] and

$$W_5(\mathbf{u}) = \bar{q}_0 \bar{g}^2 \left[ \frac{1}{\bar{g}} + \frac{1}{\alpha} \left( \frac{1}{I} - \frac{1}{\bar{m}} \right) T_c^2(\mathbf{u}) \right] \tag{7.39}$$

in [mm$^{-2}$].

**Stage 6: Spatial Integration of Secondary Quanta**

This stage describes the conversion of secondary quanta within an active element into a digital number and yields an output $d$. Assuming the dimension of an aperture is $a_x \times a_y = a_{xy}$, the quantum integration process shown in Table 7.1 with a deterministic scaling factor $k$ yields:

$$\bar{d} = k \bar{q}_0 a_{xy} \bar{g} \tag{7.40}$$

in [null] and

$$W_d(\mathbf{u}) = \frac{\bar{d}^2}{\bar{q}_0}\left[\frac{1}{\bar{g}} + \frac{1}{\alpha}\left(\frac{1}{I} - \frac{1}{\bar{m}}\right)T_c^2(\mathbf{u})\right]\text{sinc}^2(a_x u)\text{sinc}^2(a_y v) \tag{7.41}$$

in [mm²].

### Stage 7: Sampling

As shown in Table 7.1, the sampling process is the summation of the fundamental presampling NPS as well as its aliases appearing as harmonics at multiples of the sampling frequency $(1/p_x, 1/p_y)$, where $p_x$ and $p_y$ are the side dimensions of a pixel having a size of $p_{xy}$; hence,

$$W_{\text{dig}}(\mathbf{u}) = W_d(\mathbf{u}) + \sum_{i=1}^{\infty}\sum_{j=1}^{\infty} W_d(\mathbf{u} \pm \mathbf{u}_{ij}) \tag{7.42}$$

in [mm²], where $\mathbf{u}_{ij} = (i/p_x, j/p_y)$.

### Stage 8: Addition of Electronic Noise Quanta

Considering a typical dark-field or offset correction, $\bar{d}$ remains unchanged, whereas the detector output noise includes the readout electronic noise. Assuming that the readout noise is independent between pixels, we obtain:

$$W_{\text{out}}(\mathbf{u}) = W_{\text{dig}}(\mathbf{u}) + W_{\text{add}}(\mathbf{u}) = W_{\text{dig}}(\mathbf{u}) + p_{xy}\sigma_{\text{add}}^2 \tag{7.43}$$

in [mm²], where $\sigma_{\text{add}}$ is the root-mean-square readout noise.

### DQE

The DQE, as given by Equation (7.5), is applied to digital systems. Considering the presampling MTF, $T_{\text{pre}}(\mathbf{u}) = T_c(\mathbf{u})|\text{sinc}(a_x u)\text{sinc}(a_y v)|$, and the potential noise aliasing, the digital DQE for a given example is given by:

$$\begin{aligned}\text{DQE}(\mathbf{u}) &= \frac{T_{\text{pre}}^2(\mathbf{u})}{\bar{q}_0\left[W_{\text{out}}(\mathbf{u})/\bar{d}^2\right]} \\ &= \frac{T_{\text{pre}}^2(\mathbf{u})}{\dfrac{1}{\gamma_{xy}\bar{g}} + \dfrac{1}{\alpha}\left(\dfrac{1}{I} - \dfrac{1}{\bar{m}}\right)\left[T_{\text{pre}}^2(\mathbf{u}) + T_{\text{alias}}^2(\mathbf{u})\right] + \dfrac{\bar{q}_0 p_{xy}\sigma_{\text{add}}^2}{\bar{d}^2}},\end{aligned} \tag{7.44}$$

where

$$T_{\text{alias}}^2(\mathbf{u}) = \sum_{i=1}^{\infty}\sum_{j=1}^{\infty} T_c^2(\mathbf{u} \pm \mathbf{u}_{ij})\text{sinc}^2\left(a_x(u \pm i/p_x)\right)\text{sinc}^2\left(a_y(v \pm j/p_y)\right), \tag{7.45}$$

which represents the squared *aliasing* MTF. While the geometric pixel fill factor is defined as $\gamma_{xy} = \gamma_x \gamma_y = (a_x / p_x)(a_y / p_y)$, its appearance in the denominator of Equation (7.44) results from the aliasing effect due to the squared aperture MTF [164]:

$$\gamma_{xy}^{-1} = \sum_{i=0}^{\infty}\sum_{j=0}^{\infty} \mathrm{sinc}^2\left(a_x(u \pm i / p_x)\right)\mathrm{sinc}^2\left(a_y(v \pm j / p_y)\right). \tag{7.46}$$

## Notes

1. The concept of two-layered detector configuration can also be applied to the hybrid photon-counting detectors; the sensor material plus pulse-processing/counting pixel-array layers.
2. This chapter is based on a previous review article [20], with some updates to the references and figures. Section 7.1.3, describing the cascaded-systems analysis of detector performance, has been expanded to show the relationship between image quality and the detector design parameters in depth. The limitation of conventional energy-integrating detectors is discussed.
3. Do not be confused with the linear attenuation coefficient, which has the same symbol $\mu$.
4. In mammography applications, *a*-Se exhibits a higher x-ray quantum efficiency than CsI:Tl, despite its lower $Z$ and density; this is because of probable interactions of x-ray photons with its $K$-shell (binding energy $E_K = \sim 12.7$ keV) for the conventional mammographic x-ray spectra whose average energies are around 20 keV or less.
5. $u$ and $v$ in the frequency domain are the Fourier conjugates of $x$ and $y$ in the space domain, respectively, in terms of the Cartesian coordinates expression.
6. The same phenomenon is observed in direct-conversion detector materials [171], hence not avoidable in photon-counting detectors [172].
7. It is assumed that the quantum-noise-limited exposure occurs when the DQE is reduced by 50%.
8. Commercial products are provided by KA Imaging Inc. (Waterloo, ON Canada).

## References

1. Sonoda, M., Takano, M., Miyahara, J., Kato, H.: Computed radiography utilizing scanning laser stimulated luminescence. Radiology. (1983), doi: 10.1148/radiology.148.3.6878707.
2. Rowlands, J.A.: The physics of computed radiography. Phys. Med. Biol. (2002), doi: 10.1088/0031-9155/47/23/201.

3. Rowlands, J.A., Yaffe, M.J.: X-ray detectors for digital radiography. Phys. Med. Biol. (1997), doi: 10.1088/0031-9155/42/1/001.
4. Street, R.A., Nelson, S., Antonuk, L., Perez-Mendez, V.: Amorphous silicon sensor arrays for radiation imaging. Proc. Mater. Res. Soc. (1990), doi: 10.1557/PROC-192-441.
5. Floyd, C.E., Warp, R.J., Dobbins, J.T., Chotas, H.G., Baydush, A.H., Vargas-Voracek, R., Ravin, C.E.: Imaging characteristics of an amorphous silicon flat-panel detector for digital chest radiography. Radiology. (2001), doi: 10.1148/radiology.218.3.r01fe45683.
6. Kotter, E., Langer, M.: Digital radiography with large-area flat-panel detectors. Eur. Radiol. (2002), doi: 10.1007/s00330-002-1350-1.
7. Yorkston, J.: Recent developments in digital radiography detectors. Nucl. Instrum. Methods Phys. Res. A. (2007), doi: 10.1016/j.nima.2007.06.041.
8. Zhao, W., Rowlands, J.A.: X-ray imaging using amorphous selenium: feasibility of a flat panel self-scanned detector for digital radiology. Med. Phys. (1995), doi: 10.1118/1.597628.
9. Lee, D.L.Y., Cheung, L.K., Jeromin, L.S.: New digital detector for projection radiography. Proc. SPIE (1995), doi: 10.1117/12.208342.
10. Antonuk, L.E., Boudry, J., Huang, W., McShan, D.L., Morton, E.J., Yorkston, J., Longo, M.J., Street, R.A.: Demonstration of megavoltage and diagnostic x-ray imaging with hydrogenated amorphous silicon arrays. Med. Phys. (1992), doi: 10.1118/1.596802.
11. James, J.J., Davies, A.G., Cowen, A.R., O'Connor, P.J.: Developments in digital radiography: an equipment update. Eur. Radiol. (2001), doi: 10.1007/s003300100828.
12. Spahn, M.: Flat detectors and their clinical applications. Eur. Radiol. (2005), doi: 10.1007/s00330-005-2734-9.
13. Rowlands, J., Kasap, S.: Amorphous semiconductors usher in digital x-ray imaging. Phys. Today. (1997), doi: 10.1063/1.881994.
14. Chotas, H.G., Dobbins, J.T., Ravin, C.E.: Principles of digital radiography with large-area, electronically readable detectors: a review of the basics. Radiology. (1999), doi: 10.1148/radiology.210.3.r99mr15595.
15. Rowlands, J.A., Yorkston, J.: Flat panel detectors for digital radiography. In: Van Metter, R., Beutel, J., Kundel, H.L. (eds.) Handbook of Medical Imaging: Vol. 1. Physics and Psychophysics, pp. 223–328. SPIE, Bellingham (2000).
16. Street, R.A.: Large area image sensor arrays. In: Street R.A. (ed.) Technology and Applications of Amorphous Silicon, pp. 147–221. Springer, Heidelberg (2000).
17. Moy, J.-P.: Recent developments in x-ray imaging detectors. Nucl. Instrum. Methods Phys. Res. A. (2000), doi: 10.1016/S0168-9002(99)01196-1.
18. Kasap, S.O., Rowlands, J.A.: Direct-conversion flat-panel x-ray image detectors. IEE P-Circ. Dev. Syst. (2002), doi: 10.1049/ip-cds:20020350.
19. Antonuk, L.E., El-Mohri, Y., Jee, K.-W., Zhao, Q., Sawant, A.R., Su, Z., Street, R.A.: Technological pathways for 21st century active matrix x-ray imager development. Proc. SPIE (2002), doi: 10.1117/12.465545.
20. Kim, H.K., Cunningham, I.A., Yin, Z., Cho, G.: On the development of digital radiography detectors: a review. Int. J. Precis. Eng. Manuf. **9**, 86–100 (2008).
21. Cunningham, I.A., Westmore, M.S., Fenster, A.: A spatial-frequency dependent quantum accounting diagram and detective quantum efficiency model of signal and noise propagation in cascaded imaging systems. Med. Phys. (1994), doi: 10.1118/1.597401.

22. Cunningham, I.A., Shaw, R.: Signal-to-noise optimization of medical imaging systems. J. Opt. Soc. Am. A. (1999), doi: 10.1364/JOSAA.16.000621.
23. Liu, H., Karellas, A., Harris, L.J., D'Orsi, C.J.: Methods to calculate the lens efficiency in optically coupled CCD x-ray imaging systems. Med. Phys. (1994), doi: 10.1118/1.597352.
24. Yu, T., Boone, J.M.: Lens coupling efficiency: derivation and application under differing geometrical assumptions. Med. Phys. (1997), doi: 10.1118/1.598055.
25. Gagne, R.M., Quinn, P.W., Chen, L., Myers, K.J., Doyle, R.J.: Optically coupled digital radiography: sources of inefficiency. Proc. SPIE (2001), doi: 10.1117/12.430937.
26. Kim, H.K., Cho, H., Lee, S.W., Shin, Y.H., Cho, H.S.: Development and evaluation of a digital radiographic system based on CMOS image sensor. IEEE Trans. Nucl. Sci. (2001), doi: 10.1109/23.940143.
27. Cho, M.K., Kim, H.K., Graeve, T., Yun, S.M., Lim, C.H., Cho, H., Kim, J.: Measurements of x-ray imaging performance of granular phosphors with direct-coupled CMOS sensors. IEEE Trans. Nucl. Sci. (2008), doi: 10.1109/TNS.2007.913939.
28. Bates, C.W.: Scintillation processes in thin films of CsI(Na) and CsI(Tl) due to low energy x-rays, electrons and protons. Adv. Electronics Electron Phys. (1969), doi: 10.1016/S0065-2539(08)61380-3.
29. Stevels, A.L.N., Schrama de Pauw, A.D.M.: Vapor-deposited CsI:Na layers, I. morphologic and crystallographic properties, II. screen for application in x-ray imaging devices. Philips Res. Rept. **29**, 340–362 (1974).
30. Nagarkar, V.V., Gupta, T.K., Miller, S.R., Klugerman, Y., Squillante, M.R., Entine, G.: Structured CsI(Tl) scintillators for x-ray imaging applications. IEEE Trans. Nucl. Sci. (1998), doi: 10.1109/23.682433.
31. Itoh, H., Matsubara, S., Takahashi, T., Shimada, T., Takeuchi, H.: Integrated radiation detectors with a-Si photodiodes on ceramic scintillators. Jpn. J. Appl. Phys. (1989), doi: 10.1143/jjap.28.l1476.
32. Holl, I., Lorenz, E., Mageras, G.: A measurement of the light yield of common inorganic scintillators. IEEE Trans. Nucl. Sci. (1998), doi: 10.1109/23.12684.
33. van Eijk, C.W.E.: Inorganic scintillators in medical imaging. Phys. Med. Biol. (2002), doi: 10.1088/0031-9155/47/8/201.
34. Nikl, M.: Scintillation detectors for x-rays. Meas. Sci. Technol. (2006), doi: 10.1088/0957-0233/17/4/r01.
35. Knoll, G.F.: Radiation Detection and Measurement. John Wiley & Sons, Hoboken, NJ (2000).
36. Klein, C.A.: Bandgap dependence and related features of radiation ionization energies in semiconductors. J. Appl. Phys. (1968), doi: 10.1063/1.1656484.
37. Que, W., Rowlands, J.A.: X-ray photogeneration in amorphous selenium: geminate versus columnar recombination. Phys. Rev. B. (1995), doi: 10.1103/PhysRevB.51.10500.
38. Kasap, S.O., Rowlands, J.A.: Photoconductor selection for digital flat panel x-ray image detectors based on the dark current. J. Vac. Sci. Technol. A. (2000), doi: 10.1116/1.582237.
39. Hack, M., Guha, S., Shur, M.: Photoconductivity and recombination in amorphous silicon alloys. Phys. Rev. B. (1984), doi: 10.1103/PhysRevB.30.6991.
40. Guha, S., Hack, M.: Dominant recombination process in amorphous silicon alloys. J. Appl. Phys. (1985), doi: 10.1063/1.336063.

41. Abkowitz, M.: Density of states in a-Se from combined analysis of xerographic potentials and transient transport datas. Phil. Mag. Lett. (1988), doi: 10.1080/09500838808214730.
42. Haugen, C., Kasap, S.O., Rowlands, J.A.: X-ray irradiation induced bulk space charge in stabilized a-Se x-ray photoconductors. J. Appl. Phys. (1998), doi: 10.1063/1.368859.
43. Zhao, W., DeCrescenzo, G., Rowlands, J.A.: Investigation of lag and ghosting in amorphous selenium flat-panel x-ray detectors. Proc. SPIE (2002), doi: 10.1117/12.465557.
44. Zhao, B., Zhao, W.: Temporal performance of amorphous selenium mammography detectors. Med. Phys. (2005), doi: 10.1118/1.1827791.
45. Zhao, W., DeCrescenzo, G., Kasap, S.O., Rowlands, J.A.: Ghosting caused by bulk charge trapping in direct conversion flat-panel detectors using amorphous selenium. Med. Phys. (2005), doi: 10.1118/1.1843353.
46. Rau, A.W., Bakueva, L., Rowlands, J.A.: The x-ray time of flight method for investigation of ghosting in amorphous selenium-based flat panel medical x-ray imagers. Med. Phys. (2005), doi: 10.1118/1.2042248.
47. Siewerdsen, J.H., Jaffray, D.A.: A ghost story: spatio-temporal response characteristics of an indirect-detection flat-panel imager. Med. Phys. (1999), doi: 10.1118/1.598657.
48. Fujieda, I., Nelson, S., Street, R.A., Weisfield, R.L.: Radiation imaging with 2D a-Si sensor arrays. IEEE Trans. Nucl. Sci. (1992), doi: 10.1109/23.159759.
49. Kim, H.K.: Analytical model for incomplete signal generation in semiconductor detectors. Appl. Phys. Lett. (2006), doi: 10.1063/1.2191742.
50. Bloomquist, A.K., Yaffe, M.J., Mawdsley, G.E., Hunter, D.M., Beideck, D.J.: Lag and ghosting in a clinical flat-panel selenium digital mammography system. Med. Phys. (2006), doi: 10.1118/1.2218315.
51. Street, R.A., Ready, S.E., Melekhov, L., Ho, J., Zuck, A., Breen, B.N.: Approaching the theoretical x-ray sensitivity with $HgI_2$ direct detection image sensors. Proc. SPIE (2002), doi: 10.1117/12.465584.
52. Overdick, M., Baumer, C., Engel, K.J., Fink, J., Herrmann, C., Kruger, H., Simon, M., Steadman, R., Zeitler, G.: Status of direct conversion detectors for medical imaging with x-rays. IEEE Trans. Nucl. Sci. (2009), doi: 10.1109/TNS.2009.2025041.
53. Kim, H.K., Lim, C.H., Tanguay, J., Yun, S., Cunningham, I.A.: Spectral analysis of fundamental signal and noise performances in photoconductors for mammography. Med. Phys. (2012), doi: 10.1118/1.3702455.
54. Street, R.A.: Hydrogenated Amorphous Silicon. Cambridge University Press, Cambridge (1991).
55. Zhao, W., Blevis, I., Germann, S., Rowlands, J.A., Waechter, D., Huang, Z.: Digital radiology using active matrix readout of amorphous selenium: construction and evaluation of a prototype real-time detector. Med. Phys. (1997), doi: 10.1118/1.598098.
56. Zhao, W., Waechter, D., Rowlands, J.A.: Digital radiology using active matrix readout of amorphous selenium: radiation hardness of cadmium selenide thin film transistors. Med. Phys. (1998), doi: 10.1118/1.598233.
57. Zhao, W., Law, J., Waechter, D., Huang, Z., Rowlands, J.A.: Digital radiology using active matrix readout of amorphous selenium: detectors with high voltage protection. Med. Phys. (1998), doi: 10.1118/1.598229.

58. Pang, G., Zhao, W., Rowlands, J.A.: Digital radiology using active matrix readout of amorphous selenium: geometrical and effective fill factors. Med. Phys. (1998), doi: 10.1118/1.598344.
59. Zhao, W., Rowlands, J.A.: Active Matrix X-ray Imaging Array. number = 5962856 (1999).
60. Spear, W.E., Le Comber, P.G.: Substitutional doping of amorphous silicon. Solid State Commun. (1975), doi: 10.1016/0038-1098(75)90284-7.
61. Kim, H.J., Kim, H.K., Cho, G., Choi, J.: Construction and characterization of an amorphous silicon flat-panel detector based on ion-shower doping process. Nucl. Instrum. Methods Phys. Res. A. (2003), doi: 10.1016/S0168-9002(03)01040-4.
62. Munro, P., Bouius, D.C.: X-ray quantum limited portal imaging using amorphous silicon flat-panel arrays. Med. Phys. (1998), doi: 10.1118/1.598252.
63. Kameshima, T., Kaifu, N., Takami, E., Morishita, M., Yamazaki, T.: Novel large-area MIS-type x-ray image sensor for digital radiography. Proc. SPIE (1998), doi: 10.1117/12.317045.
64. Weckler, G.P.: Operation of p-n junction photodetectors in a photon flux integrating modes. IEEE J. Solid-st. Circ. **2**, 65–73 (1967).
65. Graeve, T., Huang, W., Alexander, S.M., Li, Y.: Amorphous silicon image sensor for x-ray applications. Proc. SPIE (1995), doi: 10.1117/12.206513.
66. Chaussat, C., Chabbal, J., Ducourant, T., Spinnler, V., Vieux, G., Neyret, R.: New CsI/a-Si 17" × 17" x-ray flat-panel detector provides superior detectivity and immediate direct digital output for general radiography systems. Proc. SPIE (1998), doi: 10.1117/12.317049.
67. Yarema, R.J., Zimmerman, T., Srage, J., Antonuk, L.E., Berry, J., Huang, W., Maolinbay, M.: A programmable, low noise, multichannel ASIC for readout of pixelated amorphous silicon arrays. Nucl. Instrum. Methods Phys. Res. A. (2000), doi: 10.1016/S0168-9002(99)00900-6.
68. De Geronimo, G., O'Connor, P., Radeka, V., Yu, B.: Front-end electronics for imaging detectors. Nucl. Instrum. Methods Phys. Res. A. (2001), doi: 10.1016/S0168-9002(01)00963-9.
69. Maolinbay, M., Zimmerman, T., Yarema, R.J. Antonuk, L.E., El-Mohri, Y., Yeakey, M.: Design and performance of a low noise, 128-channel ASIC preamplifier for readout of active matrix flat-panel imaging arrays. Nucl. Instrum. Methods Phys. Res. A. (2002), doi: 10.1016/S0168-9002(01)02129-5.
70. Huang, W., Antonuk, L.E., Berry, J., Maolinbay, M., Martelli, C., Mody, P., Nassif, S., Yeakey, M.: An asynchronous, pipelined, electronic acquisition system for active matrix flat-panel imagers (AMFPIs). Nucl. Instrum. Methods Phys. Res. A. (1999), doi: 10.1016/S0168-9002(99)00275-2.
71. Vedantham, S., Karellas, A., Suryanarayanan, S., Albagli, D., Han, S., Tkaczyk, E.J., Landberg, C.E., Opsahl-Ong, B., Granfors, P.R., Levis, I., D'Orsi, C.J., Hendrick, R.E.: Full breast digital mammography with an amorphous silicon-based flat panel detector: physical characteristics of a clinical prototype. Med. Phys. (2000), doi: 10.1118/1.598895.
72. Granfors, P.R., Aufrichtig, R.: Performance of a 41×41-$cm^2$ amorphous silicon flat panel x-ray detector for radiographic imaging applications. Med. Phys. (2000), doi: 10.1118/1.599010.
73. Granfors, P.R., Albagli, D., Tkaczyk, J.E., Aufrichtig, R., Netel, H., Brunst, G., Boudry, J.M., Luo, D.: Performance of a flat-panel cardiac detector. Proc. SPIE (2001), doi: 10.1117/12.430887.

74. Yorker, J.G., Jeromin, L.S., Lee, D.L.Y., Palecki, E.F., Golden, K.P., Jing, Z.: Characterization of a full-field digital mammography detector based on direct x-ray conversion in selenium. Proc. SPIE (2002), doi: 10.1117/12.465568.
75. Cheung, L.K., Jing, Z., Bogdanovich, S., Golden, K., Robinson, S., Beliaevskaia, E., Parikh. S.: Image performance of a new amorphous selenium flat panel x-ray detector designed for digital breast tomosynthesis. Proc. SPIE (2005), doi: 10.1117/12.595907.
76. Metz, C.E., Wagner, R.F., Doi, K., Brown, D.G., Nishikawa, R.M., Myers, K.J.: Toward consensus on quantitative assessment of medical imaging systems. Med. Phys. (1995), doi: 10.1118/1.597511.
77. Samei, E., Flynn, M.J., Chotas, H.G., Dobbins III, J.T.: DQE of direct and indirect digital radiography systems. Proc. SPIE (2001), doi: 10.1117/12.430953.
78. Shaw, J., Albagli, D., Wei, C.-Y., Granfors, P.R.: Enhanced a-Si/CsI-based flat-panel x-ray detector for mammography. Proc. SPIE (2004), doi: 10.1117/12.539141.
79. Rivetti, S., Lanconelli, N., Campanini, R., Bertolini, M., Borasi, G., Nitrosi, A., Danielli, C., Angelini, L., Maggi S.: Comparison of different commercial FFDM units by means of physical characterization and contrast-detail analysis. Med. Phys. (2006), doi: 10.1118/1.2358195.
80. Lazzari, B., Belli, G., Gori, C., Rosselli Del Turco, M.: Physical characteristics of five clinical systems for digital mammography. Med. Phys. (2007), doi: 10.1118/1.2742498.
81. Monnin, P., Gutierrez, D., Bulling, S., Guntern, D., Verdun, F.R.: A comparison of the performance of digital mammography systems. Med. Phys. (2007), doi: 10.1118/1.2432072.
82. Marshall, N.W.: Detective quantum efficiency measured as a function of energy for two full-field digital mammography systems. Phys. Med. Biol. (2009), doi: 10.1088/0031-9155/54/9/017.
83. Marshall, N.W., Monnin, P., Bosmans, H., Bochud, F.O., Verdun, F.R.: Image quality assessment in digital mammography: part I. technical characterization of the systems. Phys. Med. Biol. (2011), doi: 10.1088/0031-9155/56/14/002.
84. Youn, H., Han, J.C., Yun, S., Kam, S., Cho, S., Kim, H.K.: Characterization of on-site digital mammography systems: direct versus indirect conversion detectors. J. Korean Phys. Soc. (2015), doi: 10.3938/jkps.66.1926.
85. Samei, E., Flynn, M.J.: An experimental comparison of detector performance for direct and indirect digital radiography systems. Med. Phys. (2003), doi: 10.1118/1.1561285.
86. Borasi, G., Nitrosi, A., Ferrari, P., Tassoni, D.: On site evaluation of three flat panel detectors for digital radiography. Med. Phys. (2003), doi: 10.1118/1.1569273.
87. Borasi, G., Samei, E., Bertolini, M., Nitrosi, A., Tassoni, D.: Contrast-detail analysis of three flat panel detectors for digital radiography. Med. Phys. (2006), doi: 10.1118/1.2191014.
88. Weisfield, R.L., Hartney, M.A., Schneider, R., Aflatooni, K., Lujan, R.: High-performance amorphous silicon image sensor for x-ray diagnostic medical imaging applications. Proc. SPIE (1999), doi: 10.1117/12.349505.
89. Weisfield, R.L., Yao, W., Speaker, T., Zhou, K., Colbeth, R.E., Proano, C.: Performance analysis of a 127-micron pixel large-area TFT/photodiode array with boosted fill factor. Proc. SPIE (2004), doi: 10.1117/12.535896.

90. Ducourant, T., Michel, M., Vieux, G., Peppler, T., Trochet, J.C., Schulz, R.F., Bastiaens, R.J.M., Busse, F.: Optimization of key building blocks for a large-area radiographic and fluoroscopic dynamic digital x-ray detector based on a-Si:H/CsI:Tl flat panel technology. Proc. SPIE (2000), doi: 10.1117/12.384491.
91. Ducourant, T., Couder, D., Wirth, T., Trochet, J.C., Bastiaens, R.J.M., Bruijins, T.J.C., Luijendijk, H.A., Sandkamp, B., Davies, A.G., Didier, D.: Image quality of digital subtraction angiography using flat detector technology. Proc. SPIE (2003), doi: 10.1117/12.479969.
92. Ducourant, T., Wirth, T., Bacher, G., Bosset, B., Vignolle, J.-M., Blanchon, D., Betraoui, F., Rohr, P.: Latest advancements in state-of-the-art a-Si-based x-ray flat panel detectors. Proc. SPIE (2018), doi: 10.1117/12.2291908.
93. Moy, J.-P.: Signal-to-noise ratio and spatial resolution in x-ray electronic imagers: is the MTF a relevant parameter? Med. Phys. (2007), doi: 10.1118/1.2432072.
94. Kleimann, P., Linnros, J., Fröjdh, C., Petersson, C.: An x-ray imaging pixel detector based on scintillator filled pores in a silicon matrix. Nucl. Instrum. Methods Phys. Res. A. (2001), doi: 10.1016/S0168-9002(00)01089-5.
95. Rocha, J., Correia, J.: A high-performance scintillator-silicon-well x-ray microdetector based on DRIE techniques. Sensor. Actuat. A-Phys. (2001), doi: 10.1016/S0924-4247(01)00564-7.
96. Daniel, J.H., Krusor, B., Apte, R.B., Mulato, M., Van Schuylenbergh, K., Lau, R., Do, T., Street, R.A., Goredema, A., Boils-Boissier, D.C., Kazmaier, P.M.: Microelectro-mechanical system fabrication technology applied to large area x-ray image sensor arrays. J. Vac. Sci. Technol. A. (2001), doi: 10.1116/1.1380226.
97. Badel, X., Galeckas, A., Linnros, J., Kleimann, P., Fröjdh, C., Petersson, C.: Improvement of an x-ray imaging detector based on a scintillating guides screen. Nucl. Instrum. Methods Phys. Res. A. (2002), doi: 10.1016/S0168-9002(02)00956-7.
98. Tao, S., Gu, Z.H., Nathan, A.: Fabrication of $Gd_2O_2S$:Tb based phosphor films coupled with photodetectors for x-ray imaging applications. J. Vac. Sci. Technol. A. (2002), doi: 10.1116/1.1463082.
99. Rocha, J., Ramos, N., Lanceros-Mendez S., Wolffenbuttel, R., Correia, J.: CMOS x-rays detector array based on scintillating light guides. Sensor. Actuat. A-Phys. (2004), doi: 10.1016/j.sna.2003.09.043.
100. Cha, B.K., Bae, J.H., Kim, B.-J., Jeon, H., Cho, G.: Performance studies of a monolithic scintillator-CMOS image sensor for x-ray application. Nucl. Instrum. Methods Phys. Res. A. (2008), doi: 10.1016/j.nima.2008.03.034.
101. Simon, M., Engel, K.J., Menser, B., Badel, X., Linnros., J.: Clinical implications of dysregulated cytokine production. Med. Phys. (2008), doi: 10.1118/1.2839441.
102. Yun, S.M., Lim, C.H., Kim, T.W., Kim, H.K.: Pixel-structured scintillators for digital x-ray imaging. Proc. SPIE (2009), doi: 10.1117/12.811546.
103. Jung, I.D., Cho, M.K., Lee, S.M., Bae, K.M., Jung, P.G., Lee, C.H., Lee, J.M., Yun, S., Kim, H.K., Kim, S.S., Ko, J.S.: Flexible $Gd_2O_2S$:Tb scintillators pixelated with polyethylene microstructures for digital x-ray image sensors. J. Micromech. Microeng. (2008), doi: 10.1117/12.811546.
104. Jung, I.D., Cho, M.K., Bae, K.M., Lee, S.M., Jung, P.G., Kim, H.K., Kim, S.S., Ko, J.S.: Pixel-structured scintillator with polymeric microstructures for x-ray image sensors. ETRI J. (2008), doi: 10.4218/etrij.08.0208.0166.
105. Shigeta, K., Fujioka, N., Murai, T., Hikita, I., Morinaga, T., Tanino, T., Kodama, H., Okamura, M.: High spatial resolution performance of pixelated scintillators. Proc. SPIE (2017), doi: 10.1117/12.2254037.

106. Kim, H.K., Yun, S.M., Ko, J.S., Cho, G., Graeve, T.: Cascade modeling of pixelated scintillator detectors for x-ray imaging. IEEE Trans. Nucl. Sci. (2008), doi: 10.1109/TNS.2008.919260.
107. Kim, J., Kim, J., Kim, D.W., Park, J., Kim, H.K.: Theoretical investigation of partially pixelated scintillators for high-resolution imaging with less aliasing. J. Instrum. (2019), doi: 10.1117/12.2254037.
108. Yun, S., Lim, C.H., Kim, T.W., Cunningham, I., Achterkirchen, T., Kim., H.K.: Phosphor-filled micro-well arrays for digital x-ray imaging: effects of surface treatments. Proc. SPIE (2010), doi: 10.1117/12.844167.
109. Antonuk, L.E., Jee, K.-W., El-Mohri, Y., Maolinbay, M., Nassif, S., Rong, X., Zhao, Q., Siewerdsen, J.H., Street, R.A., Shah, K.S.: Strategies to improve the signal and noise performance of active matrix, flat-panel imagers for diagnostic x-ray applications. Med. Phys. (2000), doi: 10.1118/1.598831.
110. El-Mohri, Y., Antonuk, L.E., Zhao, Q., Maolinbay, M., Rong, X. Jee, K.-W., Nassif, S., Cionca, C.: A quantitative investigation of additive noise reduction for active matrix flat-panel imagers using compensation lines. Med. Phys. (2000), doi: 10.1118/1.1287053.
111. Beuville, E., Belding, M., Costello, A., Hansen, R., Petronio, S.: A high performance, low-noise 128-channel readout integrated circuit for instrumentation and x-ray applications. *IEEE Symposium Conference Record Nuclear Science 2004.* (2004), doi: 10.1109/NSSMIC.2004.1462169.
112. Street, R.A., Ready, S.E., Van Schuylenbergh, K., Ho, J., Boyce, J.B., Nylen, P., Shah, K., Melekhov, L., Hermon, H.: Comparison of $PbI_2$ and $HgI_2$ for direct detection active matrix x-ray image sensors. J. Appl. Phys. (2002), doi: 10.1063/1.1436298.
113. Kasap, S., Zahangir Kabir, M., Rowlands, J.: Recent advances in x-ray photoconductors for direct conversion x-ray image detectors. Curr. Appl. Phys. (2006), doi: 10.1016/j.cap.2005.11.001.
114. Schieber, M., Zuck, A., Gilboa, H., Zentai, G.: Reviewing polycrystalline mercuric iodide x-ray detectors. IEEE Trans. Nucl. Sci. (2006), doi: 10.1109/TNS.2006.877043.
115. Du, H., Antonuk, L.E., El-Mohri, Y., Zhao, Q., Su, Z., Yamamoto, J., Wang, Y.: Signal behavior of polycrystalline $HgI_2$ at diagnostic energies of prototype, direct detection, active matrix, flat-panel imagers. Phys. Med. Biol. (2008), doi: 10.1088/0031-9155/53/5/011.
116. Rahn, J.T., Lemmi, F., Weisfield, R.L., Lujan, R., Mei, P., Lu, J.-P., Ho, J., Ready, S.E., Apte, R.B., Nylen, P., Boyce, J.B., Street, R.A.: High-resolution high fill factor a-Si:H sensor arrays for medical imaging. Proc. SPIE (1999), doi: 10.1117/12.349529.
117. Matsuura, N., Zhao, W., Huang, Z., Rowlands, J.A.: Digital radiology using active matrix readout: amplified pixel detector array for fluoroscopy. Med. Phys. (1999), doi: 10.1118/1.598572.
118. Karim, K.S., Nathan, A., Rowlands, J.A.: Alternate pixel architectures for large-area medical imaging. Proc. SPIE (2001), doi: 10.1117/12.430903.
119. Lu, J.P., Van Schuylenbergh, K., Ho, J., Wang, Y., Boyce, J.B., Street, R.A.: Flat panel imagers with pixel level amplifiers based on polycrystalline silicon thin-film transistor technology. Appl. Phys. Lett. (2002), doi: 10.1063/1.1481788.

120. Antonuk, L.E., Li, Y., Du, H., El-Mohri, Y., Zhao, Q., Yamamoto, J., Sawant, A., Wang, Y., Su, Z., Lu, J.-P. Street, R.A., Weisfield, R., Yao, B.: Investigation of strategies to achieve optimal DQE performance from indirect-detection active-matrix flat-panel imagers (AMFPIs) through novel pixel amplification architectures. Proc. SPIE (2005), doi: 10.1117/12.596786.
121. Antonuk, L.E., El-Mohri, Y., Zhao, Q., Koniczek, M., McDonald, J., Yeakey, M., Wang, Y., Behravan, M., Street, R.A., Lu, J.: Exploration of the potential performance of polycrystalline silicon-based active matrix flat-panel imagers incorporating active pixel sensor architectures. Proc. SPIE (2008), doi: 10.1117/12.771406.
122. Antonuk, L.E., Zhao, Q., El-Mohri, Y., Du, H., Wang, Y., Street, R.A., Ho, J., Weisfield, R., Yao, W.: An investigation of signal performance enhancements achieved through innovative pixel design across several generations of indirect detection, active matrix, flat-panel arrays. Med. Phys. (2009), doi: 10.1118/1.3049602.
123. El-Mohri, Y., Antonuk, L.E., Koniczek, M., Zhao, Q., Li, Y., Street, R.A., Lu, J.-P.: Active pixel imagers incorporating pixel-level amplifiers based on polycrystalline-silicon thin-film transistors. Med. Phys. (2009), doi: 10.1118/1.3116364.
124. Koniczek, M., Antonuk, L.E., El-Mohri, Y., Liang, A.K., Zhao, Q.: Theoretical investigation of the noise performance of active pixel imaging arrays based on polycrystalline silicon thin film transistors. Med. Phys. (2017), doi: 10.1002/mp.12257.
125. Zhao, W., Hunt, D.C., Tanioka, K., Rowlands, J.A.: Indirect flat-panel detector with avalanche gain. Proc. SPIE (2004), doi: 10.1117/12.536863.
126. Zhao, W., Hunt, D., Tanioka, K., Rowlands, J.A.: Amorphous selenium flat panel detectors for medical applications. Nucl. Instrum. Methods Phys. Res. A. (2005), doi: 10.1016/j.nima.2005.04.053.
127. Zhao, W., Li, D., Rowlands, J.A., Egami, N., Takiguchi, Y., Nanba, M., Honda, Y., Ohkawa, Y., Kubota, M., Tanioka, K., Suzuki, K., Kawai, T.: An indirect flat-panel detector with avalanche gain for low dose x-ray imaging: SAPHIRE (scintillator avalanche photoconductor with high resolution emitter readout). (2008), doi: 10.1117/12.772774.
128. Li, D., Zhao, W.: SAPHIRE (scintillator avalanche photoconductor with high resolution emitter readout) for low dose x-ray imaging: spatial resolution. Med. Phys. (2008), doi: 10.1118/1.2937652.
129. Lia, D., Zhao, W., Nanba, M., Egami, N.: Scintillator avalanche photoconductor with high resolution emitter readout for low dose x-ray imaging: lag. Med. Phys. (2009), doi: 10.1118/1.3187227.
130. Zhao, W., Li, D., Reznik, A., Lui, B.J.M., Hunt, D.C., Rowlands, J.A., Ohkawa, Y., Tanioka, K.: Indirect flat-panel detector with avalanche gain: fundamental feasibility investigation for SHARP-AMFPI (scintillator HARP active matrix flat panel imager). Med. Phys. (2005), doi: 10.1118/1.2008428.
131. Wronski, M., Zhao, W., Tanioka, K., DeCrescenzo, G., Rowlands, J.A.: Scintillator high-gain avalanche rushing photoconductor active-matrix flat panel imager: zero-spatial frequency x-ray imaging properties of the solid-state SHARP sensor structure. Med. Phys. (2012), doi: 10.1118/1.4760989.
132. Scheuermann, J.R., Goldan, A.H., Tousignant, O., Léveillé, S., Zhao, W.: Development of solid-state avalanche amorphous selenium for medical imaging. Med. Phys. (2015), doi: 10.1118/1.4907971.

133. Scheuermann, J.R., Howansky, A., Hansroul, M., Léveillé, S., Tanioka, K., Zhao, W.: Toward scintillator high-gain avalanche rushing photoconductor active matrix flat panel imager (SHARP-AMFPI): initial fabrication and characterization. Med. Phys. (2018), doi: 10.1002/mp.12693.
134. Dimitrakopoulos, C., Malenfant, P.: Organic thin film transistors for large area electronics. Adv. Mater. (2002), doi: 10.1002/1521-4095(20020116)14:2<99::AID-ADMA99>3.0.CO;2-9.
135. Blakesley, J.C., Speller, R.: Modeling the imaging performance of prototype organic x-ray imagers. Med. Phys. (2008), doi: 10.1118/1.2805479.
136. Wong, W.S., Ready, S., Matusiak, R., White, S.D., Lu, J.-P., Ho, J., Street, R.A.: Amorphous silicon thin-film transistors and arrays fabricated by jet printing. Appl. Phys. Lett. (2002), doi: 10.1063/1.1436273.
137. Paul, K.E., Wong, W.S., Ready, S.E., Street, R.A.: Additive jet printing of polymer thin-film transistors. Appl. Phys. Lett. (2003), doi: 10.1063/1.1609233.
138. Arias, A.C., Ready, S.E., Lujan, R., Wong, W.S., Paul, K.E., Salleo, A., Chabinyc, M.L., Apte, R., Street, R.A., Wu, Y., Liu, P., Ong, B.: All jet-printed polymer thin-film transistor active-matrix backplanes. Appl. Phys. Lett. (2004), doi: 10.1063/1.1801673.
139. Street, R.A., Wong, W.S., Ready, S., Lujan, R., Arias, A.C., Chabinyc, M.L., Salleo, A., Apte, R., Antonuk, L.E.: Printed active-matrix TFT arrays for x-ray imaging. Proc. SPIE (2005), doi: 10.1117/12.593521.
140. Ng, T.N., Lujan, R.A., Sambandan, S., Street, R.A., Limb, S., Wong, W.S.: Low temperature a-Si:H photodiodes and flexible image sensor arrays patterned by digital lithography. Appl. Phys. Lett. (2007), doi: 10.1063/1.2767981.
141. Rossmann, K.: Measurement of the modulation transfer function of radiographic systems containing fluorescent screens. Phys. Med. Biol. (1964), doi: 10.1088/0031-9155/9/4/312.
142. Rossmann, K.: The spatial frequency spectrum: a means for studying the quality of radiographic imaging systems. Radiology. (1968), doi: 10.1148/90.1.1.
143. Rossmann, K.: Point spread-function, line spread-function, and modulation transfer function. Radiology. (1969), doi: 10.1148/93.2.257.
144. Barrett, H.H., Swindell, W.: Radiological Imaging: The Theory of Image Formation, Detection, and Processing. Academic Press, Cambridge (1981).
145. Papoulis, A.: Probability, Random Variables, and Stochastic Processes. McGraw-Hill, New York, NY (1991).
146. Jenkins, G.M., Watts, D.G. Spectral Analysis and Its Application. Holden Day, San Francisco, CA (1968).
147. Giger, M.L., Doi, K., Metz, C.E.: Investigation of basic imaging properties in digital radiography. 2. noise Wiener spectrum. Med. Phys. (1984), doi: 10.1118/1.595583.
148. Shaw, R.: The equivalent quantum efficiency of the photographic process. J. Photogr. Sc. (1963), doi: 10.1080/00223638.1963.11736919.
149. Wagner, R.F., Brown, D.G.: Unified SNR analysis of medical imaging systems. Phys. Med. Biol. (1985), doi: 10.1088/0031-9155/30/6/001.
150. Sharp, P., Barber, D.C., Brown, D.G., Burgess, A.E., Metz, C.E., Myers, K.J., Taylor, C.J., Wagner, R.F., Brooks, R., Hill, C.R., Kuhl, D.E., Smith, M.A., Wells, P., Worthington, B.: Report 54. Journal of the International Commission on Radiation Units and Measurements. (2016), doi: 10.1093/jicru/os28.1.Report54.

151. Cunningham, I.A.: Applied linear-systems theory. In: Beutel, J., Kundel, H.L., Van Metter, R. (eds.) Handbook of Medical Imaging: Vol. 1. Physics and Psychophysics. SPIE, Bellingham, WA (2000).
152. Barrett, H.H., Meyers, K.J.: Foundations of Image Science. Wiley, Hoboken, NJ (2004).
153. Yun, S., Kim, H.K., Han, J.C., Kam, S., Youn, S., Cunningham, I.A.: Power-law analysis of nonlinear active-pixel detector responses as a function of mammographic energy. Nucl. Instrum. Methods Phys. Res. A. (2017), doi: 10.1016/j.nima.2016.11.049.
154. Kam, S., Kim, D.W., Yun, S., Kim, H.K.: Power-law analysis of nonlinear active-pixel detector responses as a function of mammographic energy. Nucl. Instrum. Methods Phys. Res. A. (2019), doi: 10.1016/j.nima.2019.162674.
155. Papoulis, A.: Systems and Transforms with Applications in Optics. McGraw-Hill, New York, NY (1968).
156. Dainty, J.C., Shaw, R.: Image Science. Academic Press, Cambridge (1974).
157. Gaskill, J.D.: Linear systems, Fourier Transforms, and Optics. John Wiley & Sons, Hoboken, NJ (1978).
158. Doi, K.: Field characteristics of geometric unsharpness due to the x-ray tube focal spot. Med. Phys. (1977), doi: 10.1118/1.594378.
159. Metz, C.E., Doi, K.: Transfer function analysis of radiographic imaging systems. Phys. Med. Biol. (1979), doi: 10.1088/0031-9155/24/6/001.
160. Rabbani, M., Shaw, R., Metter, R.V.: Detective quantum efficiency of imaging systems with amplifying and scattering mechanisms. J. Opt. Soc. Am. A. (1987), doi: 10.1364/JOSAA.4.000895.
161. Rabbani, M., Metter, R.V.: Analysis of signal and noise propagation for several imaging mechanisms. J. Opt. Soc. Am. A. (1989), doi: 10.1364/JOSAA.6.001156.
162. Barrett, H.H., Wagner, R.F., Myers, K.J.: Correlated point processes in radiological imaging. Proc. SPIE (1997), doi: 10.1117/12.273975.
163. Sattarivand, M., Cunningham, I.A.: Computational engine for development of complex cascaded models of signal and noise in x-ray imaging systems. IEEE Trans. Med. Imaging. (2005), doi: 10.1109/TMI.2004.839680.
164. Zhao, W., Rowlands, J.A.: Digital radiology using active matrix readout of amorphous selenium: theoretical analysis of detective quantum efficiency. Med. Phys. (1997), doi: 10.1118/1.598097.
165. Swank, R.K.: Absorption and noise in x-ray phosphors. J. Appl. Phys. (1973), doi: 10.1063/1.1662918.
166. Lubberts, G.: Random noise produced by x-ray fluorescent screens. J. Opt. Soc. Am. (1968), doi: 10.1364/JOSA.58.001475.
167. Weisfield, R.L., Bennett, N.R.: Electronic noise analysis of a 127-um pixel TFT/photodiode array. Proc. SPIE (2001), doi: 10.1117/12.430955.
168. Maolinbay, M., El-Mohri, Y., Antonuk, L.E., Jee, K.-W., Nassif, S., Rong, X., Zhao, Q.: Additive noise properties of active matrix flat-panel imagers. Med. Phys. (2000), doi: 10.1118/1.1286721.
169. Mainprize, J.G., Hunt, D.C., Yaffe, M.J.: Direct conversion detectors: the effect of incomplete charge collection on detective quantum efficiency. Med. Phys. (2002), doi: 10.1118/1.1477235.
170. Zhao, W., Ristic, G., Rowlands, J.A.: X-ray imaging performance of structured cesium iodide scintillators. Med. Phys. (2004), doi: 10.1118/1.1782676.

171. Hillen, W., Eckenbach, W., Quadflieg, P., Zaengel, T.T.: Signal-to-noise performance in cesium iodide x-ray fluorescent screens. Proc. SPIE (1991), doi: 10.1117/12.43435.
172. Lubinsky, A.R., Zhao, W., Ristic, G., Rowlands, J.A.: Screen optics effects on detective quantum efficiency in digital radiography: zero-frequency effects. Med. Phys. (2006), doi: 10.1118/1.2188082.
173. Howansky, A., Peng, B., Lubinsky, A.R., Zhao, W.: Deriving depth-dependent light escape efficiency and optical Swank factor from measured pulse height spectra of scintillators. Med. Phys. (2017), doi: 10.1002/mp.12083.
174. Howansky, A., Lubinsky, A.R., Suzuki, K., Ghose, S., Zhao, W.: An apparatus and method for directly measuring the depth-dependent gain and spatial resolution of turbid scintillators. Med. Phys. (2018), doi: 10.1002/mp.13177.
175. IEC: Medical electrical equipment – characteristics of digital x-ray imaging devices – Part 1-2: determination of the detective quantum efficiency – detectors used in mammography. (2008).
176. Siewerdsen, J.H., Antonuk, L.E., El-Mohri, Y., Yorkston, J., Huang, W., Boudry, J.M., Cunningham, I.A.: Empirical and theoretical investigation of the noise performance of indirect detection, active matrix flat-panel imagers (AMFPIs) for diagnostic radiology. Med. Phys. (1997), doi: 10.1118/1.597919.
177. Siewerdsen, J.H., Antonuk, L.E., El-Mohri, Y., Yorkston, J., Huang, W., Cunningham, I.A.: Signal, noise power spectrum, and detective quantum efficiency of indirect-detection flat-panel imagers for diagnostic radiology. Med. Phys. (1998), doi: 10.1118/1.598243.
178. Jee, K.-W., Antonuk, L.E., El-Mohri, Y., Zhao, Q.: System performance of a prototype flat-panel imager operated under mammographic conditions. Med. Phys. (2003), doi: 10.1118/1.1585051.
179. El-Mohri, Y., Antonuk, L.E., Zhao, Q., Wang, Y., Li, Y., Du, H., Sawant, A.: Performance of a high fill factor, indirect detection prototype flat-panel imager for mammography. Med. Phys. (2007), doi: 10.1118/1.2403967.
180. Vedantham, S., Karellas, A.: Modeling the performance characteristics of computed radiography (CR) systems. IEEE Trans. Med. Imaging. (2010), doi: 10.1109/TMI.2009.2036995.
181. Bissonnette, J.-P., Cunningham, I.A., Jaffray, D.A., Fenster, A., Munro, P.: A quantum accounting and detective quantum efficiency analysis for video-based portal imaging. Med. Phys. (1997), doi: 10.1118/1.598009.
182. El-Mohri, Y., Jee, K.-W., Antonuk, L.E., Maolinbay, M., Zhao, Q.: Determination of the detective quantum efficiency of a prototype, megavoltage indirect detection, active matrix flat-panel imager. Med. Phys. (2001), doi: 10.1118/1.1413516.
183. Lachaine, M., Fourkal, M., Fallone, B.G.: Detective quantum efficiency of a direct-detection active matrix flat panel imager at megavoltage energies. Med. Phys. (2001), doi: 10.1118/1.1380213.
184. Mainprize, J.G., Ford, N.L., Yin, S., Tümer, T., Yaffe, M.J.: A slot-scanned photodiode-array/CCD hybrid detector for digital mammography. Med. Phys. (2002), doi: 10.1118/1.1446108.
185. Mainprize, J.G., Ford, N.L., Yin, S., Gordon, E.E., Hamilton, W.J., Tümer, T.O., Yaffe, M.J.: A CdZnTe slot-scanned detector for digital mammography. Med. Phys. (2002), doi: 10.1118/1.1523932.
186. Williams, M.B., Simoni, P.U., Smilowitz, L., Stanton, M., Phillips, W., Stewart, A.: Analysis of the detective quantum efficiency of a developmental detector for digital mammography. Med. Phys. (1999), doi: 10.1118/1.598741.

187. Lai, H., Cunningham, I.A.: Noise aliasing in interline-video-based fluoroscopy systems. Med. Phys. (2002), doi: 10.1118/1.1446100.
188. Vedantham, S., Karellas, A., Suryanarayanan, S.: Solid-state fluoroscopic imager for high-resolution angiography: parallel-cascaded linear systems analysis. Med. Phys. (2002), doi: 10.1118/1.1689014.
189. Ganguly, A., Rudin, S., Bednarek, D.R., Hoffmann, K.R.: Micro-angiography for neuro-vascular imaging. II. cascade model analysis. Med. Phys. (2003), doi: 10.1118/1.1617550.
190. Jain, A., Bednarek, D.R., Ionita, C., Rudin, S.: A theoretical and experimental evaluation of the microangiographic fluoroscope: a high-resolution region-of-interest x-ray imager. Med. Phys. (2011), doi: 10.1118/1.3599751.
191. Båth, M., Sund, P., Månsson, L.G.: Evaluation of the imaging properties of two generations of a CCD-based system for digital chest radiography. Med. Phys. (2002), doi: 10.1118/1.1507781.
192. Hu, Y.-H., Scaduto, D.A., Zhao, W.: Optimization of contrast-enhanced breast imaging: analysis using a cascaded linear system model. Med. Phys. (2017), doi: 10.1002/mp.12004.
193. Zhao, B., Zhao, W.: Three-dimensional linear system analysis for breast tomosynthesis. Med. Phys. (2008), doi: 10.1118/1.2996014.
194. Richard, S., Siewerdsen, J.H., Jaffray, D.A., Moseley, D.J., Bakhtiar, B.: Generalized DQE analysis of radiographic and dual-energy imaging using flat-panel detectors. Med. Phys. (2005), doi: 10.1118/1.1901203.
195. Tward, D.J., Siewerdsen, J.H.: Cascaded systems analysis of the 3D noise transfer characteristics of flat-panel cone-beam CT. Med. Phys. (2008), doi: 10.1118/1.3002414.
196. Zbijewski, W., Jean, P.D., Prakash, P., Ding, Y., Stayman, J.W., Packard, N., Senn, R., Yang, D., Yorkston, J., Machado, A., Carrino, J.A., Siewerdsen, J.H.: A dedicated cone-beam CT system for musculoskeletal extremities imaging: design, optimization, and initial performance characterization. Med. Phys. (2011), doi: 10.1118/1.3611039.
197. Gang, G.J., Zbijewski, W., Stayman, J.W., Siewerdsen, J.H.: Cascaded systems analysis of noise and detectability in dual-energy cone-beam CT. Med. Phys. (2012), doi: 10.1118/1.4736420.
198. Hunt, D.C., Tanioka, K., Rowlands, J.A.: X-ray imaging using avalanche multiplication in amorphous selenium: investigation of depth dependent avalanche noise. Med. Phys. (2007), doi: 10.1118/1.2437097.
199. Wronski, M.M., Rowlands, J.A.: Direct-conversion flat-panel imager with avalanche gain: feasibility investigation for HARP-AMFPI. Med. Phys. (2008), doi: 10.1118/1.3002314.
200. Xu, J., Zbijewski, W., Gang, G., Stayman, J.W., Taguchi, K., Lundqvist, M., Fredenberg, E., Carrino, J.A., Siewerdsen, J.H.: Cascaded systems analysis of photon counting detectors. Med. Phys. (2014), doi: 10.1118/1.4894733.
201. Cunningham, I.A.: Linear-systems modeling of parallel cascaded stochastic processes: the NPS of radiographic screens with reabsorption of characteristic x-radiation. Proc. SPIE (1998), doi: 10.1117/12.317021.
202. Yao, J., Cunningham, I.A.: Parallel cascades: new ways to describe noise transfer in medical imaging systems. Med. Phys. (2001), doi: 10.1118/1.1405842.
203. Zhao, W., Ji, W.G., Rowlands, J.A.: Effects of characteristic x rays on the noise power spectra and detective quantum efficiency of photoconductive x-ray detectors. Med. Phys. (2001), doi: 10.1118/1.1405845.

204. Hajdok, G., Yao, J., Battista, J.J., Cunningham, I.A.: Signal and noise transfer properties of photoelectric interactions in diagnostic x-ray imaging detectors. Med. Phys. (2006), doi: 10.1118/1.2336507.
205. Yun, S., Tanguay, J., Kim, H.K., Cunningham, I.A.: Cascaded-systems analyses and the detective quantum efficiency of single-Z x-ray detectors including photoelectric, coherent and incoherent interactions. Med. Phys. (2013), doi: 10.1118/1.4794495.
206. Akbarpour, R., Friedman, S.N., Siewerdsen, J.H., Neary, J.D., Cunningham, J.D.: Signal and noise transfer in spatiotemporal quantum-based imaging systems. J. Opt. Soc. Am. A. (2007), doi: 10.1364/JOSAA.24.00B151.
207. Cunningham, I.A., Yao, J., Subotic, V.: Cascaded models and the DQE of flat-panel imagers: noise aliasing, secondary quantum noise, and reabsorption. Proc. SPIE (2002), doi: 10.1117/12.465610.
208. Kim, H.K.: Generalized cascaded model to assess noise transfer in scintillator-based x-ray imaging detectors. Appl. Phys. Lett. (2006), doi: 10.1063/1.2398926.
209. Yun, S., Kim, H.K., Lim, C.H., Cho, M.K., Achterkirchen, T., Cunningham, I.A.: Signal and noise characteristics induced by unattenuated x rays from a scintillator in indirect-conversion CMOS photodiode array detectors. J. Mol. Med. (2009), doi: 10.1007/s001090000086.
210. Siewerdsen, J.H.: Optimization of 2D and 3D radiographic imaging systems. In: Samei, E., Krupinski, E. (eds.) The Handbook of Medical Image Perception and Techniques, pp. 335–355. Cambridge University Press, Cambridge (2010).
211. Alvarez, R.E., Macovski, A.: Energy-selective reconstructions in x-ray computerised tomography. Phys. Med. Biol. (1976), doi: 10.1088/0031-9155/21/5/002.
212. Samei, E., Flynn, M.J., Eyler, W.R.: Detection of subtle lung nodules: relative influence of quantum and anatomic noise on chest radiographs. Radiology. (1999), doi: 10.1148/radiology.213.3.r99dc19727.
213. Ziedes des Plantes, B.G.: Eine neuemethodezurdifferenzierung in der röntgenographie (Planigraphies). Acta Radiol. **13**, 182–192 (1932).
214. Dobbins, J.T., McAdams, H.P.: Chest tomosynthesis: technical principles and clinical update. Eur. J. Radiol. (2009), doi: 10.1016/j.ejrad.2009.05.054.
215. Flynn, M.J., McGee, R., Blechinger, J.: Spatial resolution of x-ray tomosynthesis in relation to computed tomography for coronal/sagittal images of the knee. Proc. SPIE (2007), doi: 10.1117/12.713805.
216. Dobbins, J.T., Godfrey, D.J.: Digital x-ray tomosynthesis: current state of the art and clinical potential. Phys. Med. Biol. (2003), doi: 10.1088/0031-9155/48/19/r01.
217. Dobbins III, J.T.: Tomosynthesis imaging: at a translational crossroads. Med. Phys. (2009), doi: 10.1118/1.3120285.
218. Sechopoulos, I.: A review of breast tomosynthesis. Part I. the image acquisition process. Med. Phys. (2013), doi: 10.1118/1.4770279.
219. Sechopoulos, I.: A review of breast tomosynthesis. Part II. image reconstruction, processing and analysis, and advanced applications. Med. Phys. (2013), doi: 10.1118/1.4770281.
220. Niklason, L.T., Hickey, N.M., Chakraborty, D.P., Sabbagh, E.A., Yester, M.V., Fraser, R.G., Barnes, G.T.: Simulated pulmonary nodules: detection with dual-energy digital versus conventional radiography. Radiology. (1986), doi: 10.1148/radiology.160.3.3526398.

221. Kelcz, F., Zink, F.E., Peppler, W.W., Kruger, D.G., Ergun, D.L., Mistretta, C.A.: Conventional chest radiography vs dual-energy computed radiography in the detection and characterization of pulmonary nodules. Am. J. Roentgenol. (1994), doi: 10.2214/ajr.162.2.8310908.
222. Martini, K., Baessler, M., Baumueller, S., Frauenfelder, T.: Diagnostic accuracy and added value of dual-energy subtraction radiography compared to standard conventional radiography using computed tomography as standard of reference. PLOS ONE. (2017), doi: 10.1371/journal.pone.0174285.
223. Fischbach, F., Freund, T., Röttgen, R., Engert, U., Felix, R., Ricke, J.: Dual-energy chest radiography with a flat-panel digital detector: revealing calcified chest abnormalities. Am. J. Roentgenol. (2003), doi: 10.2214/ajr.181.6.1811519.
224. Kashani, H., Varon, C.A., Paul, N.S., Gang, G.J., Van Metter, R., Yorkston, J., Siewerdsen, J.H.: Diagnostic performance of a prototype dual-energy chest imaging system: ROC analysis. Acad. Radiol. (2010), doi: 10.1016/j.acra.2009.10.012.
225. Jacobson, B.: Dichromatic absorption radiography. Dichromography. Acta Radiol. (1953), doi: 10.1177/028418515303900601.
226. Sabol, J.M., Avinash, G.B.: Novel method for automated determination of the cancellation parameter in dual-energy imaging: evaluation using anthropomorphic phantom images. Proc. SPIE (2003), doi: 10.1117/12.480195.
227. Ricke, J., Fischbach, F., Freund, T., Teichgräber, U., Hänninen, E.L., Röttgen, R., Engert, U., Eichstädt, H., Felix, R.: Clinical results of CsI-detector-based dual-exposure dual energy in chest radiography. Eur. Radiol. (2003), doi: 10.1007/s00330-003-1913-9.
228. Kuhlman, J.E., Collins, J., Brooks, G.N., Yandow, D.R., Broderick, L.S.: Dual-energy subtraction chest radiography: what to look for beyond calcified nodules. Radiographics. (2006), doi: 10.1148/rg.261055034.
229. Sabol, J.M., Liu, R., Saunders, R., Markley, J., Moreno, N., Seamans, J., Wiese, S., Jabri, K., Gilkeson, R.C.: The impact of cardiac gating on the detection of coronary calcifications in dual-energy chest radiography: a phantom study. Proc. SPIE (2006), doi: 10.1117/12.653387.
230. Shkumat, N.A., Siewerdsen, J.H., Dhanantwari, A.C., Williams, D.B., Paul, N.S., Yorkston, J., Van Metter, R.: Cardiac gating with a pulse oximeter for dual-energy imaging. Phys. Med. Biol. (2008), doi: 10.1088/0031-9155/53/21/014.
231. Ishigaki, T., Sakuma, S., Horikawa, Y., Ikeda, M., Yamaguchi, H.: One-shot dual-energy subtraction imaging. Radiology. (1986), doi: 10.1148/radiology.161.1.3532182.
232. Ishigaki, T., Sakuma, S., Ikeda, M.: One-shot dual-energy subtraction chest imaging with computed radiography: clinical evaluation of film images. Radiology. (1988), doi: 10.1148/radiology.168.1.3289096.
233. Speller, R.D., Ensell, G.J., Wallis, C.: A system for dual-energy radiography. Br. J. Radiol. (1983), doi: 10.1259/0007-1285-56-667-461.
234. Brooks, R.A., Di Chiro, G.: Split-detector computed tomography: a preliminary report. Radiology. (1978), doi: 10.1148/126.1.255.
235. Fenster, A.: Split Xenon detector for tomochemistry in computed tomography. J. Comput. Assist. Tomogr. **2**, 243–252 (1978).
236. Barnes, G.T., Sones, R.A., Tesic, M.M., Morgan, D.R., Sanders, J.N.: Detector for dual-energy digital radiography. Radiology. (1985), doi: 10.1148/radiology.156.2.4011921.

237. Allec, N., Abbaszadeh, S., Fleck, A., Tousignant, O., Karim, K.S.: K-edge imaging using dual-layer and single-layer large area flat panel imagers. IEEE Trans. Nucl. Sci. (2012), doi: 10.1109/TNS.2012.2212250.
238. Alvarez, R.E., Seibert, J.A., Thompson, S.K.: Comparison of dual energy detector system performance. Med. Phys. (2004), doi: 10.1118/1.1645679.
239. Richard, S., Siewerdsen, J.H.: Optimization of dual-energy imaging systems using generalized NEQ and imaging task. Med. Phys. (2007), doi: 10.1118/1.2400620.
240. Ergun, D.L., Mistretta, C.A., Brown, D.E., Bystrianyk, R.T., Sze, W.K., Kelcz, F., Naidich, D.P.: Single-exposure dual-energy computed radiography: improved detection and processing. Radiology. (1990), doi: 10.1148/radiology.174.1.2294555.
241. Alvarez, R.E.: Active energy selective image detector for dual-energy computed radiography. Med. Phys. (1996), doi: 10.1118/1.597831.
242. Yun, S., Han, J.C., Kim, D.W., Youn, H., Kim, H.K., Tanguay, J., Cunningham, I.A.: Feasibility of active sandwich detectors for single-shot dual-energy imaging. Proc. SPIE (2014), doi: 10.1117/12.2043368.
243. Han, J.C., Kim, H.K., Kim, D.W., Yun, S., Youn, H., Kam, S., Tanguay, J., Cunningham, I.A.: Single-shot dual-energy x-ray imaging with a flat-panel sandwich detector for preclinical imaging. Cur. Appl. Phys. (2014), doi: 10.1016/j.cap.2014.10.012.
244. Kim, S.H., Kim, D.W., Kim, D., Youn, H., Cho, S., Kim, H.K.: Microtomography with sandwich detectors for small-animal bone imaging. J. Instrum. (2016), doi: 10.1088/1748-0221/11/10/p10011.
245. Kim, J., Kim, D.W., Kim, S.H., Yun, S., Youn, H., Jeon, H., Kim, H.K.: Linear modeling of single-shot dual-energy x-ray imaging using a sandwich detector. J. Instrum. (2017), doi: 10.1088/1748-0221/12/01/c01029.
246. Ha, S., Kim, J., Yun, J., Kim, D.W., Kim, S.H., Kim, H.K.: Linear analysis of single-shot dual-energy computed tomography with a multilayer detector. J. Instrum. (2019), doi: 10.1088/1748-0221/14/01/c01022.
247. Kim, D.W., Park, J., Kim, J., Kim, H.K.: Noise-reduction approaches to single-shot dual-energy imaging with a multilayer detector. J. Instrum. (2019), doi: 10.1088/1748-0221/14/01/c01021.
248. Kim, J., Kim, D.W., Kam, S., Park, J., Shon, C.-S., Heo, S.K., Kim, H.K.: Development of a large-area multilayer detector for single-shot dual-energy imaging. J. Instrum. (2018), doi: 10.1088/1748-0221/13/12/c12018.
249. Lehmann, L.A., Alvarez, R.E., Macovski, A., Brody, W.R., Pelc, N.J., Riederer, S.J., Hall, A.L.: Generalized image combinations in dual KVP digital radiography. Med. Phys. (1981), doi: 10.1118/1.595025.
250. Maurino, S.L., Badano, A., Cunningham, I.A., Karim, K.S.: Theoretical and Monte Carlo optimization of a stacked three-layer flat-panel x-ray imager for applications in multi-spectral diagnostic medical imaging. Proc. SPIE (2016), doi: 10.1117/12.2217085.
251. Shi, L., Lu, M., Bennett, N.R., Shapiro, E., Zhang, J., Colbeth, R., Star-Lack, J., Wang, A.S.: Characterization and potential applications of a dual-layer flat-panel detector. Med. Phys. (2020), doi: 10.1002/mp.14211.
252. Tanguay, J., Kim, H.K., Cunningham, I.A.: The role of x-ray Swank factor in energy-resolving photon-counting imaging. Med. Phys. (2010), doi: 10.1118/1.3512794.

253. Tapiovaara, M.J., Wagner, R.: SNR and DQE analysis of broad spectrum x-ray imaging. Phys. Med. Biol. (1985), doi:10.1088/0031-9155/30/6/002.
254. Cahn, R.N., Cederström, B., Danielsson, M., Hall, A., Lundqvist, M., Nygren, D.: Detective quantum efficiency dependence on x-ray energy weighting in mammography. Med. Phys. (1999), doi: 10.1118/1.598807.
255. Tlustos, L.: Spectroscopic x-ray imaging with photon counting pixel detectors. Nucl. Instrum. Methods Phys. Res. A. (2010), doi: 10.1016/j.nima.2010.02.088.
256. Ren, L., Zheng, B., Liu, H.: Tutorial on x-ray photon counting detector characterization. J. X-ray Sci. Technol. (2018), doi: 10.3233/XST-16210.
257. Shikhaliev, P.M., Fritz, S.G.: Photon counting spectral CT versus conventional CT: comparative evaluation for breast imaging application. Phys. Med. Biol. (2011), doi: 10.1088/0031-9155/56/7/001.
258. Bornefalk, H.: Task-based weights for photon counting spectral x-ray imaging. Med. Phys. (2011), doi: 10.1118/1.3653195.
259. Marchal, J.P.: Extension of x-ray imaging linear systems analysis to detectors with energy discrimination capability. Med. Phys. (2005), doi: 10.1118/1.1951041.
260. Tanguay, J., Yun, S., Kim, H.K., Cunningham, I.A.: The detective quantum efficiency of photon-counting x-ray detectors using cascaded-systems analyses. Med. Phys. (2013), doi: 10.1118/1.4794499.
261. Llopart, X., Campbell, M., Dinapoli, R., San Segundo, D., Pernigotti, E.: Medipix2: a 64-k pixel readout chip with 55-μm square elements working in single photon counting mode. IEEE Trans. Nucl. Sci. (2002), doi: 10.1109/TNS.2002.803788.
262. Mettivier, G., Montesi, M.C., Russo, P.: First images of a digital autoradiography system based on a Medipix2 hybrid silicon pixel detector. Phys. Med. Biol. (2003), doi: 10.1088/0031-9155/48/12/403.
263. Mikulec, B.: Development of segmented semiconductor arrays for quantum imaging. Nucl. Instrum. Methods Phys. Res. A. (2003), doi: 10.1016/S0168-9002(03)01672-3.
264. Tanguay, J., Cunningham, I.A.: Cascaded systems analysis of charge sharing in cadmium telluride photon-counting x-ray detectors. Med. Phys. (2018), doi: 10.1002/mp.12853.
265. Wang, A.S., Harrison, D., Lobastov, V., Tkaczyk, J.E.: Pulse pileup statistics for energy discriminating photon counting x-ray detectors. Med. Phys. (2011), doi: 10.1118/1.3592932.
266. Ballabriga, R., Campbell, M., Heijne, E., Llopart, X., Tlustos, L., Wong, W.: Medipix3: a 64k pixel detector readout chip working in single photon counting mode with improved spectrometric performance. Nucl. Instrum. Methods Phys. Res. A. (2011), doi: 10.1016/j.nima.2010.06.108.
267. Taguchi, K., Frey, E.C., Wang, X., Iwanczyk, J.S., Barber, W.C.: An analytical model of the effects of pulse pileup on the energy spectrum recorded by energy resolved photon counting x-ray detectors. Med. Phys. (2010), doi: 10.1118/1.3429056.
268. Tanguay, J., Yun, S., Kim, H.K., Cunningham, I.A.: Detective quantum efficiency of photon-counting x-ray detectors. Med. Phys. (2015), doi: 10.1118/1.4903503.
269. Taguchi, K., Polster, C., Lee, O., Stierstorfer, K., Kappler, S.: Spatio-energetic cross talk in photon counting detectors: detector model and correlated Poisson data generator. Med. Phys. (2016), doi: 10.1118/1.4966699.

270. Taguchi, K., Stierstorfer, K., Polster, C., Lee, O., Kappler, S.: Spatio-energetic cross-talk in photon counting detectors: numerical detector model (PcTK) and workflow for CT image quality assessment. Med. Phys. (2018), doi: 10.1002/mp.12863.
271. Taguchi, K., Stierstorfer, K., Polster, C., Lee, O., Kappler, S.: Spatio-energetic cross-talk in photon counting detectors: N × N binning and sub-pixel masking. Med. Phys. (2018), doi: 10.1002/mp.13146.
272. Stierstorfer, K.: Modeling the frequency-dependent detective quantum efficiency of photon-counting x-ray detectors. Med. Phys. (2018), doi: 10.1002/mp.12667.
273. Persson, M., Rajbhandary, P.L., Pelc, N.J.: A framework for performance characterization of energy-resolving photon-counting detectors. Med. Phys. (2018), doi: 10.1002/mp.13172.
274. Tanguay, J., Kim, J., Kim, H.K., Iniewski, K., Cunningham, I.A.: Frequency-dependent signal and noise in spectroscopic x-ray imaging. Med. Phys. (2020), doi: 10.1002/mp.14160.
275. Rajbhandary, P.L., Persson, M., Pelc, N.J.: Detective efficiency of photon counting detectors with spectral degradation and crosstalk. Med. Phys. (2020), doi: 10.1002/mp.13889.
276. Hajdok, G., Battista, J.J., Cunningham, I.A.: Fundamental x-ray interaction limits in diagnostic imaging detectors: frequency-dependent Swank noise. Med. Phys. (2008), doi: 10.1118/1.2936412.

# 8

# *Silicon Photomultipliers in Scintillation Detectors for Nuclear Medicine*

**Martyna Grodzicka-Kobylka, Marek Moszyński, and Tomasz Szczęśniak**

**CONTENTS**



## 8.1 Introduction

Scintillation detectors are one of the most commonly used type of a radiation detector in nuclear medicine. These detectors consist of a photodetector and dense crystalline scintillation material which absorbs gamma quanta and emits light as a result. The number of photons detected or "seen" by a photodetector is the main parameter used to define its performance. These photons are the "information carriers" about the detected gamma radiation. The higher is their number, the more detailed information about the radiation is possible.

DOI: 10.1201/9781003219446-8

The detection of weak light signals, such as scintillation light, can be achieved by means of several types of photodetectors; however, photomultiplier tubes (PMTs) are the most commonly used and have been for over 70 years (Knoll 2010). The photodetector is also one of the key components defining the final performance of a scintillation detector. The main advantages of photomultipliers are: good quantum efficiency (up to 40%), high gain ($10^3$–$10^8$), low dark current (dark noise), the ability to detect single photons, and the availability of a wide range of sizes (from several millimeters to tens of centimeters in diameter). However, PMTs also have some drawbacks: they are very sensitive to magnetic fields, require high supply voltages (often above 1000 V), are susceptible to mechanical damage from excessive mechanical shock or vibration, and are expensive because the complicated mechanical structure inside the vacuum container is mostly made manually.

Currently, a silicon photomultiplier (SiPM) shows promising results in the field of positron emission tomography (PET) (Otte et al. 2005; Grodzicka et al. 2015a; Bisogni et al. 2019). A SiPM is a relatively new kind of a photodetector that is made up of a matrix of parallel connected micro avalanche photodiodes (APD cells), working in a Geiger mode (Antich et al. 1997; Buzhan 2003; Golovin and Saveliev 2004). Each APD cell generates a constant signal after detection of a single photon, and a sum of the signals from all the APD cells gives a SiPM output pulse that is proportional to the number of detected photons. However, this proportionality is disrupted because of such phenomena as cross-talks and after-pulses, which give additional, false pulses that are not directly related to the detected photon. In consequence, the number of fired APD cells is larger than the number of detected photons. Moreover, each APD cell possesses an internal recovery time that lasts from a few to tens of nanoseconds after detection of a single photon (this value depends on the size of APD cells in a SiPM as well as its production process). Therefore, if at the same time two or more photons incident on one APD cell, the second signal will be lost and linearity of the photon detection will be deteriorated. This is because the number of incident photons that hit the photosensitive area and could be potentially detected is larger than the number of fired APD cells. Summarizing, the sum of signals from all of the fired APD cells of the entire array is an output signal of a SiPM. However, the proportionality of the total number of the fired APD cells to the number of incident photons is not ideally preserved due to the earlier mentioned SiPM features (finite number of APD cells with internal recovery time, cross-talks, and after-pulses).

A SiPM's response depends on its total number of APD cells, its effective recovery time and on the duration of the detected pulse or decay time of the light pulses in the scintillators (Musienko et al. 2006; Dolgoshein et al. 2006; Renker 2006; Renker and Lorenz 2009a; Finocchiaro et al. 2009; Du and Retiere 2008; Grodzicka et al. 2015b; Grodzicka-Kobyłka et al. 2019).

SiPMs are widely tested in high energy physics, neutrino physics, and also in commercial applications such as nuclear medicine or border monitoring devices (Otte et al. 2005; Renker 2006; Renker and Lorenz 2009a; Yokoyama et al. 2010; Musienko 2011; Andreev et al. 2005; Danilov 2007; Korpar et al. 2008; Schaart et al. 2009; Otte et al. 2006; Yamamoto et al. 2005; Raylman et al. 2006; Achenbach et al. 2009; Grodzicka et al. 2017c). Until recently, the small total active area of the individual detector (from $1 \times 1$ mm$^2$ to $2 \times 2$ mm$^2$) limited their potentiality of being used in gamma spectrometry with scintillators. Nowadays, a slightly larger total active area of single detectors ($3 \times 3$ mm$^2$ and $6 \times 6$ mm$^2$) and their matrices (from $6 \times 6$ mm$^2$ to $57 \times 57$ mm$^2$), also on a common substrate, allow for efficient use of SiPMs in scintillation detectors for gamma spectroscopy (Grodzicka-Kobyłka et al. 2019).

The rapid development of SiPMs has facilitated construction of compact, efficient, and magnetic resonance (MR) compatible PET scanners (Yamamoto et al. 2005; Raylman et al. 2006). Moreover, the use of TOF information in PET has been demonstrated to enable significant improvement in image noise properties and improved lesion detection, especially in heavier patients (Allemand et al 1980; Laval et al. 1982; Mullani et al. 1981). It has recently been shown that very good timing resolution can be achieved with SiPM-based scintillation detectors (Schaart et al. 2010; Gundacker et al. 2016) opening a new way for TOF PET scanners.

## 8.2 Operation Principles

SiPMs are manufactured by several companies: Hamamatsu Photonics (Japan), ON Semiconductor (Phoenix, AZ) (technology purchased from SensL, Ireland), Zecotek Photonics (Singapore), FBK (Fondazione Bruno Kessler, Trento, Italy), Broadcom (San Jose, CA) (technology partially purchased from FBK), ST–Microelectronics (Catania, Italy), Amplification Technologies (Linden, NJ), Ketek GmbH (Munich, Germany), and others. The name "SiPM" is the most commonly used, but in the literature this type of a photodetector can be also referred to as: a multi pixel photon counter (MPPC), a micro-pixel avalanche photo diode (MAPD), a multi pixel Geiger-mode avalanche photodiode (GM–APD or G–APD), a solid-state photomultiplier (SSPM), a single-photon avalanche diode (SPAD) array, or a pixelated Geiger-mode avalanche photon detector (PPD).

A SiPM consists of many small cells of APD that are fabricated on a common Si substrate. Nowadays, depending on the manufacturer's technology, commercially available SiPMs can be a single device with a total active area from $0.18 \times 0.18$ mm$^2$ up to $6 \times 6$ mm$^2$. Detectors with a larger active area are based on SiPM arrays of different format composed of these single elements. The array elements can be built on a single substrate or multiple, separate, small devices can be stacked together to form the array. The available single SiPM devices consist of many of APD cells from ~100 to

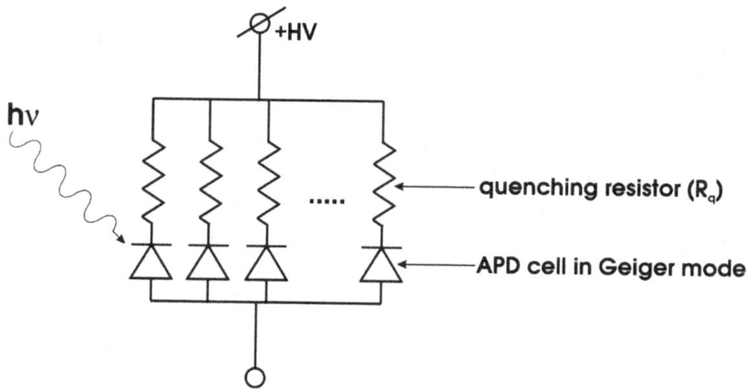


**FIGURE 8.1**
A simplified electric structure of a SiPM composed of several APD cells in a series with a quenching resistor.

15,000 per mm$^2$. The size of the individual cell varies from $15 \times 15$ μm$^2$ up to a maximum of $100 \times 100$ μm$^2$ and is constant for a given device.

Each APD cell operates in a Geiger mode, which means that a single APD cell is reverse-biased above the electrical breakdown voltage ($V_{bd}$). In these conditions the electric field in the depletion region of the APD cell is high enough for free carriers, i.e., the electrons and holes (which were produced by light absorption), to produce additional carriers by impact ionization, thus resulting in a self-sustaining avalanche (Vacheret et al. 2011). This Geiger discharge is stopped due to a drop in the voltage below the breakdown value, either passively by an external resistor connected in a series with the diode ($R_q$, quenching resistor, typically from about 100 kΩ to several MΩ, see Figure 8.1) or actively by the special quenching electronics (Bondarenko et al. 2000). After a certain effective recovery time, which lasts several tens of nanoseconds, the voltage is restored again to the operating value and the cell is ready to detect the arrival of another photon (Spinelli and Lacaita 1997). Before restoring the full voltage at a given cell, this cell can still generate signals but with reduced gain (or amplitude).

The independently operating APD cells are connected in parallel to the same readout line; therefore, the combined output signal corresponds to the sum of all fired APD cells, which is a measure of the light flux (Dolgoshein et al. 2006).

Due to the device operation principle, the avalanche can be triggered not only by photo-generated carriers, but also by carriers that are thermally generated or emitted as a result of phenomena such as after-pulses and cross-talks (discussed in Sections 8.3.4 and 8.3.5, respectively).

The most important parameters that characterize all photodetectors used in gamma-ray spectroscopy and nuclear medicine are as follows:

quantum efficiency (or photon detection efficiency, PDE), dark noise, linearity (dynamic range), excess noise factor (ENF), after-pulses and cross-talks (if exist), and gain.

## 8.3 Basic Parameters of a SiPM

One of the key parameters of a SiPM is the size of the single APD cell (also known as a microcell). It affects both the gain and the dynamic range. For geometrical considerations, for a given border technology, the smaller the microcell size, the smaller the fill factor (the percentage of the sensitive area in respect to the total area of the device). Thus, from the efficiency point of view, large cells are desirable. On the other hand, the main positive aspects of a small microcell (e.g., ≤ 30 μm) are: a higher dynamic range (due to a high number of microcells on the same substrate area), a shorter recovery time, and a smaller correlated noise (optical cross-talks and after-pulsing, both proportional to the gain of the detector) (Piemonte et al. 2013).

In Table 8.1 the main parameters of basic $3 \times 3$ mm$^2$ or $6 \times 6$ mm$^2$ Hamamatsu MPPC, ON Semi, FBK, and Ketek SiPMs are compared.

### 8.3.1 Gain

The gain of one APD cell (G) is defined as the ratio between the charge produced in a single avalanche and the elementary charge, and it can be expressed as (van Dam et al. 2010) (Vacheret et al. 2011) (Dolgoshein et al. 2006):

$$G = \frac{Q}{e} = \frac{C_{APDcell} \cdot V_{ob}}{e} \tag{8.1}$$

where:

$Q$ – is the charge of a single avalanche in an APD cell, typically created by a single carrier (unit charge) and can be triggered either by a photon or by thermal noise,

$e$ – is the elementary charge, equal to $1.602 \cdot 10^{-19}$C,

$C_{APDcell}$ – is the APD cell capacitance,

$V_{ob}$ – is the voltage-over-breakdown also referred to as overvoltage, defined as:

$$V_{ob} = (V_b - V_{bd}) \tag{8.2}$$

where:

$V_b$ – is the operating voltage or bias voltage,

$V_{bd}$ – is the breakdown voltage of a SiPM.

**TABLE 8.1**

Comparison of Parameters of the Basic SiPMs Produced by Different Manufacturers

| Parameter | Hamamatsu (Hamamatsu 2017) | ON Semi (SensL) | FBK (Cozzi 2017a) | Ketek (Ketek) |
|---|---|---|---|---|
| Type | S14160/S14161 –6050HS | J-Series 60035 | NUV-HD | PM3315-WB/ PM3325-WB |
| Active area [mm] | 6.0 × 6.0 or in array | 6.0 × 6.0 or in array | 6.0 × 6.0 or in array | 3 × 3 or in array |
| APD cell size [um] | 50 | 35 | 30/40 | 15/25 |
| Number of APD cells | 14331 | 22292 | 39900/22500 | 38800/13920 |
| Fill factor [%] | 74 | 75 | 78/83 | 80 |
| Spectral resp. range [nm] | 270–900 | 200–900 | 290–700 | 300–900 |
| PDE [%] | 50 at 2.7 V of overvoltage | 38 at 2.5 V 50 at 5 V of overvoltage | 46/56 at 4 V of overvoltage | 31/43 at 5 V of overvoltage |
| Breakdown voltage [V] | 37 | 24.5 | 26 | 27.0 |
| Gain | $2.5 \times 10^6$ | $6.3 \times 10^6$ | $1.8/3.4 \times 10^6$ | $0.3/0.87 \times 10^6$ |
| Dark current [uA] | 2.5 at 2.7 V of overvoltage | 0.9 at 2.5 V of overvoltage | 0.78 at 4 V of overvoltage | – |
| Dark count rate [kHz/mm²] | – | 50 at 2.5 V of overvoltage | 135/100 at 4.0 V of overvoltage | 50 at 2.5 V of overvoltage |
| Crosstalk prob. [%] | 7 at 2.7 V of overvoltage | 8 at 2.5 V of overvoltage | 8.8/14 at 4V of overvoltage | 7/15 at 2.5 V of overvoltage |
| Capacitance [pF] | 2000 | 4140 | 3000/3150 | 750/800 |
| Temperature coefficient [mV/K] | 34 | 21.5 | 25.0/25.0 | 22 |

The gain of a single APD cell increases linearly with the overvoltage, as opposed to the exponential voltage dependence of the gain in standard APDs. The gain is independent of the temperature if the overvoltage ($V_{ob}$) is constant, although the breakdown voltage ($V_{bd}$) is dependent on the temperature. It means that when the bias voltage is constant and the temperature has changed, the gain will also change due to a shift of the breakdown voltage. Temperature coefficient ($T_c$) is usually from 20 mV/K to 61 mV/K and strongly depends on the production process (or manufacturer). $T_c$ value does not depend on the size of the SiPM microcell.

### 8.3.2 Photon Detection Efficiency

The expected quantum efficiency (QE) of a SiPM, which is dependent on the wavelength of the incident photons ($\lambda$, nm), is more than 80% at 500 nm (Vacheret et al. 2011) (Dolgoshein et al. 2006), being the value characteristic for Si photodetectors. However, the overall PDE for the present-state SiPMs is smaller due to two additional contributions apart from the QE. The PDE can be defined as (Renker and Lorenz 2009b) (Dolgoshein et al. 2006):

$$PDE = QE \cdot FF \cdot THR \tag{8.3}$$

where:

*FF* – is a geometrical fill-factor, i.e., the ratio of the active area to the total area of the device,

*QE* – is the quantum efficiency, i.e., the probability that an incident photon will generate an electron–hole pair in a region in which carriers can produce an avalanche,

*THR* – is the combined probability of electrons and holes to initiate a Geiger discharge, which is a strong function of the electric field.

Not every carrier will succeed in generating a Geiger avalanche. In fact there is a non-zero probability that the seed carrier will lose energy to photon interactions and recombine such that the chain of ionization stops before the entire junction goes into the Geiger breakdown. The overvoltage has an influence on the electric field in the SiPM and, consequently, on the kinetic energy of the carriers, which is higher in electrons rather than in holes for a fixed electromagnetic field. The trigger probability is almost linear for a low value of overvoltage.

Typical curves of the PDE (as a function of a wavelength) for Hamamatsu MPPCs with three different APD cell sizes: 25 μm, 50 μm, and 100 μm are presented in Figure 8.2, following the manufacturer data. The curves include such effects as cross-talks and after-pulses. It is worth noting that this figure shows only an approximate curve because the manufacturer does not provide the voltage at which the curve was measured. It can only be assumed that the curves were measured for the recommended voltage suggested by the producer.

Because FF is constant for a given SiPM and QE of the device is constant for a given scintillator or a given wavelength of the laser diode pulser, the PDE of a single device depends only on the THR and in consequence is a function of the SiPM bias voltage (*V*). The photon detection efficiency (without effects of cross-talks and after-pulses) for MPPC with an APD cell size of 50 μm, as a function of overvoltage at three temperatures, is shown in Figure 8.4, according to (van Dam et al. 2010). PDE increases with the voltage-over-breakdown ($V_{ob}$); however, for a fixed overvoltage, there is no observable dependence on the temperature (van Dam et al. 2010).

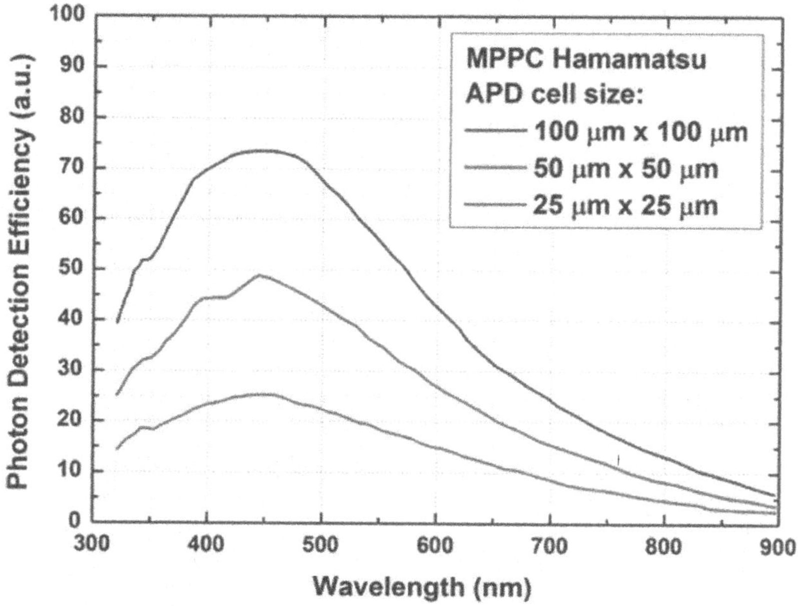


**FIGURE 8.2**
Photon detection efficiency (including effects of cross-talks and after-pulses) according to the manufacturer data (Hamamatsu 2009).

### 8.3.3 Excess Noise Factor

The ENF, affecting the performance of photodetectors in pulse height spectroscopy, is described by a pulse-to-pulse fluctuation of the charge of the output signal. The ENF for photomultipliers and APD is a result of fluctuation of the multiplication process. In the case of the SiPM, the relative gain fluctuation has only little impact on the ENF and the ENF is mainly affected by cross-talks and after-pulses.

Figure 8.3 presents the dependence of the ENF on the overvoltage at different temperatures. Measurements included in this figure were done for 2.2 × 2.2 mm SiPM (15 × 15 µm) from FBK. The ENF increases with the overvoltage due to a larger number of cross-talks and after-pulses, which introduce additional avalanches in a stochastic manner. In the example shown the ENF increases exponentially and at room temperature reaches about 2 at $V_{ob}$ of about 13 V. ENF of SiPMs can be significantly larger comparing to PMTs, where the ENF is equal to about 1.1–1.2. It is worth to note that the ENF is practically independent on the temperature for overvoltages lower than 9 V. Above this value, at 20°C, the quenching mechanism starts to be inefficient, determining a larger fluctuation of the charge produced by a single avalanche. This, in turn, produces a rapid increase of the ENF. At lower temperatures, the quenching resistance increases, postponing the

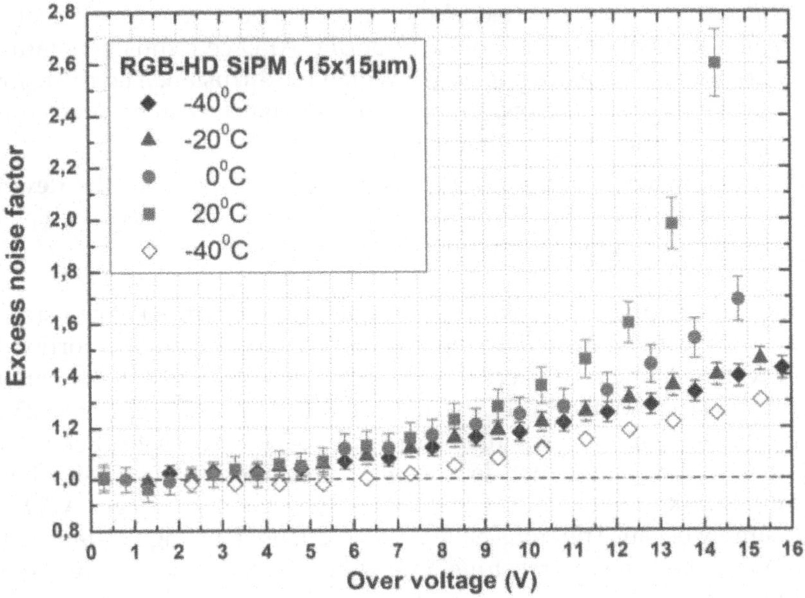


**FIGURE 8.3**
The excess noise factor vs. overvoltage measured for FBK 2.2 × 2.2 mm RGB-HD SiPM, adapted from (Grodzicka et al. 2014).

onset of this phenomenon. For a more detailed discussion see (Grodzicka et al. 2014; Piemonte et al. 2013).

### 8.3.4 Optical Cross-Talks

The optical cross-talks in SiPMs are false pulses, which are caused by optical photons emitted during an avalanche discharge in a single APD cell. These photons can trigger another Geiger discharge in the same or neighboring APD cell (Musienko et al. 2006). The earlier-mentioned photons are emitted mainly because of spontaneous, direct carrier relaxation in the conduction band and this effect is known as hot-carrier luminescence (Renker and Lorenz 2009a) (Eckert et al. 2010) (Acerbi et al. 2015). In accordance with (Chmill et al. 2017) $2.9 \times 10^{-5}$ photons with energy higher than 1.14 eV and with a wavelength less than 1 μm are emitted per carrier crossing the junction. Optical cross-talks can manifest itself in three different ways (Grodzicka et al. 2015a; Piemonte et al. 2013):

a. *Direct,* when the emitted photon generates a carrier in the active region of a neighboring APD cell, thus producing a second avalanche in coincidence with the first one.

b. *Delayed,* when the emitted photon is absorbed in the non-depleted region beneath the same or neighboring APD cell, thus generating a carrier able to reach the active region by diffusion. The diffusion process is relatively slow, so the second avalanche can be delayed in time with respect to the first one.

c. *External,* when the emitted photon tries to escape from the device but is reflected by structures placed on top of the device, such as a scintillator wrapped with reflecting/diffusing material.

The number of optical cross-talks mainly depends on an APD cell size, the distance between the high-field regions and the SiPM gain (proportional to the bias voltage). The characteristics of cross-talk probability as a function of gain for the Hamamatsu MPPC with three different APD cell sizes: 25 μm, 50 μm, and 100 μm are presented in Figure 8.4, according to (Eckert et al. 2010). It is worth noting that devices with a smaller APD cell size have a larger cross-talk probability as compared to devices with a larger APD cells. This is simply because photons have to travel a longer average distance in the case of larger cells before reaching a neighboring APD cell, where they can cause a second avalanche.

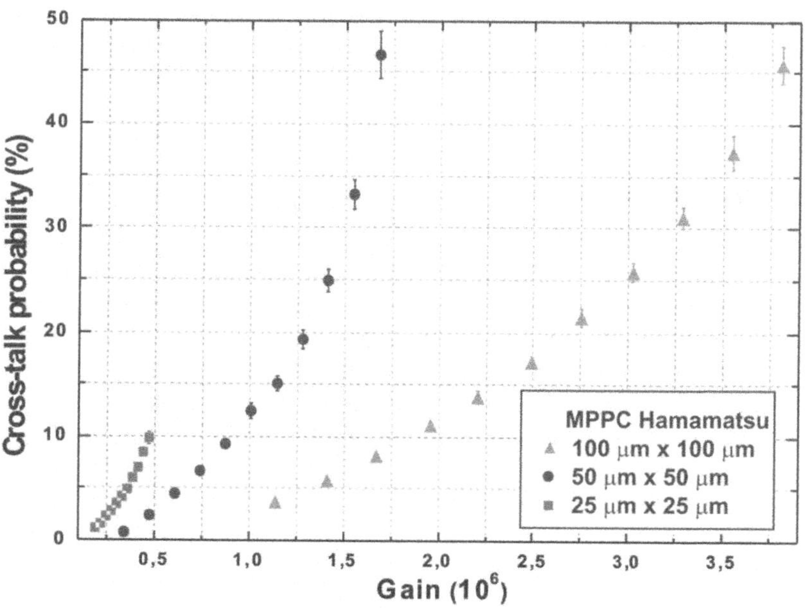


**FIGURE 8.4**
Cross-talk probability as a function of the SiPM gain for three different sensor types, adapted from (Eckert et al. 2010).

### 8.3.5 After-Pulses

The after-pulses of the SiPM are false pulses which are caused by the trapping of charge carriers created during an avalanche, by defects or impurities in the silicon lattice. The trapped carriers are released after a certain time, triggering a new avalanche inside the same APD cell as the original avalanche. The after-pulse avalanche is delayed by the trapping time. Because the traps in the silicon may be of various types with a different trap lifetime, the after-pulsing can exhibit more than one time constants and can be different for various technologies (or fabrication processes). For example, two time components: 15 ns (short) and 83 ns (long) for an MPPC with an APD cell size of 50 × 50 μm were observed by (Du 2008). For comparison, with a similar MPPC (Vacheret et al. 2011) observed three time components: 17 ns (first), 70 ns (second), and 373 ns (third). They also observed that both, the first and the second component decreased as the temperature increased. Operation at low temperatures elongates the release delay because of the elongated trapping center de-excitation time (Renker and Lorenz 2009a). Moreover, the after-pulse probability for the first component decreases with temperature, in contrast to the probability of the after-pulse for the second component, which increases with temperature.

After-pulsing may also be partially suppressed due to the presence of the effective dead time (also called the recovery time) of an APD cell. If a carrier is released while the APD cell voltage has not reached the nominal value, then the charge produced in the avalanche will be lower comparing to the nominal avalanches (see Figure 8.5). Only if the after-pulse delay is larger than the APD cell's effective recovery time, a standard avalanche signal can be triggered (Yokoyama et al. 2010; Vacheret et al. 2011; Piemonte 2006; Eckert et al. 2010).

During an avalanche only a very small fraction of the trapping level is filled. Therefore, the trap population is always well below saturation and the carrier trapping probability remains constant during all of the avalanche pulses. The probability that an after-pulse will occur increases with the amount of charge that flows through the diode during a Geiger discharge. Thus, the after-pulsing probability increases exponentially with the increasing bias voltage and, consequently, with the overvoltage. The after-pulse probability is also a function of the APD cell size and increases for a larger APD cells. The characteristics of the after-pulse probability as a function of the overvoltage for a Hamamatsu MPPC with three different APD cell sizes: 25 μm, 50 μm, and 100 μm are presented in Figure 8.6 (Eckert et al. 2010).

### 8.3.6 Dark Noise

The main sources of the dark noise in SiPMs are charge carriers generated thermally within the depletion region, which subsequently enter the Geiger

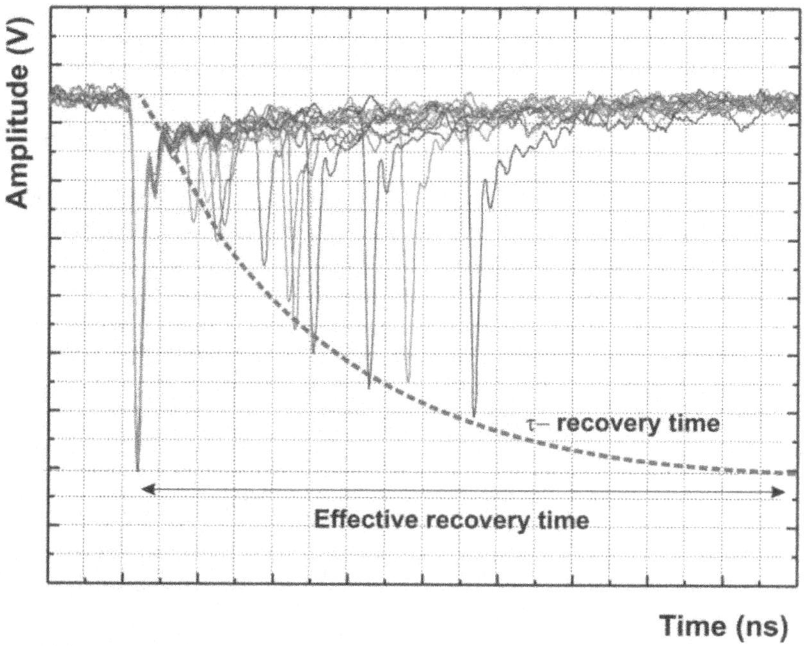


**FIGURE 8.5**
Examples of after-pulses with different delays, adapted from (Piemonte, 2006).

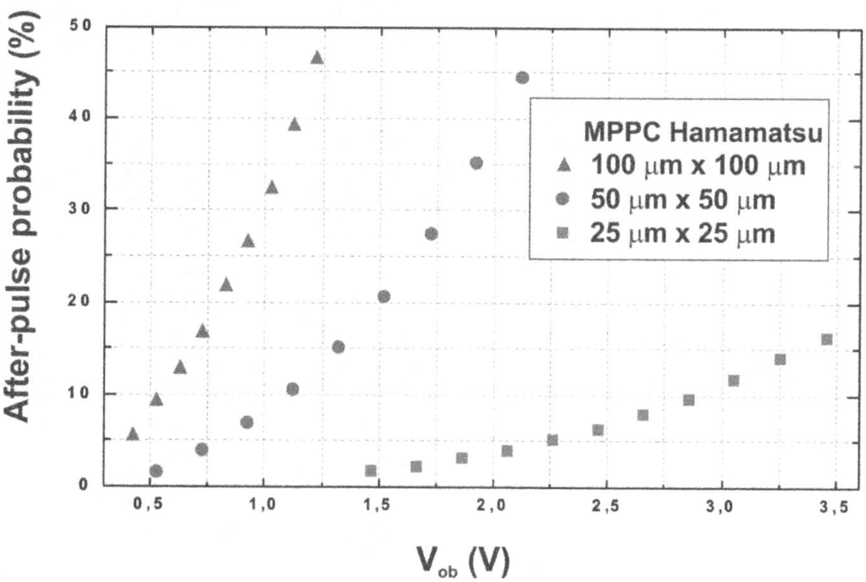


**FIGURE 8.6**
After-pulse probability as a function of the overvoltage for three different sensor types, adapted from (Eckert et al. 2010).

multiplication area and trigger avalanches. It is worth remembering that any avalanche can, in turn, initiate secondary avalanches through after-pulsing and cross-talks (Vacheret et al. 2011).

The dark noise rate of a SiPM is mainly dependent on the capacitance of an APD cell and is proportional to its total active area. The noise increases almost linearly with increasing overvoltage and exponentially with increasing temperature. Nowadays, depending on the manufacturer, commercially available SiPMs possess a dark noise rate from a few to several hundred of $kHz/mm^2$ at room temperature and gain of ~$10^6$. Such a high-dark noise rate limits the SiPM performance at room temperatures, especially for a large sensitive area (~100 $mm^2$ and above); however, only in detection of small light intensities (one, a few or a dozen of photoelectrons). It does not strongly affect the measurements in the case of larger light signals (Yokoyama et al. 2010; Buzhan et al. 2001).

### 8.3.7 Linearity of a SiPM Response

In order to define the basic parameters of a detector used for gamma spectrometry (such as energy resolution or number of photoelectrons per MeV), its response should be linear in the studied energy range or the characteristics of its nonlinear response should be known in order to correct the nonlinear data. Construction and principle of operation of SiPMs makes this type of devices nonlinear by definition. The detector consists of finite number of sensitive elements (APD cells) and this number limits the number of photons that can be detected. Moreover, even when the number of incident photons is much lower than the total number of APD cells, two photons may interact with the same cell. In such a case the second photon is lost and "invisible" for a SiPM. As a result the response of the detector (number of fired APD cells) stops to be proportional to the number of incident photons. The linear range of the SiPM response mainly depends on the total number of APD cells, the number of illuminating photons, and the effective dead time of APD cells in relation to the width (or decay time) of the light pulse (Musienko et al. 2006; Grodzicka et al. 2015b; Grodzicka-Kobyłka et al. 2019).

For light pulses shorter than the effective recovery time and for an "ideal" SiPM (without phenomena such as cross-talks and after-pulsing), the response of the SiPM is described by the following equation (Renker and Lorenz 2009b):

$$\mathrm{N_{fired}} = \mathrm{N_{total}} \cdot \left[1 - \exp\left(\frac{-(\mathrm{N_{photon}} \cdot PDE)}{\mathrm{N_{total}}}\right)\right] \mathrm{P_W} \le t_d \tag{8.4}$$

where:

$N_{fired}$ – number of excited APD cells,

$N_{total}$ – total number of APD cells,
$N_{photon}$ – number of incident photons,
*PDE* – photon detection efficiency,
$P_w$ – pulse width,
$t_d$ – effective recovery time.

The product of the number of incident photons ($N_{photon}$) and the PDE can be considered as the number of photons having the potential to be detected ($N_{pd}$).

$$N_{pd} = N_{photon} \cdot PDE \tag{8.5}$$

Figure 8.7 presents the theoretical response of three types of 1 × 1 mm² MPPCs, according to Equation (8.4), where $N_{total}$ equals to 100, 400, and 1600 APD cells and the APD cell size is equal to 100 × 100 μm², 50 × 50 μm², and 25 × 25 μm², respectively.

Phenomena such as optical cross-talks and after-pulses produce optical photons and electrons resulting from additional discharges of APD cells. Thus, for the same number of incident photons the number of fired

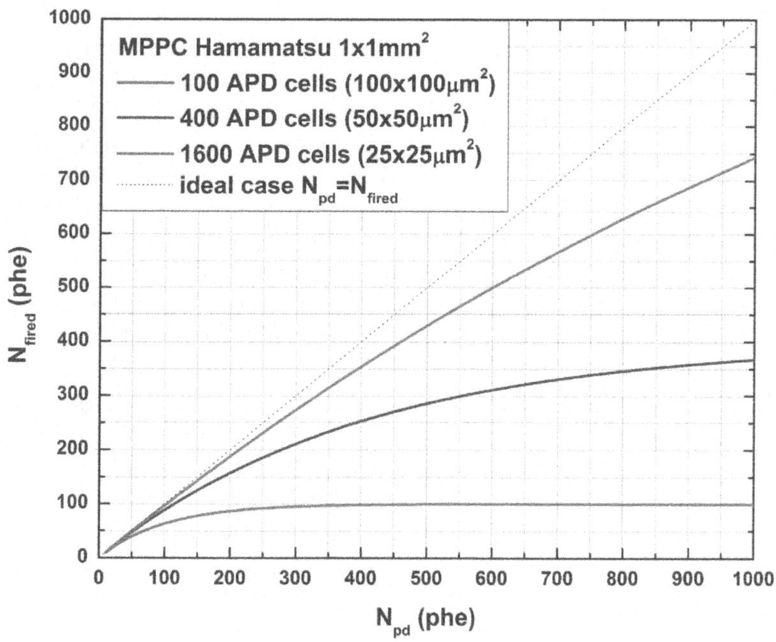


**FIGURE 8.7**
Theoretical lines – the number of fired APD cells vs. the number of photons having the potential to be detected on the MPPC for three different sensor types.

APD cells is higher than is apparent from Equation (8.4). This difference is related to the probability of cross-talks and after-pulses. These probabilities become higher with an increase of the bias voltage and depend on the type and production technology of the SiPM used. For a constant bias voltage this probability is constant and independent on the number of incident photons. Taking into account the discussed considerations, the response of a SiPM, for light pulses shorter than the effective recovery time, can be described by the equation:

$$\mathrm{N}_{\mathrm{fired}} = \mathrm{N}_{\mathrm{total}} \cdot \left[1 - \exp\left(\frac{-\left((1+P)\cdot \mathrm{N}_{\mathrm{photon}} \cdot \mathrm{PDE}\right)}{\mathrm{N}_{\mathrm{total}}}\right)\right] \mathrm{P_W} \le t_d \tag{8.6}$$

where:

$P_w$ – width of a rectangular pulse
$t_d$ – effective recovery time
$P$ – probability of cross-talks and after-pulses

The total number of events having potential to be detected ($N_{ted}$) includes optical cross-talks photons and electrons causing after-pulses and is given by:

$$\mathrm{N}_{\mathrm{ted}} = \left((1+P)\cdot \mathrm{N}_{\mathrm{photon}} \cdot \mathrm{PDE}\right) \tag{8.7}$$

For a more detailed discussion see (Grodzicka et al. 2015b). It has been experimentally proven that for pulses longer than the effective recovery time, the number of $N_{fired}$ increases and consequently the linear SiPM range is extended.

## 8.4 Transit Time Spread

The time resolution capabilities of any photodetector are described mainly by three factors: number of photoelectrons due to detected light flux, ENF, and time jitter (Moszyński and Bengtson 1979). Time jitter or transit time spread (TTS) is defined by the distribution of arrival times of pulses induced by single photons illuminating a whole surface of a photodetector. Full width at half maximum (FWHM) of this distribution describes an intrinsic timing resolution of a photodetector. Since the active layer of silicon in SiPMs is very thin (2–4 μm) and the process of the breakdown development is fast, very good timing properties, even for single photons, can be expected. Fluctuations in the avalanche development are mainly due to a

**TABLE 8.2**

Transit Time Spread of Various SiPMs

| SiPM | APD Cell Size | TTS | References |
|---|---|---|---|
| Ham 1 × 1 mm$^2$ | 50 × 50 μm$^2$ | 280 ps | Ronzhin (2010) |
| Ham 1 × 1 mm$^2$ | 50 × 50 μm$^2$ | 300 ps | Szczesniak et al. (2012) |
| FBK 1 × 1 mm$^2$ | 40 × 40 μm$^2$ | 94 ± 5 ps | Nemallapudi et al.(2016) |
| Ham 3 × 3 mm$^2$ | 50 × 50 μm$^2$ | 520 ps | Ronzhin (2010) |
| Ham 3 × 3 mm$^2$ | 50 × 50 μm$^2$ | 410 ps | Szczesniak et al. (2012) |
| Ham 3 × 3 mm$^2$ | 50 × 50 μm$^2$ | 340 ps | Mazzillo (2010) |
| Ham 3 × 3 mm$^2$ | 50 × 50 μm$^2$ | 190 ps | Gundacker et al. (2013) |
| Ham TSV 3 × 3 mm$^2$ | 50 × 50 μm$^2$ | 290 ± 7ps | Nemallapudi et al. (2016) |
| Ham LCT2 3 × 3 mm$^2$ | 50 × 50 μm$^2$ | 220 ± 7 ps | Nemallapudi et al. (2016) |
| FBK NUV, 3 × 3 mm$^2$ | 40 × 40 μm$^2$ | 175 ± 7 ps | Nemallapudi et al. (2016) |
| SensL JD0, 3 × 3 mm$^2$ | 35 × 35 μm$^2$ | 290 ± 7 ps | Nemallapudi et al. (2016) |
| SensL JD4, 3 × 3 mm$^2$ | 20 × 20 μm$^2$ | 270 ± 7 ps | Nemallapudi et al. (2016) |
| Ketek Optim., 3 × 3 mm$^2$ | 50 × 50 μm$^2$ | 330 ± 7 ps | Nemallapudi et al. (2016) |

lateral spreading by diffusion and by the photons emitted in the avalanche (Buzhan et al. 2001; Lacaita et al. 1990; Lacaita et al. 1993). Large spread in the experimental data concerning time jitter of SiPMs is observed. For 3 × 3 mm$^2$ devices the values vary between 200 ps and 500 ps at FWHM, see Table 8.2. In the case of 1 mm devices situation is similar (Ronzhin et al. 2010; Mazzillo et al. 2010; Gundacker et al. 2013; Szczesniak et al. 2012; Nemallapudi et al. 2016). Moreover, in most cases the TTS is lower for red-light photons than for blue-light ones (Ronzhin et al. 2010).

A large spread of the TTS reported in different papers is the effect of an influence of different factors on the measured quantity. The measurements are mostly carried out with illumination of SiPMs by fast light pulses of 50 ps wide or less from a laser pulser. During the measurements it is important to attenuate the detected light by at least 10–20 times to assure detection of only single photons. In (Szczesniak et al. 2012) the single photon regime was checked in two ways. First, by a comparison of average light pulses recorded by the digital oscilloscope with laser on and off – in both cases the pulses were indistinguishable. Second, by setting the number of coincidences between the laser photons (emitted with 10 kHz rate) and the SiPM response at the level below 5%. The single photon regime is crucial for this type of timing measurements since contribution of double (or multiple) photons in the detected light artificially improves the measured time jitter. Table 8.2 summarizes the recent measurements of TTS in 1 × 1 mm$^2$ and 3 × 3 mm$^2$ SiPMs.

## 8.5 SiPMs in Gamma Spectroscopy with Scintillators

The broad studies of SiPMs in gamma spectroscopy with scintillators were summarized in (Grodzicka-Kobyłka et al. 2019). The studies, started with a small 3 × 3 mm MPPC (Nassalski et al. 20107; Grodzicka et al. 2012; Grodzicka et al. 2013a; Grodzicka et al. 2013b; Grodzicka et al. 2017a; Grodzicka et al. 2017b and were later extended up to 52 × 52 mm MPPC arrays used even with 3 × 3 inch NaI(Tl) and BGO crystals (Szczesniak et al. 2017). During the studies, a number of different scintillators, like CsI:Tl, BGO, LSO, $LaBr_3$, $CeBr_3$, NaI:Tl, CsI:Na, $LaCl_3$, $CaF_2$:Eu, and $CdWO_4$ were tested. Another important studies were carried out by C. Fiorini group at Politecnico di Milano, with $LaBr_3$ and $LaBr_3$(Ce,Sr) crystals coupled to the FBK SiPM arrays (Cozzi et al. 2017; Cozzi 2018; Cozzi et al. 2016; Cozzi 2017a; Montagnani et al. 2018; Montagnani 2018a; Di Vita et al. 2019). Moreover, some other papers presenting different designs of scintillation detectors with SiPM light readout were reported in (Kim et al. 2015; Regazzoni et al. 2017; Yoo et al. 2015) and some others addressed to plasma diagnostic at JET arrangement in UK (Zychor et al. 2016; Rigamonti et al. 2016; Nocente et al. 2016).

Figure 8.8a presents a photograph of a 3 × 3 inch BGO crystal coupled to the monolithic Hamamatsu TSV MPPC array S12642-1616PB-50(X) having a 50 × 50 $\mu m^2$ cell size and effective active area of 48 × 48 $mm^2$ (16 × 16 channels). In Figure 8.8b an example with bottom view of similar coupling is shown to highlight the differences in size between the crystal and the

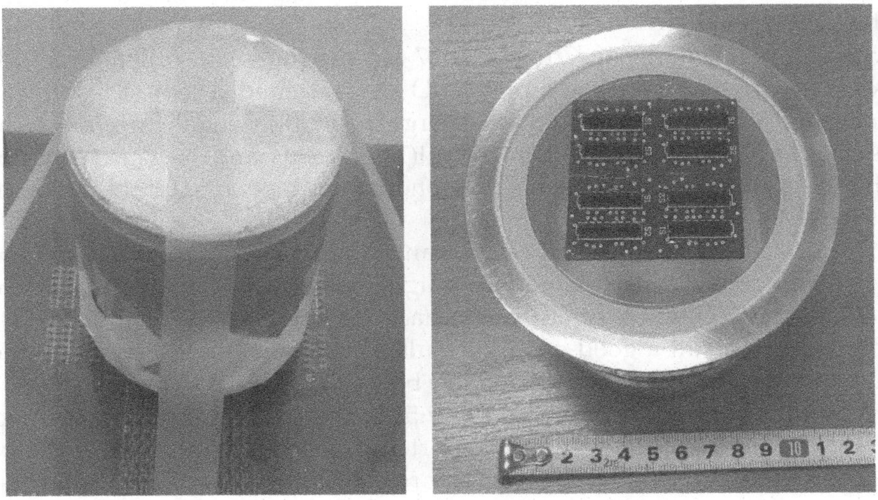

**FIGURE 8.8**
(a) Experimental setup with 3 × 3 inch BGO crystal coupled to the monolithic, 2 × 2 inch Hamamatsu TSV MPPC array S12642-1616PB-50(X) with 16 × 16 channels. (b) Example of coupling of 2 × 2 inch SiPM (composed of four 1 × 1 inch arrays) to 3 × 3 inch scintillator.

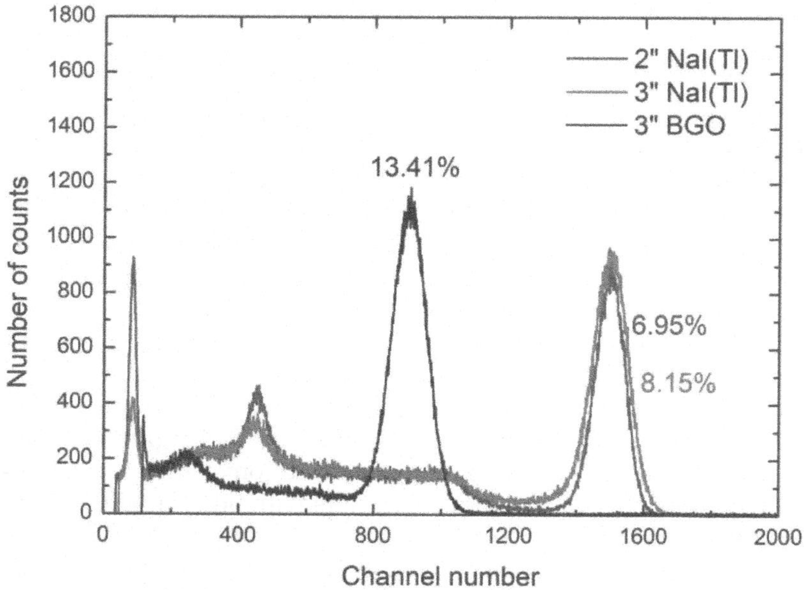


**FIGURE 8.9**
Energy spectra of Cs-137 gamma source obtained with Hamamatsu TSV MPPC array S12642-1616PB-50(X) in three detector configurations (2 and 3 inch NaI(Tl) and 3 inch BGO) for the optimal operating voltage of 67V (Szczesniak et al. 2017).

photodetector. During the measurements the open surface of the crystal (outside of the MPPC) was covered by a Teflon tape in order to maximize the light collection.

In Figure 8.9 energy spectra of Cs-137 gamma source, recorded for 2 inch NaI(Tl), 3 inch NaI(Tl), and 3 inch BGO are presented. The best result was achieved for 2 inch NaI(Tl) with energy resolution equal to 6.95% what is consistent with typical data of NaI(Tl) crystals and measurements of the same crystal with 2 inch classic photomultiplier XP5212B (Szczesniak et al. 2017).

In the case of large 3 inch NaI(Tl) and 3 inch BGO the obvious reason of energy resolution deterioration is light loss due to the area difference between MPPC and the scintillator surface. Nevertheless the recorded 8.15% and 13.4% are still a good enough result for many applications, where the compact design of the detector and low bias voltage are of importance.

The superior energy resolution of 2.58%, comparable to that measured with PMT (see Figure 8.10), was reported in (Montagnani et al. 2018) for a 3 × 3 inch $LaBr_3$(Ce,Sr) crystal. The readout of the 3 × 3 inch scintillator crystal was performed by 144 NUV-HD SiPMs from FBK arranged in a 12 × 12 matrix composed by 9 tiles, each counting 16 SiPMs, connected to the motherboard by means of 2 connectors per tile (Montagnani et al. 2018) (Montagnani 2018a) (di Vita et al. 2019).

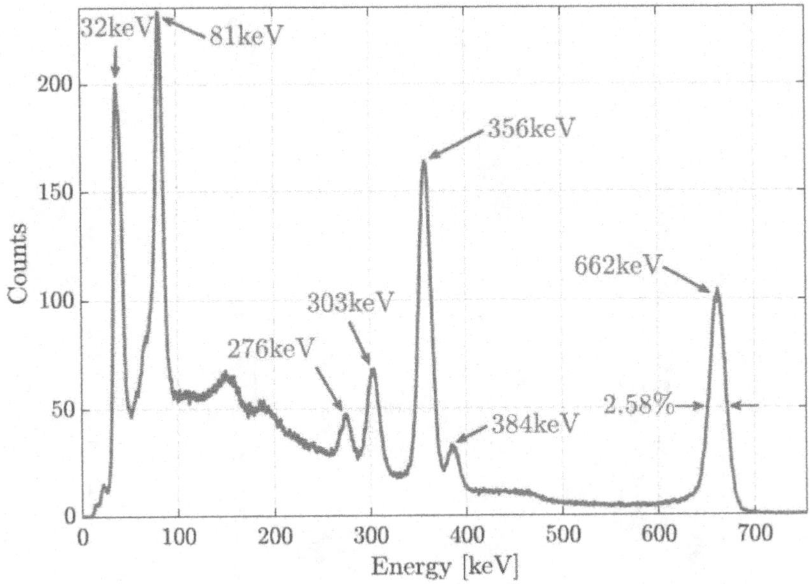


**FIGURE 8.10**
The best energy spectra acquired with SiPM and 3 inch $LaBr_3$:Ce,Sr irradiated by $^{57}$Co (122 keV), $^{133}$Ba (81 keV, 303 keV, 356 keV), $^{137}$Cs (662 keV), and $^{60}$Co (1173 keV, 1333 keV) gamma sources. Measurements were performed with an OV = 4.1 V, 800 ns of shaping time and at 30°C. An energy resolution of 2.58 ± 0.01% was achieved at 662 keV, following (Montagnani 2018a).

Energy resolution measured with small scintillators, fully fitted to the sensitive area of an MPPC array, is comparable or better than that measured with standard PMTs (Grodzicka-Kobyłka et al. 2019). It is worth pointing out that spectra measured with fast scintillators should be checked and possibly corrected for non-linear response of a SiPM array (see examples of linearity characteristics in Figure 8.11).

Good energy resolution of scintillation detectors based on SiPMs light readout is of importance in a potential application to gamma cameras and in PET scanners, limiting a contribution of false events due to scattered gamma quanta.

## 8.6 SiPMs in Fast Timing with Scintillators

The successful development of TOF PET scanners with the improved image quality, based on the photomultipliers, triggered intense studies of fast timing with SiPMs (Otte et al. 2005; Schaart et al. 2009; Yamamoto et al. 2005;

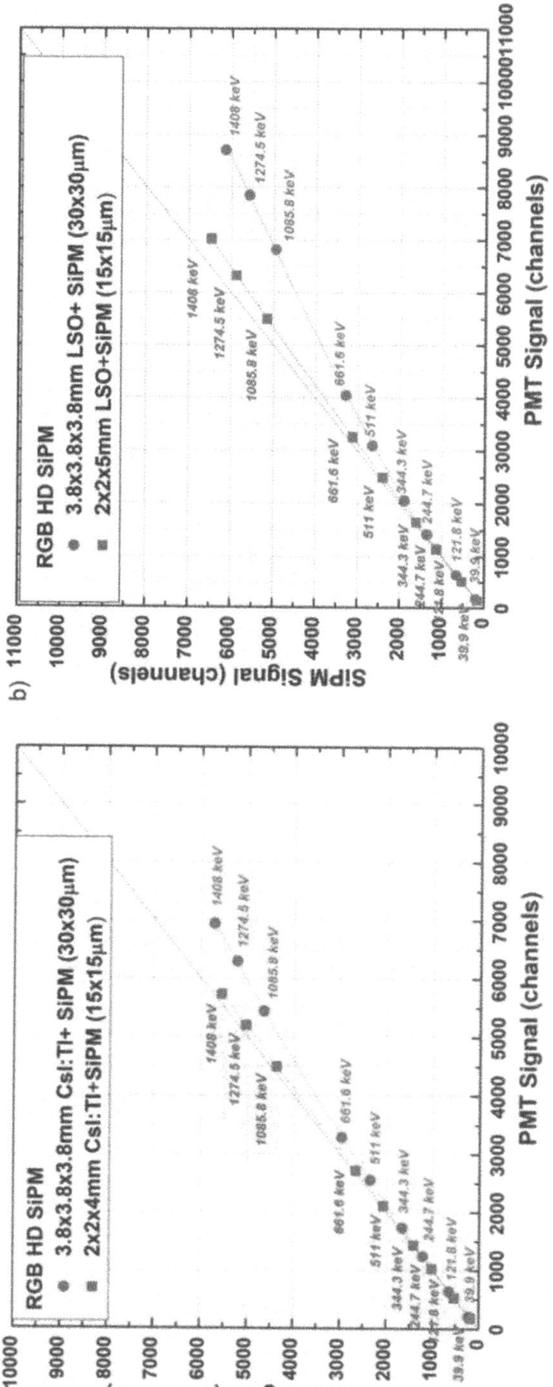


**FIGURE 8.11**
Linearity of 4 × 4 mm SiPM (30 × 30 μm) and 2 × 2 mm SiPM (15 × 15 μm) in the case of readout of different scintillation materials: (a) 3.8 × 3.8 × 3.8 mm CsI:Tl and 2 × 2 × 4 mm CsI:Tl, (b) 3.8 × 3.8 × 3.8 mm LSO and 2 × 2 × 5 mm LSO, adapted from (Grodzicka et al. 2014).

Raylman et al. 2006; Schaart et al. 2010; Kim et al. 2009; Gundacker et al. 2013; Szczesniak et al.2012) and (Gola et al. 2012; Kadrmas et al. 2009; Kolb et al. 2008; Musienko et al. 2007; Nishikido et al. 2008; Pestotnik et al. 2008; Schaart et al. 2008a; Schaart et al. 2008b; Schaart 2009; Chagani et al. 2009; Llosá et al. 2011; Roncali and Cheery 2011; Wang et al. 2012; Yamamoto et al. 2010; Szczesniak et al. 2010; Gundacker et al. 2014; Yeom et al. 2013; Llosá et al. 2010; Gundacker et al. 2012; Gundacker et al. 2013; Vinke et al. 2009; Powolny et al. 2011; Jarron et al. 2009; Seifert et al. 2012; Piemonte et al. 2011; al. 2013; Vinke et al. 2009; Powolny et al. 2011; Jarron et al. 2009; Seifert et al. 2012; Piemonte et al. 2016; Braga et al. 2012; Braga et al. 2014, Conti et al. 2009; Del Guerra et al. 2009; Lewellen et al. 2008; Moses et al. 2009; Szczesniak et al. 2009). Timing properties of SiPMs are reported to be extremely good, although the literature data are very inconsistent. The superior time resolution of 138 ps for two detectors for LYSO crystals coupled to the MPPC 050C was measured by the Delft group (Seifert et al. 2012) using the digital timing based on the Acqiris DC282 digitizer. Even better time resolution of 108 ps was reported in (Gundacker et al. 2013), as measured using NINO chips (Jarron et al. 2009) working as a leading edge discriminators. In contrast, the best time resolution of 240 ps for two detectors, measured with analog timing setups, was reported by the GE group (Kim et al. 2009). Timing resolution of 190 ps was reported by C. Piemonte group, measured by means of the Differential Leading Edge Discriminator (DLED) method, in which the arrival time of the event is not extracted directly from the signal but rather from the difference between the signal and its delayed replica (Gola et al. 2012). DLED method allows to compensate effectively the dark events allowing very low thresholds to be set in the discriminator.

In Table 8.3, the selected results of the SiPM tests in timing with LSO/LYSO crystals for 511 keV annihilation quanta are collected. Time resolution of 108 ps was measured by CERN group using Hamamatsu MPPC of $3 \times 3$ mm$^2$ size and $2 \times 2 \times 3$ mm$^3$ LSO co-doped with 0.4 mol% Ca (Powolny et al. 2011). Such good value was possible due to the fastest decay time of $30.3 \pm 1$ ns of the co-doped crystal (Powolny et al. 2011). Assuming a comparable light output of the standard LSO crystal used in (Gundacker et al. 2012) and a typical decay time of the light pulse of 40 ns, the measured time resolution follows the square root of the decay time constant. The superior timing of $73 \pm 2$ ps was reported by (Gundacker et al. 2016) with the same $2 \times 2 \times 2$ mm LSO:Ce,Ca crystal coupled to FBK NUV-HD SiPMs. For the 20 mm long crystal the time resolution was degraded to $122 \pm 3$ ps because of a contribution of the light collection time spread.

A very good time resolution of $101 \pm 2$ ps was reported by Delft group measured with $3 \times 3 \times 5$ mm$^3$ $LaBr_3$ crystal (Schaart et al. 2010). This was possible due to a larger light output and faster decay time of the $LaBr_3$ light pulse comparing to LSO/LYSO. On the other hand, the time resolution of the $LaBr_3$ detector is limited by a finite rise time of 0.93 ns of the light pulse in a typical crystal doped with 5 mol% of Ce (Levin et al. 2013).

**TABLE 8.3**

Time Resolution Measured with LSO/LYSO Crystals Coupled to SiPMs

| SiPM | Crystal | Time Resolution | Ref. |
|---|---|---|---|
| Hamamatsu MPPC –S10362-33-050C | LYSO $3 \times 3 \times 5$ mm$^3$ | $138 \pm 2$ ps | (Vinke et al. 2009) |
| Hamamatsu MPPC – S10931-050P | LSO:Ce,Ca (0.4%) $2 \times 2 \times 3$ mm$^3$ | 108 ps | (Gundacker et al. 2014) |
| Hamamatsu MPPC – S10362-33-050C | LSO $2 \times 2 \times 3$ mm$^3$ | $125 \pm 2$ ps | (Yeom et al. 2013) |
| Hamamatsu MPPC – S10362-33-050C | LYSO $3 \times 3 \times 5$ mm$^3$ | $147 \pm 3$ ps | (Yeom et al. 2013) |
| Hamamatsu MPPC – S10362-33-050C | LYSO $3 \times 3 \times 20$ mm$^3$ | $186 \pm 3$ ps | (Yeom et al. 2013) |
| Hamamatsu MPPC – S10362-33-050C | LYSO $3 \times 3 \times 10$ mm$^3$ | 240 ps | (Kim et al. 2009) |
| FBK NUV SiPM | LSO:Ce,Ca (0.4%) $2 \times 2 \times 3$ mm$^3$ | $85 \pm 4$ ps | (Nemallapudi et al. 2015) |
| FBK NUV-HD SiPM | LSO:Ce,Ca (0.4%) $2 \times 2 \times 3$ mm$^3$ | $73 \pm 2$ ps | (Gundacker et al. 2016) |
| FBK NUV-HD SiPM | LSO:Ce,Ca (0.4%) $2 \times 2 \times 20$ mm$^3$ | $122 \pm 3$ ps | (Gundacker et al. 2016) |

The data collected in Table 8.3 point out an important contribution of the light transport in the longer pixel crystals. It is the effect of the time spread of the collected light and light attenuation in a longer medium.

## 8.7 SiPMs in Medical Instrumentations

An intense study of SiPM performance in application to PET detectors has led to the development of the prototypes of TOF PET scanners (Degenhardt et al. 2012; Levin et al. 2013; Schneider et al. 2013) and the first PET/MR commercial SIGNA™ PET/MR scanner proposed by GE Healthcare (GE Healthcare's 2014).

Figure 8.12 shows a comparison of PET detectors based on PMTs and on SiPMs following developments of Siemens Healthinners (Siemens). The comparison of both detector designs confirms the overall superiority of SiPMs over PMTs, both in costeffectiveness and performance (Wagatsuma et al. 2017).

One of the first prototypes of Time-of-Flight PET systems based on SiPMs has been developed at Stanford University for simultaneous whole body PET/MR imaging (Levin et al. 2013). This PET system comprises

**FIGURE 8.12**
Picture showing the comparison of PET detectors based on PMTs (left) and on SiPMs (right) in their implementation by Siemens Healthineers.

of 5 rings of 112 detector blocks (each block is a 4 × 9 array of LYSO crystals, 3.95 × 5.3 × 25 mm³) coupled to 1 × 3 arrays of SiPM devices. Using a Ge-68 pin source, the measured PET energy resolution was 10.5% FWHM at 511 keV for both RF ON and RF OFF. The per crystal timing resolution was 390 ps with RF OFF and 399 ps with RF ON. Transaxial spatial resolution of 3.9 mm FWHM was measured at 1 cm from the isocenter using an F-18 capillary tube point source.

The Technical University of Munich (TUM) has developed a PET prototype based on digital SiPMs (dSiPMs) (Haemisch et al. 2012; Frach et al. 2009; Frach et al. 2010; Mandai et al. 2012; Schaart et al. 2011) and GAGG scintillators (Schneider et al. 2013) consisting of two facing modules of dSiPMs

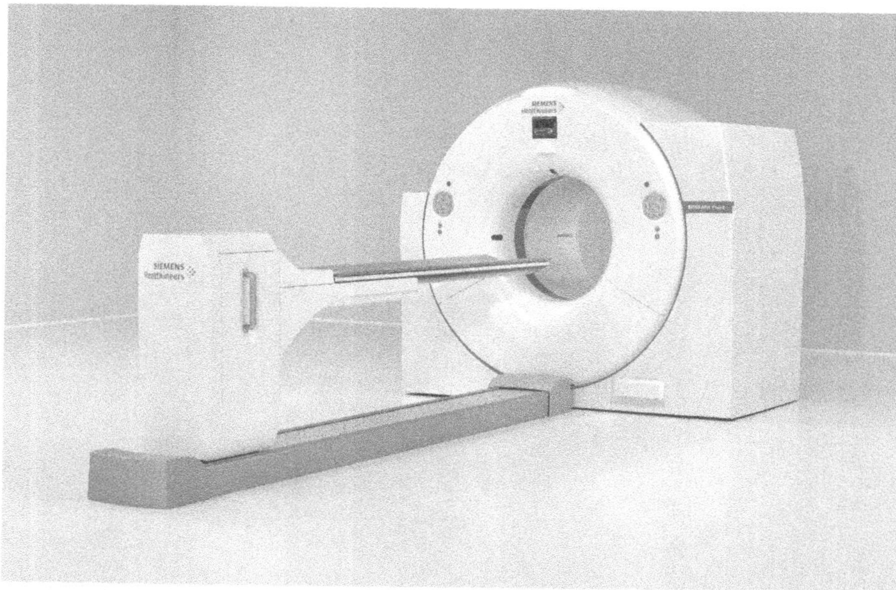

**FIGURE 8.13**
Siemens Biograph Vision PET/CT scanner. Courtesy of Siemens Healthcare, GmbH.

assembled in combination with a rotational stage for the objects. Using the GAGG scintillators, coincidence timing resolution of this system was equal to 430 ps FWHM and energy resolution was equal to 9.0% FWHM, while the LSO coupled to these dSiPMs resulted in coincidence timing resolution of 200 ps FWHM and energy resolution of 9.7% FWHM at 511 keV.

The Siemens Biograph Vision (see Figure 8.13) is the newest SiPM-based PET/CT scanner composed of detector mini-blocks, each having a 5 × 5 array of 3.2 × 3.2 × 20 $mm^3$ LSO crystals coupled to a 4 × 4 SiPM array constructed by Hamamatsu (Reddin et al. 2018) (Carlier et al. 2020). It is characterized by a high-time resolution of 214 ps and an excellent energy resolution of 10.1% at 511 keV gamma quanta. The first production model of the Biograph Vision PET/CT scanner was installed at the Hospital of Pennsylvania University in Philadelphia, in September 2018 and was put into clinical operation at the beginning of 2019. Its 1 m long, whole body version, mCT Biograph Quadra PET is characterized by 220 ps timing resolution.

The Ultra-High-Resolution Digital whole body PET scanner (uExplorer) (Badawi et al. 2019) (Spencer et al. 2020; Moses and Derenzo 1999) creates pictures of organs and tissues inside human body in the Axial FOV of 194 cm. It can perform a total body reconstruction within eight minutes compared to 20 minutes with a conventional PET scanner. The uExplorer PET scanner was developed by the University of California, Davis (UC Davis) research team, jointly led by Simon R. Cherry and Ramsey D. Badawi and its commercial collaborator, United Imaging Healthcare, headquartered in Shanghai, China. It is based on LYSO crystals and characterized by timing resolution of 400 ps.

Besides of PET detectors with the TOF capability and dual modality PET/MR scanners, SiPMs were recently proposed to be used in the handheld gamma camera for intraoperative imaging (Popovic et al. 2014). The camera incorporates a cerium doped lanthanum bromide (LaBr:Ce) plate scintillator, an array of 80 SiPM photodetectors and a two-layer parallel-hole collimator. The disk-shaped camera housing is 75 mm in diameter, approximately 40.5 mm thick and has a mass of only 1.4 kg, permitting either handheld or arm-mounted use. The field of view is circular with a 60-mm diameter. The gamma camera has an intrinsic spatial resolution of 4.2 mm FWHM, an energy resolution of 21.1% FWHM at 140 keV, and sensitivity of 481 or 73 cps/MBq when using the single or double-layer collimators, respectively.

Recent studies of SiPM applications in gamma spectrometry with scintillators, summarized in (Grodzicka-Kobyłka et al. 2019), opened a new field of SiPMs use in medical instrumentation.

## References


Acerbi, F., et al., "NUV silicon photomultipliers with high detection efficiency and reduced delayed correlated-noise." *IEEE Trans. Nucl. Sci.*, vol. 62(3), June 2015: 1318–1325.

Achenbach P., A. S. Lorente, S. S. Majos, and J. Pochodzalla. "Future use of silicon photomultipliers for KAOS at MAMI and PANDA at FAIR." *Nucl. Instrum. Methods Phys. Res. A*, (October 2009): 358–361.

Allemand R., C. Gresset, and J. Vacher, "Potential advantages of a cesium fluoride scintillator for time-of-flight positron camera." *J. Nucl.Med.* (1980): 153–155.

Andreev, V., et al., "A high-granularity scintillator calorimeter readout with silicon photomultipliers." *Nucl. Instrum. Methods Phys. Res. A*, March 2005: 368–380.

Antich P. P., E. N. Tsyganov, N. A. Malakhov, and Z. Y. Sadygov, "Avalanche photo diode with local negative feedback sensitive to UV, blue and green light." *Nucl. Instrum. Meth. Phys. Res.*, (April 1997): 491–498.

Badawi, R. D., et al., "First human imaging studies with the EXPLORER total-body PET scanner." *J. Nucl. Med.*, vol. 60(3), March 2019: 299–303.

Bisogni M. G., A. Del Guerra, and N. Belcari, "Medical applications of silicon photomultipliers." *Nucl. Instrum. Meth. Phys. Res., A*, 926 (2019): 118–128.

Bondarenko, G., et al, "Limited Geiger–mode microcell silicon photodiode: New results." *Nucl. Instr. Meth.*, vol. A, March 2000: 187–192.

Braga L. H. C., L. Gasparini, and D. Stoppa, "A time of arrival estimator based on multiple timestamps for digital PET detectors." *Proc. IEEE Nucl. Sci. Symp. Med. Imag. Conf.*, 2012.

Braga, L. H. C., et al., "A fully digital 8 16 SiPM array for PET applications with per-pixel TDCs and real-time energy output." *IEEE J. Solid-State Circuits*, 2014: 301.

Buzhan, P., et al., *"An Advanced Study of Silicon Photomultiplier." ICFA Instrumentation Bulletin*, 2001.

———. "Silicon photomultiplier and its possible applications." *Nucl. Instr. Meth. Phys. Res. A*, January 2003: 48–52.

Carlier, T., et al., "From a PMT-based to a SiPM-based PET system: A study to define matched acquisition/reconstruction parameters and NEMA performance of the Biograph Vision 450." *EJNMMI Physics*, vol. 7, 2020: 55.

Chagani, H., et al., "Tests of silicon photomultiplier PET modules." IEEE Nucl. Sci. Sympos. Med. Imag. Conf. Rec., 2009: 1518–1520.

Chmill V., E. Garutti, R. Klanner, M. Nitschke, J. Schwandt, "On the characterisation of SiPMs from pulse-height spectra." *Nucl. Instr. Meth. A*, A854 (2017): 70.

Conti M., L. Eriksson, H. Rothfuss, and C. L. Melcher, "Comparison of fast scintillators with TOF PET potential." *IEEE Trans. Nucl. Sci.*, (June 2009): 926–933.

Cozzi G., "Development of scintillation detectors based on silicon photomultipliers for high-energy gamma-ray applications", PhD thesis, Politecnico di Milano, Italy, 2017a.

———. "High-resolution gamma-ray spectroscopy with a SiPM-based detection module for 1" and 2" $LaBr_3$:Ce readout." *IEEE Trans. Nucl. Sci.*, 65(1); (Jan. 2018): 645–655.

Cozzi G., L. Buonanno, P. Busca, M. Carminati, C. Fiorini, G. L. Montagnani, F. Acerbi, A. Gola, G. Paternoster, C. Piemonte, V. Regazzoni, N. Blasi, F. Camera, B. Million, "A SiPM-based Detection Module for 1" and 2" LaBr3:Ce Readout for Nuclear Physics Applications", presented at *IEEE NSS-MIC Conference*, Atlanta, USA, 2017

Cozzi, G., et al., "Development of a SiPM-based detection module for large $LaBr_3$:Ce scintillators for nuclear physics applications." *Proc. Nuclear Science Symp. Conf. Record*, 2016.

Danilov M., "CALICE collaboration, scintillator tile hadron calorimeter with novel SiPM readout." *Nucl. Instrum. Methods Phys. Res. A*, (October 2007): 451–456.

Degenhardt, C., et al., "Performance evaluation of a prototype positron emission tomography scanner using digital photon counters (DPC)." *NSS-MIC Conference Record*, 2012: 2820.

Del Guerra A., N. Belcari, M. G. Bisogni, G. Llosá, S. Marcatili, and S. Moehrs, "Advances in position-sensitive photodetectors for PET applications." *Nucl. Instrum. Methods Phys. Res. A*, (June 2009): 319–322.

Di Vita D., L. Buonanno, G. L. Montagnani, A. Minerva, A. Giannoni, M. Carminati, F. Camera, and C. Fiorini, A high dynamic range 144-SiPM detection module for gamma spectroscopy and imaging with 3." *LaBr3, IEEE NSS-MIC Conference*, 2019, Conference Records.

Dolgoshein, B., et al., "Status report on silicon photomultiplier development and its applications." *Nucl. Instr. Meth. Phys. Res.*, vol. A, July 2006: 368–376.

Du Y., and F. Retiere, "After–pulsing and cross–talk in multi–pixel photon counter." *Nucl. Instr. Meth.*, A (November 2008): 396–401.

Eckert P., H. C. Schultz–Coulon, W. Shen, R. Stamen, and A. Tadday, "Characterisation studies of silicon photomultipliers." *Nucl. Instr. Meth.*, A (August 2010): 217–226.

Finocchiaro, P., et al., "Features of silicon photo-multipliers: Precision measurements of noise, cross-talk, after pulsing, detection efficiency." *IEEE Trans. Nucl. Sci.*, June 2009: 1033–1041.

Frach T., G. Prescher, C. Degenhardt, and B. Zwans, "The digital silicon photomultiplier – System architecture and performance evaluation." *NSS-MIC Conference Record*, 2010.

Frach T., G. Prescher, C. Degenhardt, R. de Gruyter, A. Schmitz, and R. Ballizany, "The digital silicon photomultiplier – Principle of operation and intrinsic detector performance." *Nucl. Sci. Symp. Conf. Record (NSSIMIC)*, 2009: 1959–1965.

GE Healthcare's, "New SIGNA™ PET/MR." 2014.

Gola A., C. Piemonte, and A. Tarolli. "The DLED algorithm for timing measurements on large area SiPMs coupled to scintillators." *IEEE Trans. Nucl. Sci.*, (April 2012): 358–365.

Golovin V., and V. Saveliev, "Novel type of avalanche photodetector with Geiger mode operation." *Nucl. Instrum. Methods Phys. Res. A*, A518 (February 2004): 560–564.

Grodzicka M., M. Moszyński, and T. Szczęśniak, Silicon photomultipliers in detectors for nuclear medicine, in *Radiation Detectors for Medical Imaging*, CRC Press, 2015a.

Grodzicka M., M. Moszyński, T. Szczęśniak, D. Wolski, M. Kapusta, and M. Szawłowski, "Energy resolution of small scintillation detectors with SiPM light readout." *JINST*, 8 (2013a): P02017.

Grodzicka M., M. Moszyński, T. Szczęśniak, M. Szawłowski, and J. Baszak, "Characterization of 4×4 ch MPPC array in scintillation spectrometry." *JINST*, (2013b): P09020.

Grodzicka M., M. Moszyński, T. Szczęśniak, M. Szawłowski, D. Wolski, and J. Baszak, "MPPC array in the readout of CsI(Tl), LSO:Ce(Ca), $LaBr_3$, and BGO scintillators." *IEEE Trans. Nucl. Sci.*, (2012): 3294–3303.

Grodzicka M., M. Moszyński, T. Szczęśniak, M. Szawłowski, D. Wolski, and K. Grodzicki, "Characterization of silicon photomultipliers: Effective dead time, new method of evaluating the single photoelectron response, gamma spectrometry with BGO scintillator." *Nucl. Instr. Meth. Phys. Res. A*, A783 (2015b): 58–64.

Grodzicka M., T. Szczęśniak, M. Moszyński, S. Korolczuk, J. Baszak, and M. Kapusta, "Characterization of large TSV MPPC arrays (4 × 4 ch and 8 × 8 ch) in scintillation spectrometry." *Nucl. Instr. Meth. A*, A869 (2017a) 153–162.

———. "Comparison of SensL and Hamamatsu 4_4 channel SiPM arrays in gamma spectrometry with scintillators." *Nucl. Instrum. Methods Phys. Res. A*, vol. 856, 2017b: 53–64.

———. "Performance of FBK high-density SiPMs in scintillation spectrometry." *JINST*, 2014: P08004.

———. "Silicon photomultipliers in scintillation detectors used for gamma ray energies up to 6.1 MeV." *Nucl. Instr. Meth. A*, vol. A 874, 2017c: 137–148.

Grodzicka-Kobyłka M., M. Moszyński, and T. Szczęśniak, "Silicon photomultipliers in gamma spectroscopy with scintillators." *Nucl. Instr. Meth. A*, A926 (2019): 128.

Gundacker S., A. Knapitsch, E. Auffray, P. Jarron, T. Meyer, and P. Lecoq. "Time resolution deterioration with increasing crystal length in a TOF-PET system." *Nucl. Instrum. Meth. Phys. Res. A*, (2014): 92–100.

Gundacker, S., et al., "A systematic study to optimize SiPM photodetectors for highest time resolution in PET." *IEEE Trans. Nucl. Sci.*, 2012: 1798.

———. "Sipm time resolution: From single photon to saturation." *Nucl. Instr. Meth. Phys. Res. A*, 2013: 569–572.

———. "Time of flight positron emission tomography towards 100ps resolution with L(Y)SO: An experimental and theoretical analysis." *JINST*, 2013.

———. "State of the art timing in TOF-PET detectors with LuAG,GAGG and L(Y)SO scintillators of various sizes coupled toFBK-SiPMs" *JINST, 2016*: P08008

Haemisch Y., T. Frach, C. Degenhardt, and A. Thon, "Fully digital arrays of silicon photomultipliers (dSiPM)—A scalable alternative to vacuum photomultiplier tubes (PMT)." *Physics Procedia*, (2012): 1546–1560.

Hamamatsu. "MPPC Multi–Pixel Photon Counter Technical Information." 2009.

Hamamatsu, Product flyer, 17.10.2017, MPPC for scintillation.

Jarron, P., et al., "Time based readout of a silicon photomultiplier (SiPM) for time of flight positron emission tomography (TOF-PET)." *Nucl. Sci. Symp. Conf. Record (NSS/MIC)*, 2009: 1212–1219.

Kadrmas D.J., M. E. Casey, M. Conti, B. W. Jakoby, C. Lois, and D. W. Townsend. "Impact of time-of-flight on PET tumor detection." *J. Nucl. Med.*, (2009): 1315–1323.

KETEK, Product Data Sheet SiPM – Silicon Photomultiplier PM3315-WB/PM3325-WB, https://www.ketek.net/wp-content/uploads/2016/11/KETEK-PM3315-WB-PM3325-WB-Datasheet.pdf

Kim C. L., G. C. Wang, and S. Dolinsky, "Multi-pixel photon counters for TOF PET detector and its challenges." *IEEE Trans. Nucl. Sci.*, (Octomber 2009): 2580–2585.

Kim, C., et al., "Replacement of a photomultiplier tube in a 2-inch thallium-doped sodium iodide gamma spectrometer with silicon photomultipliers and a light guide." *Nuclear Engineering and Technology*, vol. 47, 2015: 479.

Knoll Glenn F. *Radiation Detection and Measurement*. John Wiley & Sons, 2010.

Kolb A., M. S. Judenhofer, E. Lorenz, D. Renker, and B. J. Pichler, "PET block detector readout approaches using G-APDs." *IEEE Nucl. Sci. Symp. Conf. Records*, 2008.

Korpar, S., et al., "Measurement of Cherenkov photons with silicon photomultipliers." *Nucl. Instrum. Methods Phys. Res. A*, August 2008: 13–17.

Lacaita A. L., F. Zappa, S. Bigliardi, and M. Manfredi. "On the Bremsstrahlung origin of hot–carrier–induced photons in silicon devices." *IEEE Trans. Electron Devices*, (1993): 577–582.

Lacaita A. L., M. Mastrapasqua, M. Ghioni, and S. Vanoli, "Observation of avalanche propagation by multiplication assisted diffusion in p-n junctions." *Appl. Phys. Lett.*, (July 1990): 489–491.

Laval, M., et al., "Contribution of the time-of-flight information to the positron tomographic imaging." Proc. 3rd World Congress Nucl. Med. and Biol. Paris, France: Pergamon Press, Paris, 1982: 2315.

Levin C., G. Glover, T. Deller, D. McDaniel, W. Peterson, and S. H. Maramraju, "Prototype time-of-flight PET ring integrated with a 3T MRI system for simultaneous whole-body PET/MR imaging." *J Nucl Med.*, (2013): 148.

Lewellen T. K., "Recent developments in PET detector technology." *Phys. Med. Biol.*, (2008): R287–R317.

Llosá, G., et al., "Characterization of a PET detector head based on continuous LYSO crystal and monolithic 64-pixel silicon photomultiplier matrices." *Phys. Med. Biol.*, 2010.

———. "Development of a PET prototype with continuous LYSO crystals and monolithic SiPM matrices." IEEE Nucl. Sci. Sympos. Med. Imag. Conf. Rec., 2011: 3631–3634.

Mandai, S., and E. Charbon, "Multi-channel digital SiPMs: Concept, analysis and implementation." Proc. IEEE Nucl. Sci. Symp. Med. Imag. Conf., 2012: 1840–1844.

Mazzillo, M., et al., "Timing performances of large area silicon photomultipliers fabricated at STMicroelectronics." *IEEE Trans. Nucl. Sci.*, August 2010: 2273–2279.

Montagnani G. L., Development of a 3" $LaBr_3$ SiPM-Based Detection Module for High Resolution Gamma Ray Spectroscopy and Imaging, Doctoral Thesis, Politecnico di Milano, 2018a.

Montagnani G. L. L. Buonanno, D. Di Vita, A. Minerva, M. Carminati, F. Camera, A. Gola, V. Regazzoni, F. Acerbi, and C. Fiorini, Development of 3" $LaBr_3$ SiPM-Based Detection Modules for High Dynamic Range Gamma Ray Spectroscopy and Imaging, IEEE NSS-MIC 2018, Conference Records.

Moses W. W., and S. E. Derenzo, "Prospects for time-of-flight PET using LSO scintillator." *IEEE Trans. Nucl. Sci.*, (June 1999): 474–478.

Moses W. W., et al., "EXPLORER: A Total-Body PET Scanner for Biomedical Research", April 2009, https://indico.CERN.ch

Moszyński M., and B. Bengtson, "Status of timing with plastic scintillation detectors." *Nucl. Instr. Meth.*, (January 1979): 1–31.

Mullani N. A., D. C. Ficke, R. Hartz, J. Markham, and G. Wong, "System design of a fast PET scanner utilizing time-of-flight." *IEEE Trans. Nucl. Sci.*, (February 1981): 104–107.

Musienko Y., "State of the art in SiPM's." *CERN, SiPM workshop*, February 16, 2011.

Musienko Y., E. Auffray, P. Lecoq, S. Reucroft, J. Swain, and J. Trummer, "Study of multi-pixel Geiger-mode avalanche photodiodes as a read-out for PET." *Nucl. Instrum. Methods Phys. Res. A*, (2007): 362–365.

Musienko Y., S. Reucroft, and J. Sawin, "The gain, photon detection efficiency and excess noise factor of multi–pixel Geiger–mode avalanche photodiodes." *Nucl. Instr. Meth. Phys. Res., A*, (November 2006): 57–61.

Nassalski, A., et al., "Comparative study of scintillators for PET/CT detectors." *IEEE Trans. Nucl. Sci.*, Febuary 2007: 3–10.

Nemallapudi M. V., S. Gundacker, P. Lecoq, and E. Auffray, "Single photon time resolution of state of the art SiPMs." *JINST*, 11 (2016): P10016.

Nemallapudi, M. V., et al., "Sub-100 ps coincidence time resolution for positron emission tomography with LSO:Ce codoped with Ca." *Phys. Med. Biol.*, vol. 60, 2015: 4635.

Nishikido, F., et al., "Four-layer DOI-PET detector with a silicon photomultiplier array." IEEE Nucl. Sci. Symp. Conf. Record, 2008: 3923–3925.

Nocente, M., et al., "Gamma-ray spectroscopy at MHz counting rates with a compact $LaBr_3$ detector and silicon photomultipliers for fusion plasma applications." *Rev. Sci. Instr.*, vol. 87, 2016: 11E714.

Otte, A. N., et al., "A test of silicon photomultipliers as readout for PET." *Nucl. Instrum. Methods Phys. Res. A*, April 2005: 705–715.

———. "Prospects of using silicon photomultipliers for the astroparticle physics experiments EUSO and MAGIC." *IEEE Trans. Nucl. Sci.*, April 2006: 636–640.

Pestotnik R., S. Korpar, H. Chagani, R. Dolenec, P. Krizan, and A. Stanovnik. "Silicon photo-multipliers as photon detectors for PET." IEEE Nucl. Sci. Symp. Conf. Record, 2008: 3123–3127.

Piemonte C., "Development of silicon photomultipliers @ IRST." *FNAL*, 2006.

Piemonte, C., et al., "Characterization of the first FBK high–density cell silicon photomultiplier technology." *IEEE Trans. Electron Devices*, August 2013: 2567–2573.

———. "Performance of NUV-HD silicon photomultiplier technology." *IEEE Trans. Elec. Dev.*, vol. 63(3), March 2016: 1111–1116.

———. "Timing performance of large area SiPMs coupled to LYSO using dark noise compensation methods." *IEEE Nucl. Sci. Symp. Conf. Record*, 2011: 59.

Popovic, K., et al., "Development and characterization of a round hand-held silicon photomultiplier based gamma camera for intraoperative imaging." *IEEE Trans. Nucl. Sci.*, 2014: 1084.

Powolny, F., et al., "Time-based readout of a scilicon photomultiplier (SiPM) for time of flight positron emission tomography (TOF-PET)." *IEEE Trans. Nucl. Sci.*, 2011: 597–604.

Raylman, R. R., et al., "Simultaneous MRI and PET imaging of a rat brain." *Phys Med Biol.*, November 2006: 6371–6379.

Reddin, J. S., J. S. Scheuermann, D. Bharkhada, A. M. Smith, M. E. Casey, M. Conti, and J. S. Karp, "Performance evaluation of the SiPM-based Siemens biograph vision PET/CT system." 2018 *IEEE NSS-MIC Conference, Sydney, Australia, Conference Records*.

Regazzoni, V., et al., "Characterization of high density SiPM non-linearity and energy resolution for prompt gamma imaging applications." *J. Instrum.*, vol. 12(7), July 2017: P07001.

Renker, D., "Geiger–mode avalanche photodiodes, history, properties and problems." *Nucl. Instrum. Meth. Phys. Res. A*, (November 2006): 48–56.

———. "New developments on photosensors for particle physics." *Nucl. Instrum. Methods Phys. Res. A*, (January 2009b): 207–212.

Renker, D., and E. Lorenz, "Advances in solid state photon detectors." *JINST*, (2009a): P04004.

Rigamonti, D., et al., "Performance of the prototype $LaBr_3$ spectrometer developed for the JET gamma-ray camera upgrade." *Rev. Sci. Instr.* vol. 87, 2016: 11E717.

Roncali, E., and S. R. Cheery, "Application of silicon photomultipliers to positron emission tomography." *Ann. Biomed. Engin.*, (2011): 1358–1377.

Ronzhin, A., et al., "Tests of timing properties of silicon photomultipliers." *Nucl. Instr. Meth. Phys. Res. A*, April 2010: 38–44.

Schaart, D. R., H. T.van Dam, G. J. van der Lei, and S. Seifert, "The digital SiPM: Initial evaluation of a new photosensor for time-of-flight PET." *IEEE Nucl. Sci. Symp. Med. Imaging Conf.*, Valencia, Spain, October 2011: 23–29.

Schaart, D. R., et al., "A novel, SiPM-array-based, monolithic scintillator detector for PET." *Phys. Med. Biol.*, May, 2009: 3501–3512.

———. "First experiments with $LaBr_3$:Ce crystals coupled directly to silicon photomultipliers for PET applications." *IEEE Nucl. Sci. Symp. Conf. Record*, 2008a: 3991–3994.

———. "LaBr(3):Ce and SiPMs for time-of-flight PET: Achieving 100 ps coincidence resolving time." *Phys. Med. Biol.* January 2010: 79–89.

———. "SiPM-array based PET detectors with depth-of-interaction correction." *IEEE Nucl. Sci. Symp. Conf. Record*, 2008b: 3581–3585.

Schneider F., K. Shimazoe, I. S. Schweiger, K. Kamada, H. Takahashi, and S. Ziegler. "A PET prototype based on digital SiPMs and GAGG scintillators." *J Nucl Med.*, (2013): 429.

Seifert, S., et al., "A comprehensive model to predict the timing resolution of SiPM-based scintillation detectors: Theory and experimental validation." *IEEE Trans. Nucl. Sci.*, 2012: 1.

SensL, J-Series High PDE and Timing Resolution, TSV Package, http://sensl.com/downloads/ds/DS-MicroJseries.pdf

Siemens-Healthineers, Biograph Vision Quadra, https://www.siemens-healthineers.com/molecular-imaging/pet-ct/biograph-vision-quadra

Spencer, B. A., et al., "Performance evaluation of the uEXPLORER total-body PET/CT scanner based on NEMA NU 2-2018 with additional tests to characterize long axial field-of-view PET scanners." *J. Nucl. Med.*, vol. 120, 2020: 250597.

Spinelli A., and A. L. Lacaita, "Physics and numerical simulation of single photon avalanche diodes." *IEEE Trans. Electron Devices*, (November 1997): 1931–1943.

Szczesniak T., M. Grodzicka-Kobylka, M. Moszynski, M. Szawlowski, S. Mianowski, and D. Wolski, Performance of 2 inch and 3 inch scintillation detectors with SiPM light readout, presented at 2017 *IEEE NSS-MIC Conference*, Atlanta, USA, Conference Records.

Szczesniak T., M. Moszynski, L. Swiderski, A. Nassalski, P. Lavoute, and M. Kapusta, "Fast photomultipliers for TOF PET." *IEEE Trans. Nucl. Sci.*, (February 2009): 173–181.

Szczesniak, T., M. Moszynski, M. Grodzicka, M. Szawłowski, D. Wolski, and J. Baszak, "Time jitter of silicon photomultipliers." *IEEE NSS-MIC Conference Records on CD*, 2012: nr ref.: K2467.

Szczesniak, T., et al., "Time resolution of scintillation detectors based on SiPM in comparison to photomultipliers." *Nucl. Sci. Symp. Conf. Record (NSS/MIC)*, 2010.

Vacheret, A., et al., "Characterization and simulation of the response of multi–pixel photon counters to low light levels." *Nucl. Instr. Meth.*, vol. A, November 2011: 69–83.

van Dam, H. T., et al., "A comprehensive model of the response of silicon photomultipliers." *IEEE Trans. Nucl. Sci.*, August 2010: 2254–2266.

Vinke, R., et al., "Optimizing the timing resolution of SiPM sensors for use in TOF-PET detectors." *Nucl. Instrum. Meth. Phys. A*, 2009: 188.

Wagatsuma K, K. Miwa, M. Sakata, K. Oda, H. Ono, M. Kameyama, J. Toyohara, and K. Ishii, "Comparison between new-generation SiPM-based and conventional PMT-based TOF-PET/CT." *Physica Med.*, 42 (2017): 203–210.

Wang, Y., et al., "Long design and performance evaluation of a compact, large-area PET detector module based on silicon photomultipliers." *Nucl. Instrum. Meth. Phys. Res. A*, 2012: 49–54.

Yamamoto, S., S. Takamatsu, H. Murayama, and K. Minato, "A block detector for a multi-slice depth of interaction MR compatible PET." *IEEE Trans Nucl Sci.*, (February 2005): 33–37.

Yamamoto, S., et al., "Development of a Si-PM based high-resolution PET system for small animals." *Phys. Med. Biol.*, 2010: 5817–5831.

Yeom, J.Y., R. Vinke, and C. S. Levin, "Optimizing timing performance of silicon photomultiplier-based scintillation detectors." *Phys. Med. Biol.*, (2013): 1207.

Yokoyama, M., et al., "Performance of multi–pixel photon counters for the T2K near detectors." *Nucl. Instr. Meth.*, vol. A, October 2010: 567–573.

Yoo, H., et al., "Optimal design of a CsI(Tl) crystal in a SiPM based compact radiation sensor." *Rad. Meas.*, vol. 82, 2015: 102.

Zychor, I., et al., "High performance detectors for upgraded gamma ray diagnostics for JET DT campaigns." *Phys. Scr.*, vol. 91, 2016: 064003.

# 9

# *Tip Avalanche Photodiode – A New Wide Spectral Range Silicon Photomultiplier*

**Sergey Vinogradov, Elena Popova, Wolfgang Schmailzl, and Eugen Engelmann**

**CONTENTS**



## 9.1 Introduction

Silicon photomultipliers (SiPMs) are a well-established new generation of photodetectors widely recognized due to their performance in a photon-number-resolving detection of low-level light pulses. Nowadays, the

DOI: 10.1201/9781003219446-9

development of the SiPMs seems to be one of the most promising innovations toward an ideal photon detector. The majority of modern analog SiPMs are designed as an array of independent single-photon avalanche diode (SPAD) cells or pixels with individual quenching resistors connected to a common electrode. An avalanche region and adjacent photosensitive part of the cell are formed by planar p-n junctions. A periphery of the cells is typically formed by guard rings or trenches to provide independent operations of the cells. SiPM operates in a so-called Limited Geiger mode – the avalanche discharge is started as an unlimited breakdown process and then turned to extinction due to negative feedback provided by the quenching resistor. The fired cell in the Limited Geiger mode avalanche process generates a self-calibrated output charge, and the SiPM response appears to be an analog-like discrete sum of binary responses from all fired cells. Key SiPM advantages – a high gain, ultra-low excess noise of multiplication, fast time response, superior photon-number (energy), and time resolution – make SiPMs highly competitive with vacuum PMTs and MCPs, linear mode and Geiger mode APDs, and SPAD arrays in an emerging range of applications. Research and development of SiPMs have a long history starting from the mid-1970s in Russia [1, 2].

It was shown that avalanche breakdown in the metal-insulator-semiconductor (MIS) structures is quenched by an accumulation of avalanche carriers on a semiconductor-insulator interface and corresponding screening of the electric field providing efficient self-calibration of the avalanche charge. These basic studies carried out in the Lebedev Physical Institute resulted in a concept of the single-carrier avalanche with negative feedback (SC ANF) process and operating principles of solid-state photomultipliers [3, 4]. At the same time, the ANF APD designs have been evolved from planar MIS to various metal-resistor-semiconductor (MRS) APD structures as described in detail in references [5–8]. The planar approach [9, 10] appeared to be the most successful in terms of performance, reliability, reproducibility, applicability, and recognition. Since the mid-2000s, the SiPMs of the planar design have been developed by Hamamatsu [11], SensL [12], STM [13], FBK [14], Excelitas [15], and KETEK [16]. This device generation was worldwide recognized as the SiPM, a new photon-number-resolving avalanche detector of outstanding performance [17]. However, the planar analog SiPMs are approaching the physical limits of the design performance with its main inherent trade-off between photon detection efficiency (PDE) and dynamic range (DR) because of the dead space at the cell periphery [18]. This trade-off becomes challenging when developing a SiPM that should be sensitive for a wide spectral range, especially for red and near-infrared (NIR) wavelengths. The development of a "NIR-enhanced" SiPM is anticipated to be a breakthrough in many applications – first and foremost in LIDARs – as well as in medicine and health sciences, biology and physics, environmental monitoring, and quantum telecommunications. To achieve efficient absorption of photons and a high PDE for the red part of the light spectrum, a thicker photosensitive layer is required due to the low absorption coefficient of silicon at these wavelengths. But the available increase of the thickness is eliminated by the so-called "border effect". The thicker the active layer of a SiPM for efficient absorption of NIR photons, the higher the losses of a sensitive volume adjacent to trenches [19]. Further, the breakdown voltage increases proportionally to the depletion depth for planar technologies. This leads to high operating voltages in the Geiger mode with high temperature coefficients. Simultaneously, the lateral distance of the active area to the microcell edge should be increased to prevent edge breakdowns at high voltages.

Thinking about the SiPM design of ultimate performance where high PDE in a wide spectral range should be complemented by a high DR and fast time response, we turned to an idea to benefit from nonplanar configurations of p-n junctions and nonuniform electric fields. Indeed, the earliest device of this type was the "Spherical" Si APD with a radius of the p-n junction of 2 μm and a breakdown voltage of 50V operated in a photon-counting mode [20].

In the intermediate development stage, the Russian MRS ANF APDs have also evolved from planar to nonplanar designs like the "spherical" one. They were formed as an array of avalanche micro-channels – n+ diffusion dots (pixels) on a p-Si wafer – covered by a thin resistive SiC layer as quenching resistor and a thin semitransparent metal layer as a common electrode [5–8, 21, 22]. The schematic design of this device is shown in Figure 9.1 [5]. The n+ "dots" have been introduced mainly for localization of the avalanche on the "dot" to prevent its indeterminate lateral spread that may cause uncontrolled and inefficient quenching including unlimited discharge at the local defects of the silicon interface. Low capacitance of the "dots" and low breakdown voltage was identified as advantages of the MRS APD with respect to SPAD [8]. However, drawbacks of this device were low PDE, high DCR, and low production yield.

To improve geometrical efficiency and transparency of the MRS APD, the n+ "dots" were relocated from the surface in the depth of the p-Si layer.

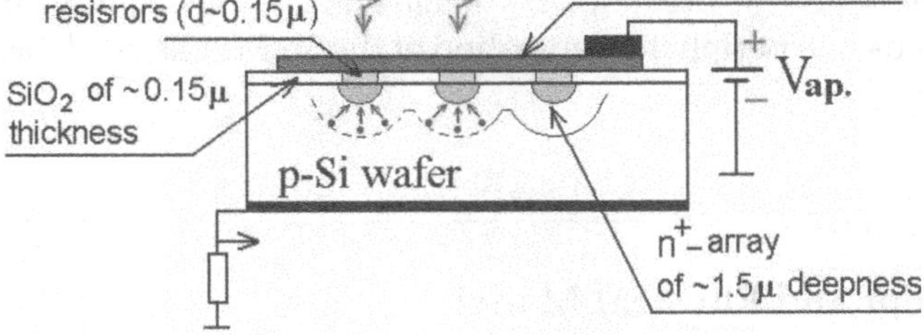


**FIGURE 9.1**
The basic version of Avalanche Micro-pixel PhotoDiode (AMPD) [5] also known as Micro-channel MRS APD [7].

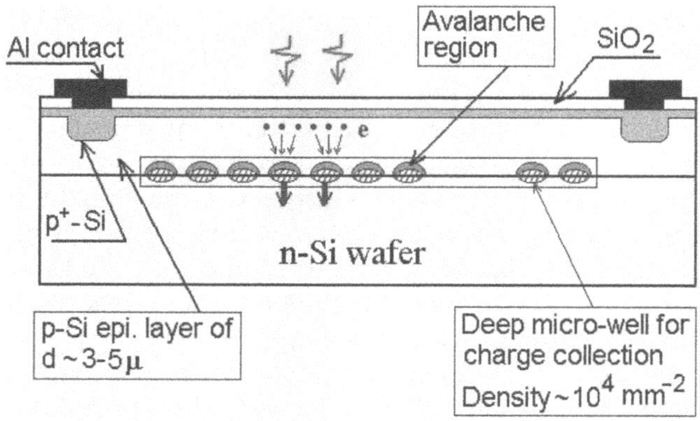


**FIGURE 9.2**
Advanced version of Avalanche Micro-pixel PhotoDiode (AMPD) with deep micro-wells [5] also known as Micro-pixel APD (MAPD) [7].

Such buried n+ "dots" are used to collect photoelectrons, enhance electric field, and localize avalanche in the same way as the surface ones. Moreover, being "micro-wells" for the multiplied electrons in the p-type silicon, they also provide quenching of the avalanche without dedicated resistors. The schematic design of the Micro-pixel APD (MAPD) based on deep n+ "micro-wells" [23, 24] is shown in Figure 9.2 [5]. The MAPDs demonstrate a unique pixel-density of up to $4 \times 10^4$ $mm^{-2}$ with the highest geometric efficiency of almost 100% and quantum efficiency of 80%. However, the peak PDE value of up to 40% appears to be about twice lower than it could be expected at given quantum efficiency. The MAPD design is also associated with a long pixel recovery time due to slow and field-dependent reemission of the electrons from "micro-wells" eliminating its DR for continuous illumination and long light pulse detection.

To overcome the above-mentioned limitations and combine the advantages of the planar SiPMs as well as MRS APDs and MAPDs, we developed a new SiPM concept – the "Tip Avalanche Photodiode" (TAPD) [25, 26]. The concept is based on the properties of tip-like electrodes to focus and enhance the electric field, to reduce the breakdown voltage and cell capacitance, and to eliminate the needs in a peripheral separation of the SiPM cells (avalanche regions).

## 9.2 Concept and Physical Model

The simplified cross-section of the TAPD is shown in Figure 9.3. A spherical tip that consists of high doped n-silicon is placed in a low-doped p-type epitaxial layer. The bias supply is provided through a quenching resistor

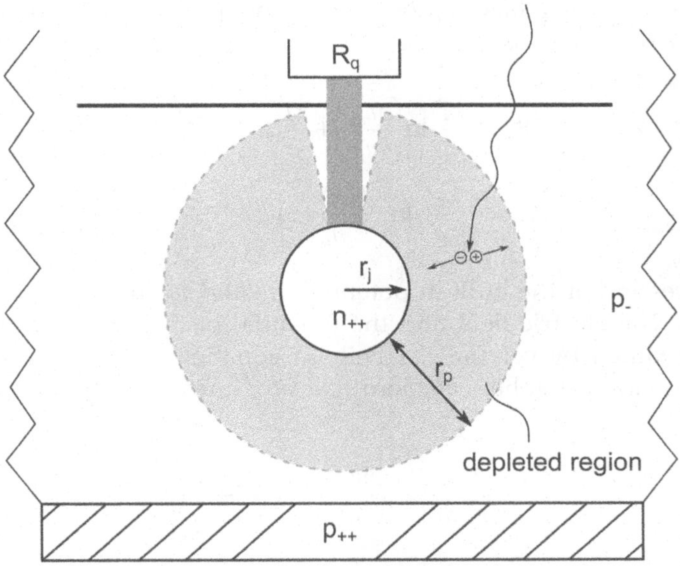


**FIGURE 9.3**
Schematic of the spherical SiPM concept [26].

$R_q$ that is placed on the surface. The connection of the tip to the quenching resistor is realized by a conductive pillar. The p-n junction on the surface of the tip causes depletion and therefore an electric field around the sphere.

### 9.2.1 Analytic Description

In this article, the basic properties of the new SiPM concept are derived using a simplified model of an n-doped sphere inside an infinite p-doped bulk. The transition from the n-doped to the p-doped region is first approximated as an abrupt junction where the depleted space charge region has a box profile (in 3D spherical shells).

In thermal equilibrium, the charge carrier currents of drift and diffusion cancel out and the Fermi level $E_F$ is constant throughout the junction:

$$\begin{aligned} J_n &= q\mu_n\left(n\varepsilon + \frac{kT}{q}\frac{dn}{dr}\right) = 0 \\ J_p &= q\mu_p\left(p\varepsilon + \frac{kT}{q}\frac{dp}{dr}\right) = 0 \end{aligned} \tag{9.1}$$

The depletion approximation considers complete impurity ionization in the n- and p-region. At thermal equilibrium (subscript '*o*') the free charge carrier concentration can be simplified to $n_{no} \approx N_D$ in the n-region and $p_{po} \approx N_A$ in the p-region. The condition of Equation (9.1) requires a constant Fermi level and therefore a built-in voltage is present in the p-n junction. The built-in potential $\Psi_{bi}$ can be written as:

$$\begin{aligned}\Psi_{bi} &= \frac{kT}{q}\ln\left(\frac{n_{no}}{n_i}\right)+\frac{kT}{q}\ln\left(\frac{p_{po}}{n_i}\right)\\ &\approx \frac{k_BT}{q}\ln\left(\frac{N_DN_A}{n_i^2}\right)\end{aligned} \tag{9.2}$$

This expression of the built-in potential is valid for planar and spherical junctions. The electric field and the potential of the space charge region can be obtained by solving the Poisson equation. The one dimensional Poisson equation in spherical coordinates for any arbitrary charge density $\rho(r)$ is [27]:

$$\frac{1}{r^2}\frac{d}{dr}\left(r^2\frac{d\Psi_i(r)}{dr}\right)=-\frac{\rho(r)}{\varepsilon\varepsilon_0} \tag{9.3}$$

In the first case of an abrupt junction, the charge carrier densities are given by the completely ionized acceptor and donor impurities, $N_A$ and $N_D$. The ionized regions and the notation of the dimensional variables are shown in Figure 9.4. The analytic solution for the electric field in the n-region and p-region ($\mathcal{E}_n$ and $\mathcal{E}_p$) can be obtained from integration of Equation (9.3) with the boundary condition of zero electric field outside the depletion region:

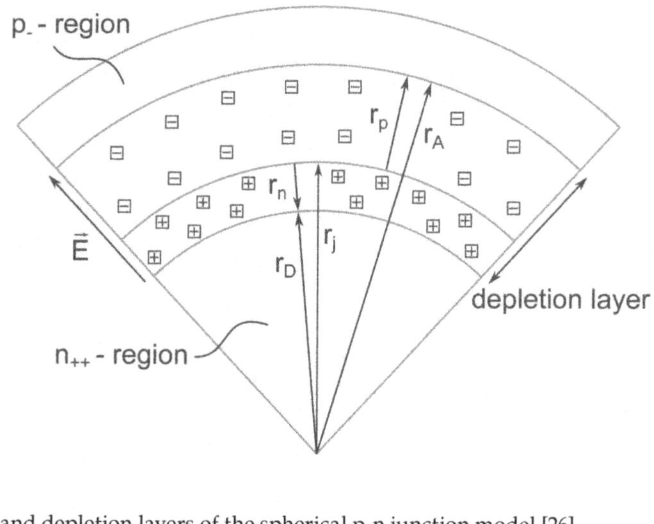


**FIGURE 9.4**
Dimensions and depletion layers of the spherical p-n junction model [26].

$$\begin{aligned} \varepsilon_n(r) &= \frac{eN_D}{3\varepsilon\varepsilon_0}\frac{\left(r^3-r_D^3\right)}{r^2}, \\ \varepsilon_p(r) &= \frac{eN_D}{3\varepsilon\varepsilon_0}\frac{\left(r_j^3-r_D^3\right)}{r^2}-\frac{eN_A}{3\varepsilon\varepsilon_0}\frac{\left(r^3-r_j^3\right)}{r^2} \end{aligned} \tag{9.4}$$

The electric field in the n-region is zero at the inner depletion edge $r_D$ and will increase linearly if the term $r_D^3/r^2$ is small. In the p-region, the expression of the electric field has two terms. The first term is dominant if $N_D \gg N_A$ and the electric field will decrease proportionally to $1/r^2$. The second term becomes dominant if the first term is small enough with increasing radius and the electric field will equal to zero at the outer depletion edge $r_A$.

In this concept, the donor concentration in the sphere is much higher than the acceptor concentration in the epitaxial layer. An exemplary electric field distribution according to Equation (9.4) for different metallurgical junction radii is presented in Figure 9.5. Here, the donor and acceptor concentrations were arbitrarily set to $N_A = 1\times10^{14}\,\mathrm{cm}^{-1}$, $N_D = 1\times10^{18}\,\mathrm{cm}^{-}$. The depletion width of each sphere size was adapted to create a depletion potential of 40V.

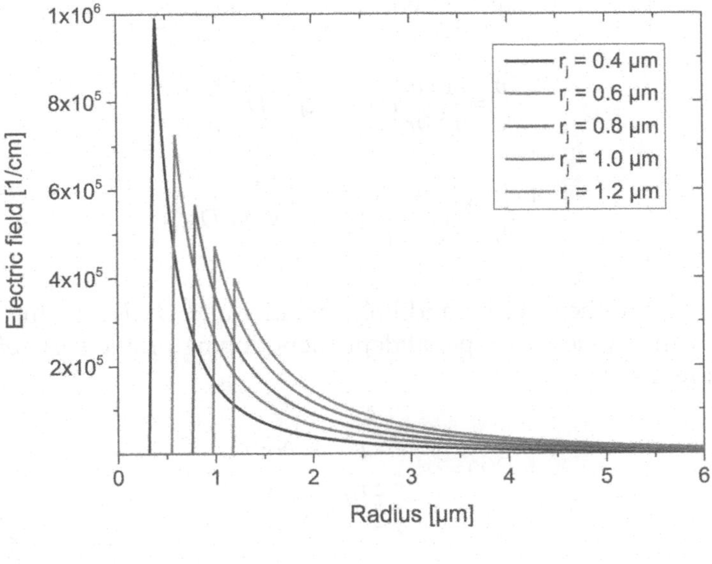


**FIGURE 9.5**
Electric field distribution obtained from the analytic solution for different sphere radii at a depletion potential of 40V [26].

### 9.2.2 Numerical Simulation

The electric field in the previous section is derived using an approximation of the depletion regions as box profiles. In a processed device, the dopant concentration will vary along with the radius of the sphere. A dopant distribution is typically created by ion implantation and thermal annealing during different process steps. The diffusion of impurities is thermally activated and occurs in direction of the concentration gradient. In the case of the TAPD, especially the donor impurities of the tip will start to diffuse into the p-region. Three different structure sizes were processed and tested. The nominal junction radii of these devices are 0.6 µm, 0.8 µm, and 1.0 µm. The measured dopant profile of these devices is used for the following simulations.

The complete impurity ionization (as assumed in Equation (9.4)) is prevented by diffusion of free charge carriers into the depleted regions [28]. While Equation (9.1) deals with the steady-state, the continuity equations describe the net current flowing in and out of a region of interest. Here, the generation and recombination in the depletion region are neglected. The divergence operator has to be adapted for one-dimensional spherical coordinates. The simplified continuity equations and the differential equation of the electric field from the system of the coupled equation that has to be solved:

$$\frac{\partial n}{\partial t} = \frac{1}{r^2}\frac{\partial}{\partial r} r^2 \left( \mu_n \varepsilon n + D_n \frac{\partial n}{\partial r} \right) \tag{9.5}$$

$$\frac{\partial p}{\partial t} = \frac{1}{r^2}\frac{\partial}{\partial r} r^2 \left( -\mu_p \varepsilon p + D_p \frac{\partial p}{\partial r} \right) \tag{9.6}$$

$$\frac{1}{r^2}\frac{d}{dr} r^2 \varepsilon(r) = \frac{q}{\varepsilon_s}\left(p(r) - n(r) + D(r)\right) \tag{9.7}$$

where $\mu_n$, $\mu_p$ are the electron and hole mobility and $D_n$, $D_p$ are the electron and hole diffusivities. The spatial dependent doping profile $D(r)$ defines the p- and n-region as:

$$D(r) = \left( \begin{array}{ll} r \le r_j; & N_D(r) \\ r \ge r_j; & -N_A(r) \end{array} \right. \tag{9.8}$$

The presented governing system of equations is considered in a spherical domain. The boundary conditions in the center and on the spherical shell of the region of interest were of the Dirichlet type. We used the finite difference

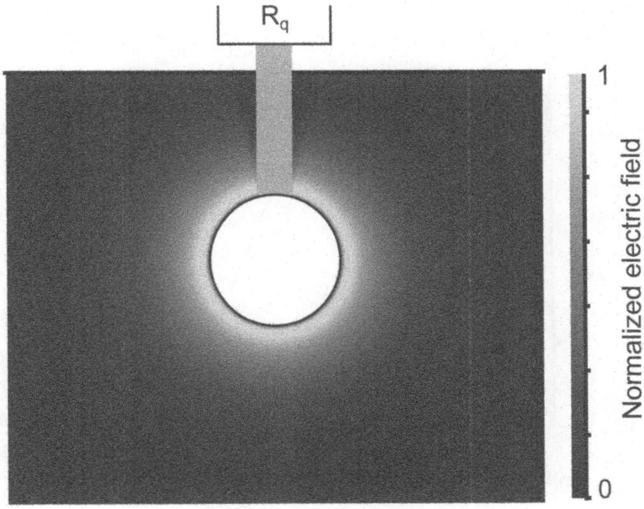


**FIGURE 9.6**
Schematic 2D electric field distribution due to the spherical depletion around the tip [26].

method with a centred difference stencil for the spatial numerical approximation. The approximation in time of the continuity equations was done using the Crank Nicolson method [29]. All solutions were verified with the commercial TCAD tool Silvaco [30].

In Figure 9.6, a schematic 2D distribution of the normalized electric field around the tip is illustrated. The highest field strength is located close to the tip surface and decreases with increasing distance to the junction. The electric field is present up to the passivated surface and a spherical active volume is available for electron attraction in direction of the n-region. However, impact ionization multiplication due to a high-electric field occurs just close to tip surface. The solution of the numerical approximation at the breakdown voltage (here 43.4V) is presented in Figure 9.7 for a nominal junction radius of 0.6 μm. The ionization rates $\alpha_n$ and $\alpha_p$ for electrons and holes depend on the electric field:

$$\alpha(\varepsilon) = \alpha_\infty \exp\left(\frac{-b}{\varepsilon}\right) \tag{9.9}$$

with $\alpha_\infty$ and $b$ at room temperature taken from [31] for electrons and holes, respectively. To estimate the breakdown voltage of different sphere sizes, the multiplication of charge carriers due to impact ionization has to be evaluated using the spatial dependent ionization rates at different bias voltages. A breakdown occurs if the impact ionization multiplication becomes

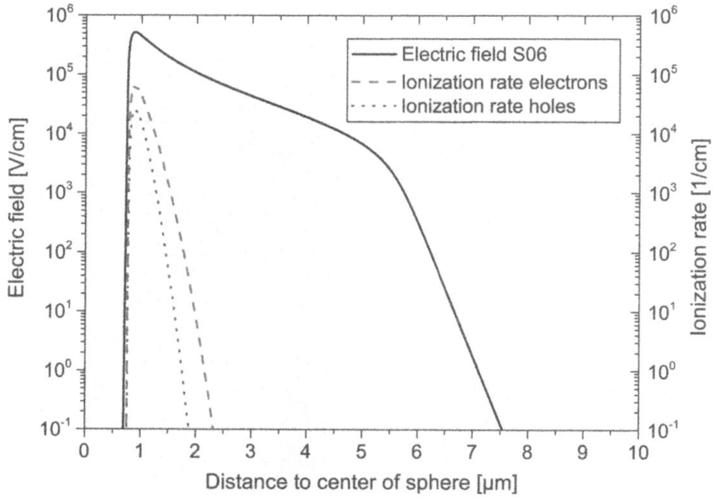


**FIGURE 9.7**
Spatial distribution of the electric field and the ionization rates of electrons and holes for the smallest structure S06 [26].

infinite which is equivalent to the condition that the ionization integral [32] equals one:

$$\int_{r_D}^{r_A} \alpha_n \exp\left[-\int_{r}^{r_A} (\alpha_n - \alpha_p) dr'\right] dr = 1 \tag{9.10}$$

The results of this evaluation are presented in Table 9.1. The breakdown voltage increases with the size of the tip due to a lower electric field at the same bias voltage. The range of the electric field directly affects the photon detection capability regarding light with increasing wavelength. The absorption of a photon inside of a material at a certain depth $x$ can be described by the Beer-Lambert law. The probability for the generation of charge carriers at a certain depth inside the epitaxial layer decreases exponentially with increasing distance to the surface. Additionally, in the range of visible light, the

**TABLE 9.1**
Tested Device Structures and Nominal Radii of the Metallurgical Junction

| Structure Name | Nominal Radius ($r_j$) | Breakdown Voltage |
|---|---|---|
| S06 | 0.6 µm | 43.4V |
| S08 | 0.8 µm | 50.7V |
| S10 | 1.0 µm | 53.9V |

absorption coefficient decreases with increasing wavelength. A SiPM with a high-charge collection efficiency aims to absorb as much of the incoming photon flux as possible. Consequently, the electric field has to be present as deep as possible. Photoelectrons generated at a distance to the tip are first accelerated due to the drift field in direction of the multiplication region (high electric field) and a Geiger discharge is triggered.

The simulated range of the electric field into the epitaxial layer for the three structure sizes at different bias voltages is shown in Figure 9.9. All structures reach at least 8 μm at their respective breakdown voltage. The maximal active volume is reached placing the center of the sphere at a depth of the maximal depletion range. In this configuration, photoelectrons that are generated close to the surface and up to the maximal active depth are detected. This advantage of the new concept allows for a high PDE in a wide spectral range. Theoretically, it is possible to reach a total active depth of 20 μm with the current technology (see Figure 9.9) and consequently a PDE above 30% at 905 nm. Our prototypes were processed in an epitaxial layer of 12 μm, which limits the maximal depleted volume for all structure types.

### 9.2.3 Cell Placement

The operation of a SiPM above the breakdown voltage in Geiger mode requires a serially connected quenching resistor to keep the device in a quasi-stable state. The potential of the epitaxial layer during operation is set to ground while the tip is biased through the resistor. The spherical shape of the active volume allows a high-density placement of multiple single cells in an array. In the top view, the cells are placed in a hexagonal grid achieving a theoretical packing density of $\eta \approx 90.7\%$. The bias voltage is supplied through a metal grid connected to the quenching resistor of each cell. The schematic layout is shown in Figure 9.8. The aim of the layout is to cover

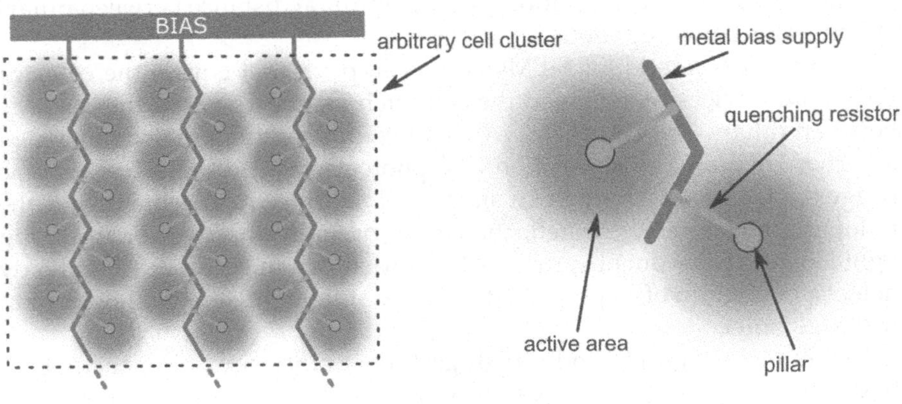


**FIGURE 9.8**
High-density layout and bias supply of a TAPD array [26].

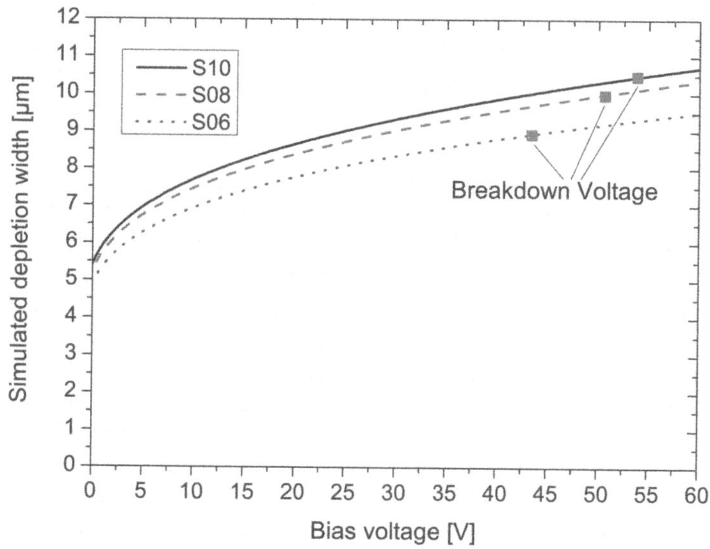


**FIGURE 9.9**
Range of the electric field for the three structure sizes [26].

the smallest possible area with metal. The metal on the surface blocks visible light therefore reducing the active area of the SiPM for light detection. The presented placement offers the advantage to use just one metal line for two rows of SiPM cells.

The ratio between the uncovered area and the total area is called geometric efficiency (see also Section 9.3.2). Compared to planar devices, the new concept offers the advantage of a frameless layout. The TAPD cells are biased through the center and can be placed close to each other without losing active area. The limitations for the minimal distance between pillars are still under investigations.

A typical parameter for SiPMs is the cell pitch that equals the cell size for planar devices. We defined the cell pitch for the TAPD as the distance between pillars (see Figure 9.8). A small cell pitch offers the advantage of a higher DR [18]. The absorption of a photoelectron in the active volume triggers a Geiger-discharge of a single cell. During the discharge and the following recharge of the cell, the cell is partially inactive for incoming light [33, 34]. Consequently, the SPAD array shown in Figure 9.8 could only detect a maximum of 24 photons during a fast light pulse (e.g., shorter than recharge time).

The aim of a high DR and a high-geometric efficiency is in general contradictory. Regarding the results of Figure 9.9, we chose a cell pitch of 15 µm for the processed devices. The active areas of two neighboring cells for the smallest structure S06 are just in contact. The geometric efficiency of our device is only determined by the opaque materials like metal lines and semi-transparent materials like the quenching resistors, which are located on top of the active area. For the prototypes with a 15-µm pitch, the nominal geometric efficiency of 83% was realized. The value was calculated from the layout, accounting for metal lines and quenching resistors as opaque.

### 9.2.4 Depletion Layer Capacitance

The depletion layer around the tip creates a certain amount of charge on each side of the junction whereas the total charge is zero. The incremental charge $dQ_D$ of one side of the junction upon an incremental change of the bias voltage defines the depletion layer capacitance $C_D = dQ_D/dV$. The spatial depletion of the respective structures was obtained with the solution of the continuity equations (see Equation (9.7)). The simulated depletion layer capacitance dependent on the bias voltage is presented in Figure 9.18. All three structures show a similar curve progression while the increased tip size leads to a higher capacitance.

The simulated depletion layer capacitance is part of the total cell capacitance $C_{cell}$, which includes additionally the parasitic capacitance due to the pillar and the connection to the quenching resistor. The main goal for the presented new SiPM design was to achieve a high PDE over a wide spectral range. The recovery time (recharge time) should be short to provide a fast device with a high DR. Consequently, a low-cell capacitance is beneficial. In respect to the simulations of this section, we chose the smallest device S06 as the most promising structure.

## 9.3 Metrological Characterization

### 9.3.1 Single Electron Response

To achieve a high-photon count rate and a strong ambient light immunity, a fast recovery of the microcells is required. In Figure 9.10, the single electron response (SER) of the TAPD is shown at an excess bias voltage of $\Delta V = 4\,\text{V}$. It was measured as the voltage drop across a 25 Ω load resistor. The decay part of the pulse consists of two exponential components with time constants of $\tau_1 \approx 0.5\,\text{ns}$ and $\tau_2 \approx 4.3\,\text{ns}$. After approximately 9.5 ns, the micro-cells are recovered to 90% of their maximum charge. Using the double-light-pulse method, as proposed in [35], comparable results were achieved. This confirms that the SER shape reflects the true recovery process in this case. With this result, the TAPD has the fastest recovery time with respect to state-of-the-art planar SiPMs with enhanced red-sensitivity [36, 37].

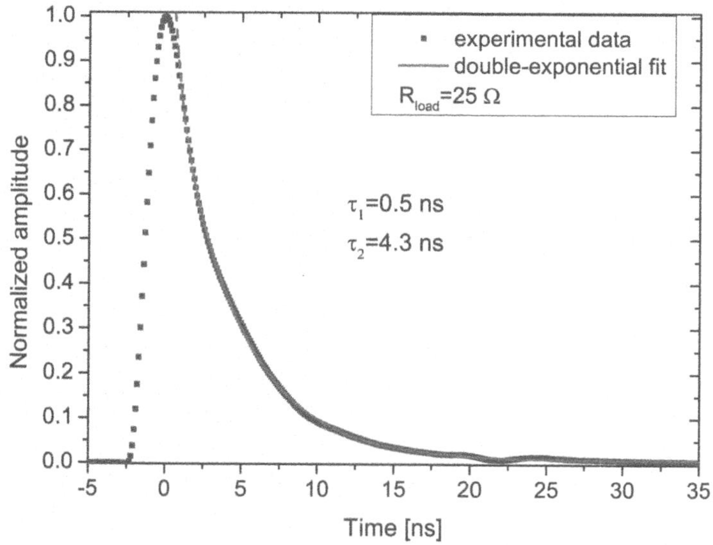


**FIGURE 9.10**
Normalized pulse shape of the S06 structure at an excess bias voltage of 4V at 21°C [26].

Figure 9.11 provides an example of a single photoelectron charge spectrum. For this spectrum, we used a pulsed laser illumination with a pulse width of 70 ps. The acquisition was synchronous with the light pulses. The charge was integrated within a time window of 10 ns. The peaks up to

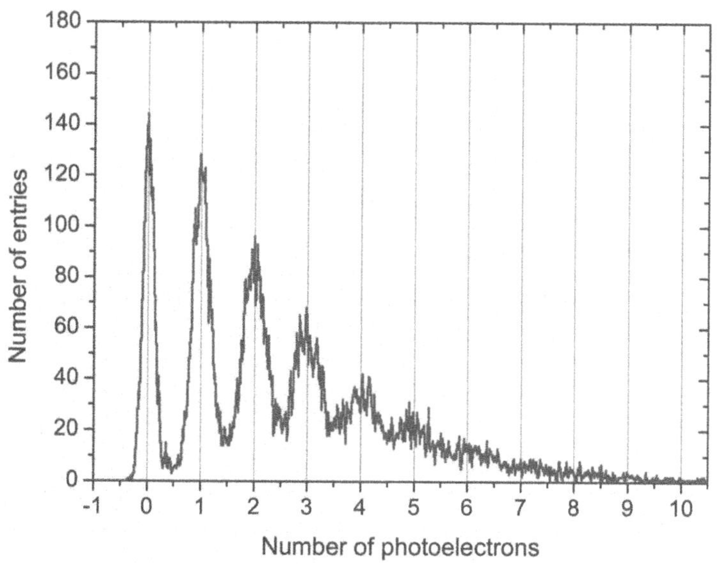


**FIGURE 9.11**
Single photoelectron charge spectrum of the S06 structure at an excess bias voltage of 4V at 21°C [26].

4 photoelectrons are well separated, which makes precise single photon counting possible.

### 9.3.2 Photon Detection Efficiency

The PDE describes the capability of the sensor to detect light as the ratio of the average number of detected and the average number of incident photons. In Equation (9.11), the PDE is described as a product of three quantities:

$$PDE = \varepsilon \cdot QE \cdot P_{trigg} \tag{9.11}$$

i. The geometric efficiency $\varepsilon$ describes the fraction of the SiPM area, which is able to detect photons. The area of the SiPM which is not sensitive to light is mainly due to the metal lines for signal readout, the quenching resistors, guard rings for electric field attenuation toward the micro-cell edges, and trenches for the suppression of optical crosstalk.
ii. The quantum efficiency $QE$ describes the efficiency to collect a fraction of charge carriers that a photon generated within the active volume of a micro-cell. To maximize the PDE for blue light, the depletion region has to be extended as close to the surface as possible. Lower energetic photons are also absorbed in larger depths. To reach an enhanced detection of red and NIR light, the active region has to reach deeper inside the silicon. With our devices, we reach a spherical depletion volumes with radii of about 8–9 µm (see Figure 9.9).
iii. The avalanche triggering probability $P_{trigg}$ describes the probability that a generated e-h pair will successfully initiate an avalanche breakdown by impact ionization.

In this work, the PDE was measured by using the continuous low-level light method, as reported in [38–41]. The incident photon rate was determined by a calibrated reference PIN-diode [42]. Both, the SiPM and the reference diode were homogeneously illuminated. The homogeneous part of the light spot was significantly larger than the active area of the photosensors. The photon rate was determined as the difference of the SiPM pulse count rate during illumination and in dark conditions. The count rates were measured according to the method described in Section 9.3.3. In Figure 9.12, the PDE is shown as a function of the excess bias voltage for several wavelengths from 460 µm to 905 µm. In this spectral range, the PDE reaches 90% of its saturation value at excess bias voltages between 4V and 5V. This is comparable to state-of-the-art planar SiPMs that are optimized for the detection of blue light, and hence have smaller depletion volumes [43].

In Figure 9.13, the PDE is shown at different wavelengths. This result was obtained by converting the spectral response measurement with a monochromator into an absolute PDE measurement as described in reference [43]. The TAPD demonstrates the highest peak PDE value of 73% at 600 nm

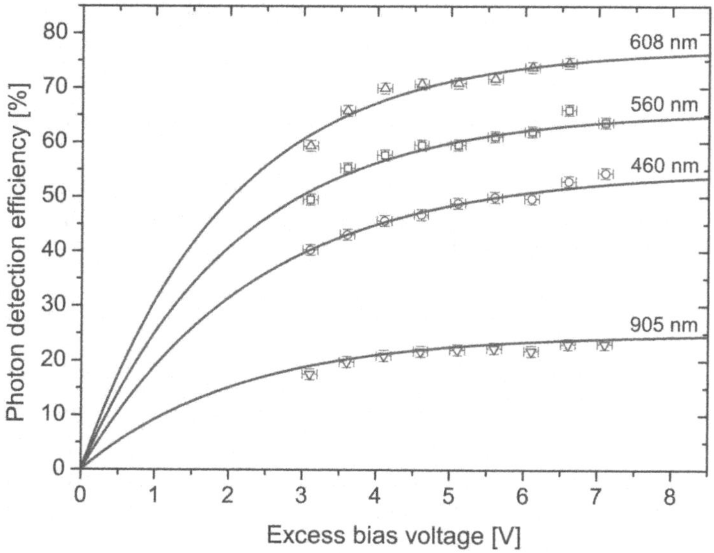


**FIGURE 9.12**
Photon detection efficiency vs. the excess bias voltage for the S06 structure at 21°C [26]. The lines are drawn as eye guides.

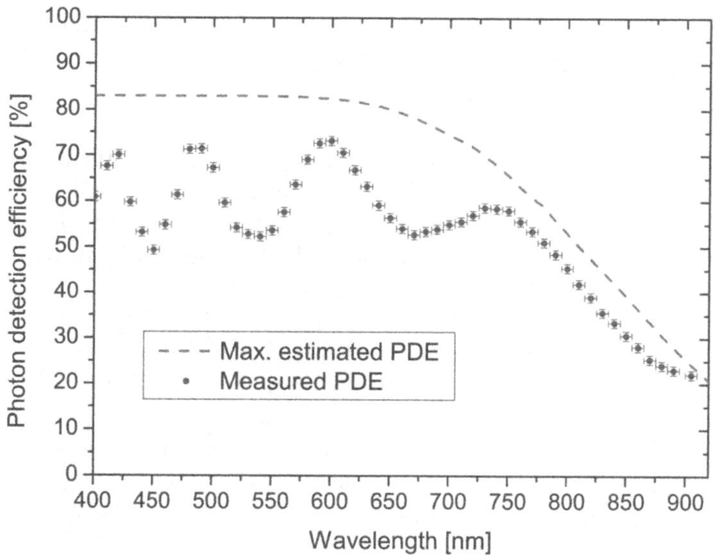


**FIGURE 9.13**
Photon detection efficiency vs. wavelength for the S06 structure at an excess bias voltage of 5V at 21°C [26].

compared with the state-of-the-art devices with a 15-μm pitch size [44, 45]. Additionally, the PDE curve does not show the typical fast decrease with increasing wavelengths and remains above a value of 45% up to a wavelength of 800 nm. The measured PDE is in good agreement with the expected values when assuming a Geiger-efficiency of 90% (see Equation (9.11) and dashed curve in Figure 9.13). For wavelengths in the NIR-regime, we expect the geometric efficiency to increase approximately to 90% due to the decreasing absorption by the semi-transparent quenching resistors. The oscillations in the PDE curve are caused by destructive interference due to the protective $Si / SiO_2$ stack on top of the entrance window of the prototype devices. The results of the applied method are in agreement with the well-known method based on pulsed laser illumination [43, 46]. The applied method offers the possibility to directly measure the absolute PDE for a larger number of wavelengths due to the easy access to LEDs of different wavelengths.

### 9.3.3 Dark Count Rate and Delayed Correlated Pulses

In this work, the dark count rate and the probability of delayed correlated pulses are determined by using the method proposed in reference [40]. The method is based on the analysis of the complementary cumulative distribution function (CCDF) of pulses subsequent to a primary dark pulse. Contrary to the pulse counting approach, the applied method provides the benefit that delayed correlated pulses that exceed the detection threshold do not contribute to the dark count rate of the device [41]. This advantage is of special significance for SiPMs with a fast recovery time, like the TAPD, where delayed correlated pulses reach the maximum charge after a few ns. The CCDF-method is also suited to measure the count rate of a continuous Poissonian photon output from a light source. In this way, the PDE can be measured (cf. Section 9.3.2).

As delayed correlated pulses, we understand pulses that may be caused by three kinds of effects:

> The first one is afterpulsing. Here, trapping centers present as the energy states within the bandgap may capture electrons or holes from the conduction or valence band and re-emit them after a certain delay-time $\Delta t$ into the same band. If the trapping center is located in the active region of a micro-cell, the re-emitted charge carrier has a finite probability to trigger a subsequent avalanche in the same micro-cell. The delay-times depend on the respective trap type and may vary by many orders of magnitude.
>
> The second effect is the so-called delayed optical crosstalk. Here, one or more photons that are emitted during the avalanche breakdown are absorbed outside the high field region. The generated minority charge carriers then diffuse towards the active region and can trigger consecutive breakdowns of a neighboring micro-cell. This process is significantly slower compared to the prompt optical crosstalk.

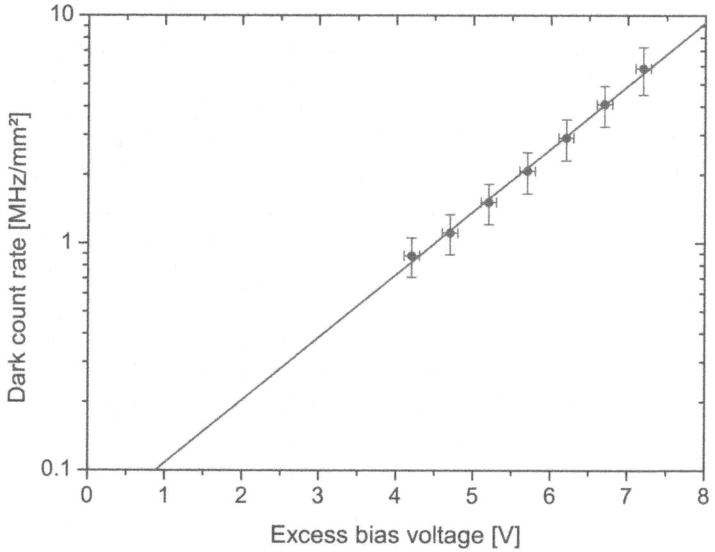


**FIGURE 9.14**
Dark count rate vs. the excess bias voltage for the S06 structure at 21°C [26]. The error bars represent the standard deviation of five samples. The line is drawn as an eye guide.

> In the third case the diffusing charge carrier triggers the original micro-cell, and this effect is called optically-induced afterpulsing [47].

The applied method is only valid under the assumption that the time constants of the delayed correlated effects are smaller with respect to the reciprocal of the dark count rate. In reference [48], the applied measurement procedure is described in detail.

In Figure 9.14, the dark count rate is shown as a function of the excess bias voltage at 21°C. At recommended excess bias voltages between $\Delta V = 4\,\text{V}$ and $\Delta V = 5\,\text{V}$, the prototypes show a DCR between $700\,\text{kHz/mm}^2$ and $1.3\,\text{MHz/mm}^2$. In comparison, KETEK's planar 15 μm pitch SiPM shows a DCR of typically $125\text{kHz}/\text{mm}^2$ at $\Delta V = 5\,\text{V}$ [49].

This result matches our expectations since the active volume of the TAPD is about a factor 10 larger with respect to the planar structures. The state of the art red-sensitive SiPMs from other manufacturers show dark count rates between $600\,\text{kHz/mm}^2$ [36] and $3.5\,\text{MHz/mm}^2$ [37] at the recommended operation voltages. In Figure 9.15, the probability of delayed correlated pulses is plotted as a function of the excess bias voltage. It is below 3% at excess bias voltages up to $\Delta V = 5\,\text{V}$.

### 9.3.4 Prompt Optical Crosstalk

During an avalanche breakdown, optical photons are generated by a variety of processes. These photons are able to propagate to neighboring

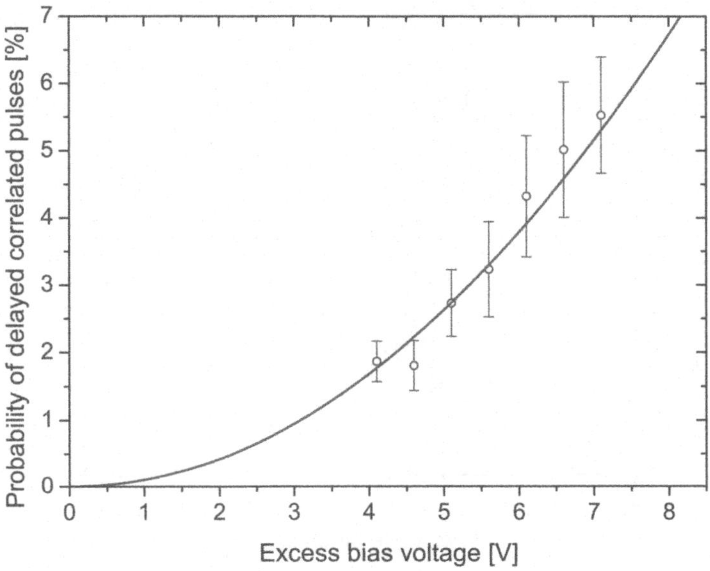


**FIGURE 9.15**
Probability of delayed correlated pulses vs. the excess bias voltage for the S06 structure at 21°C [26]. The error bars represent the standard deviation of five samples.

micro-cells and initiate further avalanche breakdowns. The propagation may occur by a direct path or by several reflections at the top and bottom sides of the device. In either case, the time difference between the first and the consecutive pulse is not sufficient for a distinction between the two pulses. For this reason, only one pulse with an amplitude of multiple photoelectron equivalent (p.e.) is registered. This effect is called "prompt optical crosstalk" (CT). The prompt optical crosstalk probability scales with the number of generated photons during an avalanche breakdown, the geometric cross-section for the interaction between two micro-cells, and the avalanche triggering probability. The prompt optical crosstalk probability $P_{CT}$ is estimated as shown in Equation (9.12). Here, $DCR_n$ is the dark count rate measured by the pulse counting method with a discriminator threshold of $n$ p.e. [34, 43, 46, 50]:

$$P_{CT} \approx \frac{DCR_{1.5}}{DCR_{0.5}} \tag{9.12}$$

At typical excess bias voltages between 4V and 5V, we measure crosstalk probabilities between 27% and 35% as shown in Figure 9.16. For comparison: the reported values from other state-of-the-art devices with a red-enhanced sensitivity are between 20% and 43% [36, 37].

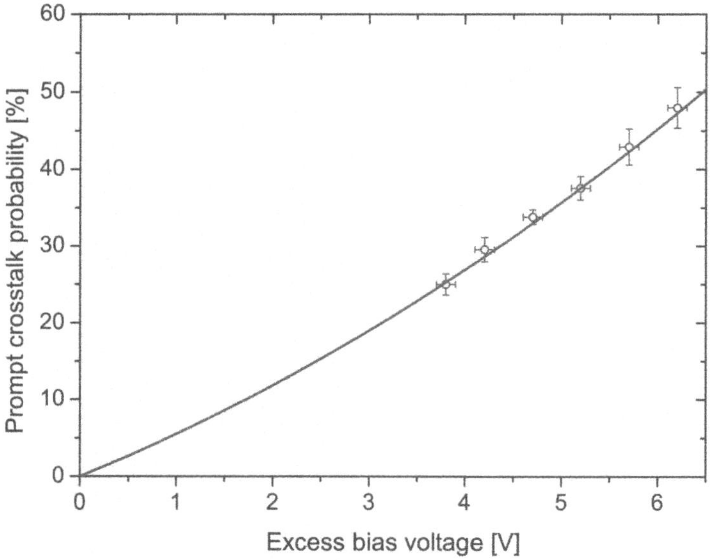


**FIGURE 9.16**
Prompt optical crosstalk probability of the S06 structure at 21°C [26].

### 9.3.5 Gain

The fast recovery time, described in Section 9.3.1, is realized by a strongly reduced micro-cell capacity. As a consequence, the intrinsic gain is about a factor of 20 lower with respect to the KETEK's planar SiPM products [49]. In Figure 9.17, the absolute gain is shown as a function of the excess bias voltage for two different structures. The higher gain of the S10 structure can be attributed to the larger ball radius and the larger contribution from the parasitic capacitance of the quenching resistor. To measure the gain, we applied a combination of the single photoelectron charge spectrum and the dark current [48, 51]. In this procedure, the single photoelectron spectrum was recorded in a dark environment with a trigger threshold set to 0.5 pe. To receive the probability density function (PDF) of the primary and secondary events, the spectrum was normalized to the total number of detected events. The expected number of firing micro-cells per initial photoelectron is described by the mean value of the PDF which is equivalent to the excess charge factor (ECF) [52]. The dark count rate was determined as described in Section 9.3.3. In this case, the contribution from afterpulses is neglected. Since the investigated devices have a low-afterpulsing probability (see Figure 9.15), this approach is reasonable. The gain $G$ is then determined by using Equation (9.13). $I_{dark}$ is the dark current at the respective operation voltage and the same temperature:

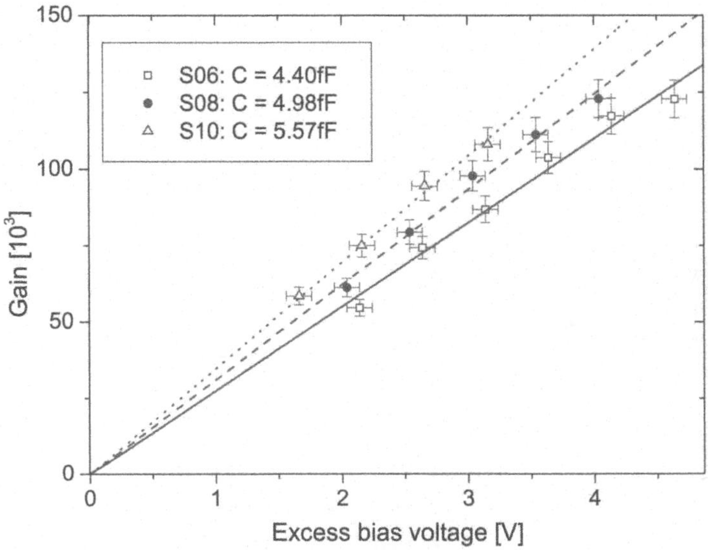


**FIGURE 9.17**
Gain vs. the excess bias voltage for the S06, S08, and the S10 structure at 21°C [26].

$$G(V) = \frac{I_{dark}(V)}{q \cdot DCR(V) \cdot ECF(V)} \tag{9.13}$$

Typically, the gain of a single cell is assumed to be proportional to the cell capacitance and can be expressed as:

$$G = \frac{Q}{e} = \frac{\Delta V \cdot C_{\text{cell}}}{e} \tag{9.14}$$

where $\Delta V$ stands for the excess bias voltage. According to (see Equation (9.14)), we extracted the micro-cell capacitances from the slopes of the linear fits in Figure 9.17. The experimentally determined micro-cell capacitances differ from the simulated ones by about 3.3 tF to 3.8 tF (see Figure 9.18). This discrepancy is not understood, yet. One possible explanation is to attribute this discrepancy to the parasitic capacitance of the quenching resistor and the variation of the real geometry from the simulated spherical one. Here, we would like to point out that we expect the parasitic capacitance of the quenching resistor to be of the same order of magnitude as the micro-cell capacitance, and hence significantly contribute to the charge output of the TAPD. Another possible explanation is that the gain cannot be described by a simple product of the micro-cell capacitance and the excess bias voltage for our specific field distribution [7, 53].

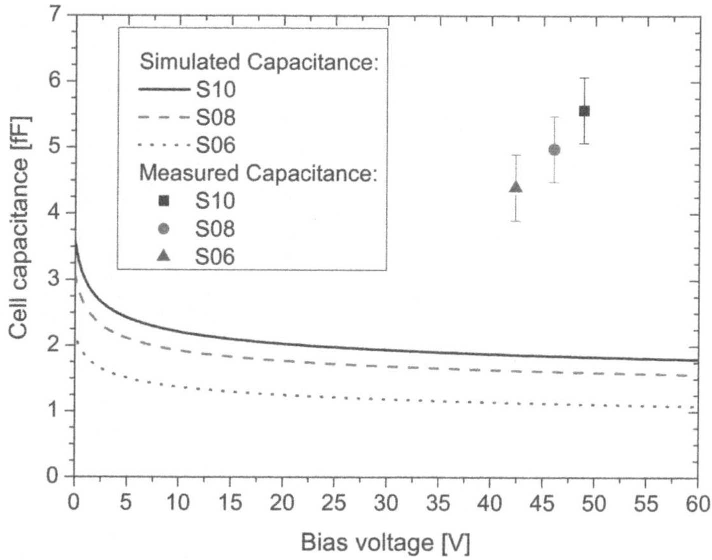


**FIGURE 9.18**
Simulated depletion layer capacitance and measured total cell capacitance (plotted at $V_{bd}$) [26].

### 9.3.6 Temperature Dependence of the Breakdown Voltage

Especially in automotive applications, the systems must operate in a wide temperature range. Independent of whether the photosensor has a temperature stabilization/compensation or not, a low-temperature coefficient of the breakdown voltage ($V_{bd}$) is beneficial. For our TAPD, we measure a linear decrease of $V_{bd}$ with temperature (see Figure 9.19). The slopes increase with increasing pillar diameter from 26 mV/°C for the S06 structure to 31.5 mV/°C for the S10 structure. These values are comparable to 22 mV/°C for planar KETEK SiPMs [49], despite the larger depletion width and the increased breakdown voltage. The breakdown voltage was determined from the inverse logarithmic derivative (ILD) of the reverse current-voltage-characteristic with low-level light illumination, as described in reference [54]. The simulated breakdown voltages (see Table 9.1) overestimate the experimental values at room temperature. The discrepancy decreases from 5V for the S10, to 4.7V for the S08, and 1V for the S06 structure. We attribute this behavior to the fact that the deviation of the processed structures from a perfect sphere increases with increasing tip size. For the S10 structure, the shape of the tip is closer to the one of an ellipse than a sphere. Here, the breakdown voltage is defined by the point with the lowest curvature.

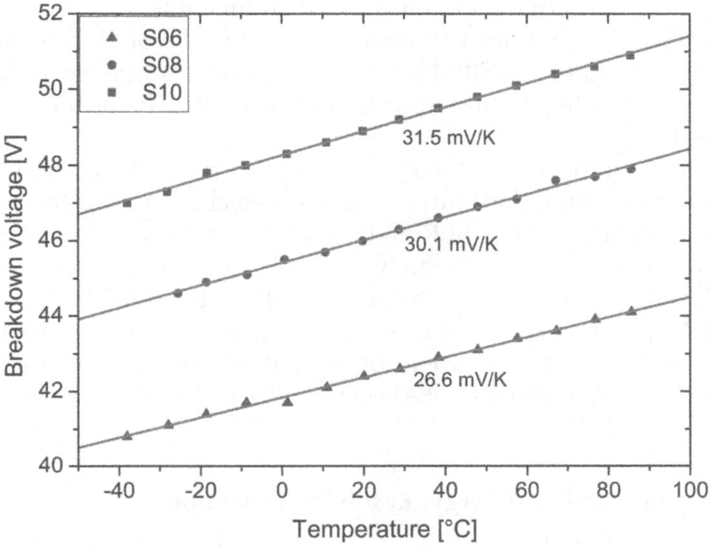


**FIGURE 9.19**
Breakdown voltage vs. temperature [26]. The error bars are within the symbols.

## 9.4 Discussion on the Applicability of TAPD

Let us consider the most general application areas of the SiPM as a photon-number resolving detector, namely:

1. To count photons of low-intensity light flux as a single photon counter;
2. To measure the number of photons in a light pulse as a multi-pixel photon counter that output charge is determined by a number of fired pixels;
3. To measure the arrival time of a light pulse as a fast analog photodetector;
4. To measure a waveform of an incident light signal as a fast analog photodetector.

### 9.4.1 Photon Counting and Current Integration

The SiPM performance in photon counting is considered as hardly competitive with dedicated single-element SPAD counters due to its lower geometric efficiency (periphery of cells) and higher dark noise (millimeter-scale area). However, at high light intensities, when time intervals between detection events approach a dead time of the counter, the SPAD becomes saturated while the multipixel SiPM still provides a linear response in the output current (current integration) measurements even if SERs of fired pixels are overlapped [55].

The TAPD should be very competitive with the conventional SiPM in both photon counting and output current detection modes for high light intensities due to very fast SER and very fast recovery time of 4 ns and a high-pixel density of approximately $10^4$ mm$^{-2}$. Low-intensity photon counting could hardly be a proper primary goal for the TAPD due to a high DCR of 1 Mcps, i.e., an order of magnitude higher comparing with the blue-sensitive SiPM and a few orders of magnitude higher than the DCR of wide-spectral single-element SPAD counters, e.g., Excelitas (former Perkin-Elmer) SPCM-AQR.

### 9.4.2 Photon-Number (Energy) Resolving Detection

SiPMs have initially been recognized and widely implemented as scintillation detectors due to superior photon number resolution at room temperature and corresponding energy resolution in calorimetry of high energy particles [56–58]. Performance in the energy resolution ER is defined by detective quantum efficiency (DQE) as well as by total excess noise factor (ENF) of the detection process [18]. The key contributions to the DQE in a linear detection mode above readout and dark noises and below saturation are PDE and ENF of correlated events (crosstalk and afterpulsing), namely:

$$ER = \sqrt{\frac{ENF_{tot}}{N_{ph}}} \tag{9.15}$$

$$DQE = \frac{1}{ENF_{tot}} \sim \frac{PDE}{ENF_{corr}} \tag{9.16}$$

Record-high PDE of ~70% makes the TAPD one of the best energy-resolving SiPMs in terms of DQE as expressed by Equations. (9.15, 9.16). Despite rather high optical crosstalk of ~30% (ENF ~1.3), the TAPD DQE of ~54% is almost at the top of modern SiPM performance as discussed in reference [59]. Moreover, in contrast with the record-high DQE SiPM of 100 μm pixels developed by MEPhI/Zelenograd with DQE of 55%, the high DQE of the TAPD is also associated with a very high DR.

### 9.4.3 Time-of-Flight Resolving Detection

Recent advances in R&D toward ultimate time resolution of scintillation detectors, first and foremost for TOF PET applications, resulted in a clear understanding of the key factors affecting performance of the detectors [60].

SiPMs demonstrated considerable progress and very promising results in the time resolution (TR), first, as a direct consequence of improvements in energy resolution and its key components (*PDE* and $ENF_{corr}$ as expressed in Equation (9.16)). The influence of single photon time resolution (SPTR, $\sigma_{sprt}$) of the photodetector on TR has been figured out later because that happens often been masked by slow scintillator timing (decay time $\tau_d$ and rise time $\tau_r$, optical transient time spread $\sigma_{otts}$) as well as nonoptimal read-out AQC electronics. Recently, the TR of bi-exponential scintillation detection with SiPM has been approximately expressed in an analytical form [61] as:

$$TR \approx \sqrt{\frac{ENF_{tot}}{N_{ph}} \cdot \tau_d \cdot \left(a \cdot \tau_r + b \cdot \sqrt{\sigma_{otts}^2 + \sigma_{sprt}^2}\right)} \quad (9.17)$$

with $a \approx 1.4$ and $b \approx 1.1$.

Indeed, SPTR of SiPM affects multiphoton TR only in case of negligible contributions from scintillator timing ($\sigma_{sprt} \gg \sigma_{otts}, \tau_r$). A similar result has been obtained for the Gaussian-shape light pulse detection. The influence of the SER pulse shape on TR is also typically masked by the slow decay time of the scintillator, but the fast SER is an obvious advantage for the leading edge discrimination of a few first photon detection events. Overall, the fast and very efficient TAPD is expected to be rather competitive in the time-of-flight resolving detection, especially in the case of relatively slow multiphoton pulses.

### 9.4.4 Waveform Measurement

Waveform measurement, especially with an objective of the arbitrary waveform reconstruction, is the most challenging task for SiPM, and it almost undeveloped area except for a few initial studies, for example, related to beam loss monitoring for particle accelerators [62]. The transient SiPM response on the arbitrary transient signal reveals a rather complex history-dependent dynamic behavior with a strong dependence on the mean number of detected photons per pixel per recovery time $\tau_{rec}$. The key figure of merit in such a case could be defined as the intensity of SiPM recovery. If the intensity of the photon detection events $I_{det}$ approaches recovery $I_{rec}$, the SiPM becomes nonlinear and approaches saturation with dramatic losses of information from the incident optical signal. Therefore, the DR of the SiPM is limited by the relation

$$I_{det} \sim I_{rec} = \frac{N_{pix}}{\tau_{rec}} \quad (9.18)$$

## 9.5 Conclusion

In this work, we presented a new SiPM of a nonplanar design, the TAPD. The TAPD concept is based on the properties of tip-like electrodes to focus and enhance the electric field, to reduce the breakdown voltage and cell capacitance, and to eliminate the needs in the borders for a peripheral separation of the APD cells. The tip-like electrodes are realized as conductive pillars with quasi-spherical tips located in the depth of an epitaxial layer. We gave a theoretical overview of the physical models and presented the metrological characterization of the existing prototypes. Our simulations are in good agreement with the obtained experimental data. With the TAPD of a small micro-cell pitch of 15 μm, we achieved a record PDE over a wide spectral range from 400 nm to 905 nm with the peak PDE of 73% at 600 nm in comparison with the state-of-the-art SiPMs of any pitch. In combination with a high DR for a short-light pulse detection due to high micro-cell density, a fast micro-cell recovery time of 4 ns provides a record-high DR for a long-light pulse and CW flux detection. A low-breakdown voltage of about 50V and a low-temperature coefficient of the breakdown voltage of 26 mV/K allows using the TAPD as conveniently as other modern SiPMs. However, the TAPD drawbacks are high DCR of about 1 Mcps and prompt optical crosstalk of about 30% but this level of noise is comparable with competing NIR-enhanced SiPMs. Certainly, we see a lot of space for considerable improvements of the TAPD design and performance in a near future and will be working hard on this promising wide-spectral high-dynamic-range and fast-timing SiPM.

## References


1. N. I. Golbraikh, A. F. Plotnikov, and V. E. Shubin, "Pulse avalanche photodetector based on a metal-insulator-semiconductor structure," *Kvantovaia Elektronika Moscow*, vol. 2, pp. 2624–2626, Dec. 1975.
2. A. B. Kravchenko, A. F. Plotnikov, and V. É. Shubin, "Feasibility of construction of a pulsed avalanche photodetector based on an MIS structure with stable internal amplification," *Soviet Journal of Quantum Electronics*, vol. 8, no. 9, pp. 1086–1089, Sep. 1978. [Online]. Available: https://doi.org/10.1070/qe1978v008n09abeh010725
3. D. A. Shushakov and V. E. Shubin, "New solid state photomultiplier," in *Optoelectronic Integrated Circuit Materials, Physics, and Devices*, M. Razeghi, Y.-S. Park, and G. L. Witt, Eds., vol. 2397, International Society for Optics and Photonics. SPIE, 1995, pp. 544–554. [Online]. Available: https://doi.org/10.1117/12.206900

4. V. E. Shubin and D. A. Shushakov, "Avalanche photodetectors," *Encyclopedia of Optical and Photonic Engineering*, pp. 121–141, 2003. [Online]. Available: https://doi.org/10.1081/E-EOE2-120047097
5. Z. Sadygov, "Three advanced designs of micro-pixel avalanche photodiodes: Their present status, maximum possibilities and limitations," in *Talk given at the 4th International Conference on New Developments in Photodetection (NDIP05)*, Beaune, France, 06 2005.
6. D. McNally and V. Golovin, "Review of solid state photomultiplier developments by CPTA and photonique sa," in *Talk given at the 5th International Conference on New Developments in Photodetection (NDIP08)*, Aix-le-Bains, France, 06 2008.
7. Z. Sadygov, A. Sadigov, and S. Khorev, "Silicon photomultipliers: Status and prospects," *Physics of Particles and Nuclei Letters*, vol. 17, no. 2, pp. 160–176, 2020. [Online]. Available: http://link.springer.com/10.1134/S154747712002017X
8. G. Bondarenko, P. Buzhan, B. Dolgoshein, V. Golovin, E. Guschin, A. Ilyin, V. Kaplin, A. Karakash, R. Klanner, V. Pokachalov, E. Popova, and K. Smirnov, "Limited Geiger-mode microcell silicon photodiode: New results," *Nuclear Instruments and Methods in Physics Research Section A: Accelerators, Spectrometers, Detectors and Associated Equipment*, vol. 442, no. 1, pp. 187–192, 2000. [Online]. Available: http://www.sciencedirect.com/science/article/pii/S016890029901219X
9. P. Buzhan, B. Dolgoshein, A. Ilyin, V. Kantserov, V. Kaplin, A. Karakash, A. Pleshko, E. Popova, S. Smirnov, Y. Volkov, L. Filatov, and S. Klemin, "An advanced study of silicon photomultiplier," *ICFA Instrumentation Bulletin*, vol. 23, pp. 28–41, 2001.
10. B. Dolgoshein, V. Balagura, P. Buzhan, M. Danilov, L. Filatov, E. Garutti, M. Groll, A. Ilyin, V. Kantserov, V. Kaplin, A. Karakash, F. Kayumov, S. Klemin, V. Korbel, H. Meyer, R. Mizuk, V. Morgunov, E. Novikov, P. Pakhlov, and I. Tikhomirov, "Status report on silicon photomultiplier development and its applications," *Nuclear Instruments and Methods in Physics Research Section A: Accelerators, Spectrometers, Detectors and Associated Equipment*, vol. 563, pp. 368–376, 06 2006. [Online]. Available: https://doi.org/10.1016/j.nima.2006.02.193
11. K. Sato, K. Yamamoto, K. Yamamura, S. Kamakura, and S. Ohsuka, "Application oriented development of multi-pixel photon counter (mppc)," in *IEEE Nuclear Science Symposuim Medical Imaging Conference*, 2010, pp. 243–245. [Online]. Available: http://ieeexplore.ieee.org/document/5873756/
12. A. G. Stewart, V. Saveliev, S. J. Bellis, D. J. Herbert, P. J. Hughes, and J. C. Jackson, "Performance of 1-mm$^2$ silicon photomultiplier," *IEEE Journal of Quantum Electronics*, vol. 44, no. 2, pp. 157–164, 2008. [Online]. Available: https://doi.org/10.1109/JQE.2007.910940
13. M. Mazzillo, G. Condorelli, D. Sanfilippo, G. Valvo, B. Carbone, G. Fallica, S. Billotta, M. Belluso, G. Bonanno, L. Cosentino, A. Pappalardo, and P. Finocchiaro, "Silicon photomultiplier technology at stmicroelectronics," *IEEE Transactions on Nuclear Science*, vol. 56, no. 4, pp. 2434–2442, 2009. [Online]. Available: https://doi.org/10.1109/TNS.2009.2024418
14. C. Piemonte, "A new silicon photomultiplier structure for blue light detection," *Nuclear Instruments and Methods in Physics Research Section A: Accelerators, Spectrometers, Detectors and Associated Equipment*, vol. 568, no. 1, pp. 224–232, 2006. [Online]. Available: http://www.sciencedirect.com/science/article/pii/S016890020601271X

15. P. Bérard, M. Couture, P. Deschamps, F. Laforce, and H. Dautet, "Characterization study of a new uv-sipm with low dark count rate," *Nuclear Instruments and Methods in Physics Research Section A: Accelerators, Spectrometers, Detectors and Associated Equipment*, vol. 695, pp. 35–39, 2012. [Online]. Available: http://www.sciencedirect.com/science/article/pii/S0168900211020821
16. C. Dietzinger, T. Ganka, W. Gebauer, N. Miyakawa, P. Iskra, and F. Wiest, "Silicon photomultipliers with enhanced blue-light sensitivity," in Talk given at PhotoDet, Seoul, Korea (South), 2012. [Online]. Available: https://indico.cern.ch/event/164917/session/0/contribution/18/material/slides/0.pdf
17. R. Mirzoyan, P. Buzhan, B. Dolgoshein, V. Kaplin, E. Popova, and M. Teshima, "SiPM: On the way at becoming an ideal low light level sensor," in Talk given at IEEE Nucl. Sci. Symp. Med. Imaging Conf., Knoxville, USA, 2010.
18. S. Vinogradov, T. Vinogradova, V. Shubin, D. Shushakov, and C. Sitarsky, "Efficiency of solid state photomultipliers in photon number resolution," *IEEE Transactions on Nuclear Science*, vol. 58, no. 1, PART 1, pp. 9–16, 2011. [Online]. Available: http://ieeexplore.ieee.org/document/5692134/
19. F. Acerbi, G. Paternoster, A. Gola, N. Zorzi, and C. Piemonte, "Silicon photomultipliers and single-photon avalanche diodes with enhanced NIR detection efficiency at FBK," *Nuclear Instruments and Methods in Physics Research Section A: Accelerators, Spectrometers, Detectors and Associated Equipment*, vol. 912, pp. 309–314, 2018, new developments in photodetection 2017. [Online]. Available: http://www.sciencedirect.com/science/article/pii/S0168900217313542
20. H. Sigmund, "Photoelectrical properties of spherical avalanche diodes in silicon," *Infrared Physics*, vol. 8, no. 4, pp. 259–264, 1968. [Online]. Available: http://www.sciencedirect.com/science/article/pii/0020089168900341
21. A. Gasanov, V. Golovin, Z. Sadygov, N. Yusipov, ""Avalanche semiconductor radiation detector", ru patent no. 1702831 (1989)." [Online]. Available: https://www1.fips.ru/fips_servl/fips_servlet?DB=RUPAT&DocNumber=1702831
22. Z. Y. Sadygov, I. M. Zheleznykh, N. A. Malakhov, V. N. Jejer, and T. A. Kirillova, "Avalanche semiconductor radiation detectors," *IEEE Transactions on Nuclear Science*, vol. 43, no. 3, pp. 1009–1013, 1996. [Online]. Available: http://ieeexplore.ieee.org/document/510748/
23. Z. Ya. Sadygov, ""Microchannel avalanche photodiode," ru patent no. 2316848 (2008)." [Online]. Available: http://www1.fips.ru/fips_servl/fips_servlet?DB=RUPAT&DocNumber=2316848
24. N. Anfimov, I. Chirikov-Zorin, A. Dovlatov, O. Gavrishchuk, A. Guskov, N. Khovanskiy, Z. Krumshtein, R. Leitner, G. Meshcheryakov, A. Nagaytsev, A. Olchevski, T. Rezinko, A. Sadovskiy, Z. Sadygov, I. Savin, V. Tchalyshev, I. Tyapkin, G. Yarygin, and F. Zerrouk, "Novel micropixel avalanche photodiodes (MAPD) with super high pixel density," *Nuclear Instruments and Methods in Physics Research Section A: Accelerators, Spectrometers, Detectors and Associated Equipment*, vol. 628, no. 1, pp. 369–371, 2011, vCI 2010. [Online]. Available: http://www.sciencedirect.com/science/article/pii/S0168900210015457
25. P. Iskra, S. Vinogradov, "Radiaton detector, method for producing a radiation detector and method for operating a radiation detector," European Patent Application EP3640682A1. [Online]. Available: https://patents.google.com/patent/EP3640682A1

26. E. Engelmann, W. Schmailzl, P. Iskra, F. Wiest, E. Popova, and S. Vinogradov, "Tip avalanche photodiode—a new generation silicon photomultiplier based on non-planar technology," *IEEE Sensors Journal*, vol. 21, no. 5, pp. 6024–6034, 2021. [Online]. Available: https://ieeexplore.ieee.org/document/9274364/
27. B. Jayant Baliga and S. K. Ghandhi, "Analytical solutions for the breakdown voltage of abrupt cylindrical and spherical junctions," *Solid-State Electronics*, vol. 19, pp. 739–744, 1976. [Online]. Available: https://doi.org/10.1016/0038-1101(76)90152-0
28. S. M. Sze and K. K. NG, *Physics of Semiconductor Devices*, 3rd ed. John Wiley & Sons, Inc, Hoboken, NJ, USA, 2007. [Online]. Available: https://doi.org/10.1002/0470068329
29. J. Crank and P. Nicolson, "A practical method for numerical evaluation of solutions of partial differential equations of the heat-conduction type," *Mathematical Proceedings of the Cambridge Philosophical Society*, vol. 43, no. 1, pp. 50–67, 1947. [Online]. Available: https://doi.org/10.1017/S0305004100023197
30. Silvaco, https://silvaco.com/tcad/
31. R. Van Overstraeten and H. De Man, "Measurement of the ionization rates in diffused silicon p-n junctions," *Solid-State Electronics*, vol. 13, no. 5, pp. 583–608, 1970. [Online]. Available: https://doi.org/10.1016/0038-1101(70)90139-5
32. R. J. McIntyre, "Multiplication noise in uniform avalanche diodes," *IEEE Transactions on Electron Devices*, vol. ED-13, no. 1, pp. 164–168, 1966. [Online]. Available: https://doi.org/10.1109/T-ED.1966.15651
33. H. Oide, H. Otono, S. Yamashita, T. Yoshioka, H. Hano, and T. Suehiro, "Study of afterpulsing of MPPC with waveform analysis," Proceedings of Science, 2007, international Workshop on New Photon-Detectors, PD 2007; Conference date: 27-06-2007 through 29-06-2007.
34. D. Renker and E. Lorenz, "Advances in solid state photon detectors," *Journal of Instrumentation*, vol. 4, no. 04, P04 004, 54 pages, Apr. 2009. [Online]. Available: https://doi.org/10.1088/1748-0221/4/04/P04004
35. E. Popova, "Evaluation of high UV sensitive SiPMs from MEPhI/MPI for use in liquid argon," PoS, vol. PhotoDet 2012, p. 034, 2013.
36. Product announcement NIR-SiPM, Broadcom, "Broadcom's nir sipm technology sets new performance standards for lidar." [Online]. Available: https://www.broadcom.com/products/optical-sensors/silicon-photomultiplier-sipm
37. Product Data Sheet RB-Series SiPM, on Semiconductor. [Online]. Available at: https://www.onsemi.com/pub/Collateral/MICRORB-SERIES-D.PDF
38. C. Piemonte, A. Ferri, A. Gola, A. Picciotto, T. Pro, N. Serra, A. Tarolli, and N. Zorzi, "Development of an automatic procedure for the characterization of silicon photomultipliers," in 2012 IEEE Nuclear Science Symposium and Medical Imaging Conference Record (NSS/MIC), 2012, pp. 428–432.
39. V. Chaumat, "Sipm PDE measurement with continuous and pulsed light," *PhotoDet 2012*, p. 058, 2013.
40. S. Vinogradov, "Precise metrology of sipm: Measurement and reconstruction of time distributions of single photon detections and correlated events," in 2016 IEEE Nuclear Science Symposium, Medical Imaging Conference and Room-Temperature Semiconductor Detector Workshop (NSS/MIC/RTSD), 2016, pp. 1–4.

41. E. Engelmann, "Sipm noise measurement with waveform analysis," in *Talk Given at International Conference on the Advancement of Silicon Photomultipliers (ICASiPM), Schwetzingen, Germany*, 2018.
42. Calibrated Si Photodiode - FDS100-CAL from Thorlabs. [Online]. Available at: https://www.thorlabs.com/thorproduct.cfm?partnumber=FDS100-CAL
43. A. N. Otte, D. Garcia, T. Nguyen, and D. Purushotham, "Characterization of three high efficiency and blue sensitive silicon photomultipliers," *Nuclear Instruments and Methods in Physics Research Section A: Accelerators, Spectrometers, Detectors and Associated Equipment*, vol. 846, pp. 106–125, 2017. [Online]. Available: http://www.sciencedirect.com/science/article/pii/S0168900216309901
44. A. Gola, F. Acerbi, M. Capasso, M. Marcante, A. Mazzi, G. Paternoster, C. Piemonte, V. Regazzoni, and N. Zorzi, "Nuv-sensitive silicon photomultiplier technologies developed at fondazione bruno kessler," *Sensors*, vol. 19, p. 308, 01 2019.
45. K. Yamamoto, "Recent development of mppc at hamamatsu for photon counting applications," in Talk Given at 5th International Workshop on New Photon Detectors (PD18), Tokyo, Japan, 2018.
46. P. Eckert, H.-C. Schultz-Coulon, W. Shen, R. Stamen, and A. Tadday, "Characterisation studies of silicon photomultipliers," *Nuclear Instruments and Methods in Physics Research Section A: Accelerators, Spectrometers, Detectors and Associated Equipment*, vol. 620, no. 2, pp. 217–226, 2010.
47. A. Gola, "Noise sources in silicon photomultipliers," in Talk Given at ICASIPM, Schwetzingen, Germany, 2018. [Online]. Available: https://indico.gsi.de/event/6990/contributions/31520/attachments/22639/28400/ICASiPM_2018_-_A_Gola_-_Noise_Sources_in_SiPMs_-_v5.pdf
48. E. Engelmann, *"Dark Count Rate of Silicon Photomultipliers - Metrological Characterization and Suppression,"* Cuvillier, Göttingen, 2018.
49. Product Data Sheet PM1125-WB-B0. [Online]. Available at: https://www.ketek.net/wp-content/uploads/KETEK-PM1125-WB-B0-Datasheet.pdf
50. K. O'Neill and C. Jackson, "SensL B-Series and C-Series silicon photomultipliers for time-of-flight positron emission tomography," *Nuclear Instruments and Methods in Physics Research, Section A: Accelerators, Spectrometers, Detectors and Associated Equipment*, vol. 787, pp. 169–172, 2015.
51. R. Pagano, D. Corso, S. Lombardo, G. Valvo, D. N. Sanfilippo, G. Fallica, and S. Libertino, "Dark current in silicon photomultiplier pixels: Data and model," *IEEE Transactions on Electron Devices*, vol. 59, no. 9, pp. 2410–2416, 2012.
52. R. Klanner, "Characterisation of SIPMS," *Nuclear Instruments and Methods in Physics Research Section A: Accelerators, Spectrometers, Detectors and Associated Equipment*, vol. 926, pp. 36–56, 2019, silicon photomultipliers: Technology, Characterisation and Applications. [Online]. Available: http://www.sciencedirect.com/science/article/pii/S0168900218317091
53. G. Kawata, K. Sasaki, and R. Hasegawa, "Avalanche-area dependence of gain in passive-quenched single-photon avalanche diodes by multiple-photon injection," *IEEE Transactions on Electron Devices*, vol. 65, no. 6, pp. 2525–2530, 2018.
54. V. Chmill, E. Garutti, R. Klanner, M. Nitschke, and J. Schwandt, "Study of the breakdown voltage of SIPMS," *Nuclear Instruments and Methods in Physics Research Section A: Accelerators, Spectrometers, Detectors and Associated Equipment*,

vol. 845, pp. 56–59, 2017, *proceedings of the Vienna Conference on Instrumentation 2016*. [Online]. Available: http://www.sciencedirect.com/science/article/pii/S016890021630256X

55. S. Vinogradov, A. Arodzero, R. Lanza, and C. Welsch, "SIPM response to long and intense light pulses," *Nuclear Instruments and Methods in Physics Research Section A: Accelerators, Spectrometers, Detectors and Associated Equipment*, vol. 787, pp. 148–152, 2015, new developments in photodetection NDIP14. [Online]. Available: http://www.sciencedirect.com/science/article/pii/S0168900214013850
56. E. Garutti, "Silicon photomultipliers for high energy physics detectors," *Journal of Instrumentation*, vol. 6, no. 10, C10003, 16 pages, Oct. 2011. [Online]. Available: http://dx.doi.org/10.1088/1748-0221/6/10/C10003
57. E. Popova, "Silicon photomultiplier: application for high-granularity scintillator calorimeters," *IEEE Nuclear Science Symposium Conference Record*, vol. 1, pp. 102–106, 2004. [Online]. Available: http://ieeexplore.ieee.org/document/1462077/
58. S. Gomi, T. Nakaya, M. Yokoyama, H. Kawamuko, T. Nakadaira, and T. Murakami, "Research and development of MPPC for T2K experiment," *Proceedings of Science*, vol. PD07, p. 015, 2006.
59. S. Vinogradov and E. Popova, "Status and perspectives of solid state photon detectors," *Nuclear Instruments and Methods in Physics Research Section A: Accelerators, Spectrometers, Detectors and Associated Equipment*, vol. 952, p. 161752, 2020, 10th International Workshop on Ring Imaging Cherenkov Detectors (RICH 2018). [Online]. Available: http://www.sciencedirect.com/science/article/pii/S0168900218318813
60. P. Lecoq, "Pushing the limits in time-of-flight pet imaging," *IEEE Transactions on Radiation and Plasma Medical Sciences*, vol. 1, no. 6, pp. 473–485, 2017. [Online]. Available: https://ieeexplore.ieee.org/document/8049484/
61. S. Vinogradov, "Approximations of coincidence time resolution models of scintillator detectors with leading edge discriminator," *Nuclear Instruments and Methods in Physics Research Section A: Accelerators, Spectrometers, Detectors and Associated Equipment*, vol. 912, pp. 149–153, 2018, new developments in photodetection 2017. [Online]. Available: http://www.sciencedirect.com/science/article/pii/S0168900217311956
62. S. Vinogradov, L. Devlin, E. Nebot del Busto, M. Kastriotou, and C. Welsch, "Challenges of arbitrary waveform signal detection by SIPM in beam loss monitoring systems with cherenkov fibre readout," *Proceedings of Science*, vol. 6-9-July-2015, no 70, pp. 4–8, 2015. [Online]. Available: http://pos.sissa.it/archive/conferences/252/070/PhotoDet2015_070.pdf

# 10

# *Silicon Photomultiplier-Based Scintillation Detectors for Photon-Counting X-Ray Imaging*

**Stefan J. van der Sar and Dennis R. Schaart**

**CONTENTS**



## 10.1 General Introduction

Medical imaging using X-rays comes in two flavors: projection radiography and computed tomography (CT). In both modalities, two-dimensional projection images are obtained by sending a poly-energetic X-ray beam (diagnostic energy range: 25–150 keV) through a patient and measuring the attenuation of the X-ray beam intensity using a pixelated, energy-integrating detector. Whereas projection radiography is based on a single projection

DOI: 10.1201/9781003219446-10

image, CT relies on projection images obtained from many angles around the patient, which are subsequently used to reconstruct cross-sectional slices of the patient. This enables three-dimensional imaging (Prince and Links 2015).

Image contrast is based on differences in the linear attenuation coefficient among tissues in the human body. As shown in Figure 10.1, air (e.g., in the lungs), bone tissue, and soft tissue can be particularly well distinguished in this way. However, the differences in the linear attenuation coefficient among various soft tissues are much smaller. Therefore, contrast agents based on elements with a relatively high atomic number, such as iodine and barium, are often used to make certain anatomical structures more visible (Prince and Links 2015). While X-ray imaging is one of the most widely used medical

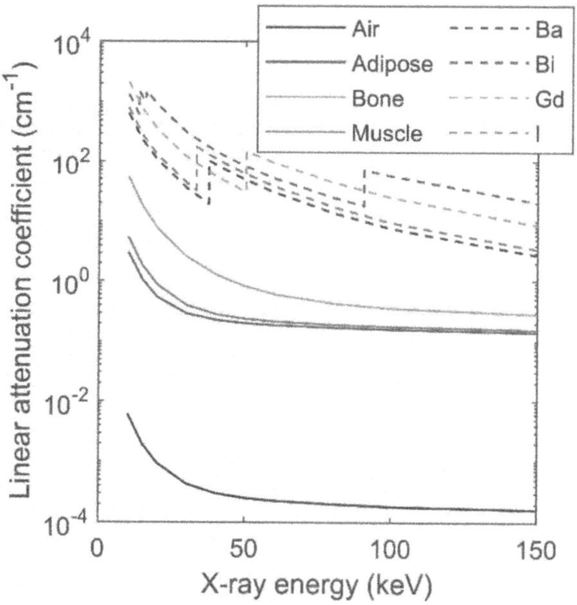


**FIGURE 10.1**
(1) Contrast in medical X-ray imaging is based on differences in the linear attenuation coefficients of tissues. Therefore, air (e.g., in the lungs), soft tissues (e.g., adipose and muscle tissue), and bone can be well distinguished from each other. The contrast between two soft tissues is limited, however. Therefore, contrast agents based on iodine (I) and barium (Ba) are commonly administered to enhance contrast. (2) The differences in the linear attenuation coefficients tend to increase for decreasing X-ray energy, which gives photon-counting detectors an advantage over energy-integrating detectors in terms of achievable image contrast for a given radiation dose or contrast agent load. (3) Each contrast agent adds a K-edge discontinuity to the overall linear attenuation coefficient. Therefore, material decomposition in the presence of $n$ contrast agents requires photon-counting detectors that group the measured X-ray photons into $2+n$ energy bins. (4) Since this allows to quantify multiple contrast agents at the same time, new contrast agents for X-ray imaging, e.g., based on gadolinium (Gd) and bismuth (Bi), are under investigation (Tao et al. 2019). Data were obtained from the NIST databases 8 and 126.

imaging modalities, it has not reached its full clinical potential yet, in particular due to the use of energy-integrating detectors (Flohr et al. 2020).

## 10.2 Energy-Integrating Detectors

In an X-ray detector, the energy deposited by an X-ray photon is ultimately converted into a number of electric charges, together forming a small current pulse. For imaging purposes, detectors must be pixelated. Operation in energy-integrating mode means that each pixel has its own read-out channel in which all current pulses are integrated during a certain exposure time. At the end of this time window, the integrated signal is digitized. This output signal is related to the total charge generated in the pixel during the exposure time and therefore to the total energy deposition by the X-ray photons. Dividing the signal measured with a patient in the beam by the signal measured without patient yields the attenuation of the X-ray beam intensity (Prince and Links 2015).

Higher energy X-ray photons typically generate more charge and stronger current pulses than lower energy X-ray photons. Therefore, they contribute more to the output signal. This leads to suboptimal image contrast, because the differences in linear attenuation coefficients among tissues and contrast agents are more pronounced at lower X-ray energies, as can be appreciated from Figure 10.1. It also leads to a reduction of the signal-to-noise ratio (SNR) as described by the Swank factor (Swank 1973). In addition, energy-integrating detectors integrate all noise during the exposure time, further reducing the SNR. As a result, the contrast-to-noise ratio (CNR), i.e., the product of image contrast and SNR, of X-ray imaging systems equipped with energy-integrating detectors is suboptimal. This also implies that unnecessarily high radiation doses and/or contrast agent loads are delivered to patients in order to obtain the desired image quality.

## 10.3 Photon-Counting Detectors

The opposite of an energy-integrating detector is a photon-counting detector, which measures the total number of detected X-ray photons in each pixel during the exposure time instead of the total energy deposited. The number of counts is incremented by one every time the train of detector pulses crosses a certain threshold. In this operating mode, all X-ray photons equally contribute to the output signal, regardless of their energy. In addition, noise does not contribute to the output signal, provided that the threshold is set above the noise level. These two facts increase both the contrast and the SNR of the resulting image. Consequently, photon-counting detectors offer superior CNR compared to energy-integrating detectors. This feature can be used to improve image quality on the one hand, or to reduce radiation dose and/or contrast agent load on the other hand.

### 10.3.1 Dual-Energy Photon-Counting Detectors

If a second, higher threshold is implemented in the photon-counting detector, it will be possible to discriminate the lower- and higher-energy X-ray photons from each other. In this way, all the opportunities of dual-energy X-ray imaging become available for photon-counting X-ray imaging, as well (McCollough et al. 2015). However, photon-counting systems offer the additional advantages of a simple system design, i.e., a single, conventional X-ray tube and a single detector, and a truly simultaneous acquisition of the lower- and the higher-energy projection data. Furthermore, the degree of spectral separation between both data sets is only limited by the spectral response of the detector.

### 10.3.2 Multi-Energy Photon-Counting Detectors

If more than two thresholds are implemented in the photon-counting detector, the detected X-ray photons can be assigned to more than two energy bins and multi-energy X-ray imaging becomes available. This finds application in generating quantitative, material-specific images, which is commonly called material decomposition. Since the dimensionality of the linear attenuation coefficient of materials naturally occurring in the human body is likely two in the diagnostic energy range (photoelectric and Compton interactions), a measurement at two energies (i.e., dual-energy imaging) is usually sufficient for material decomposition (Alvarez and Macovski 1976). However, contrast agents typically have a K-edge discontinuity in their linear attenuation coefficient that falls within the diagnostic energy range (see Figure 10.1). This increases the dimensionality of the linear attenuation coefficient to $2+n$ with $n$ the number of such contrast agents present in the field-of-view. Consequently, measurements at $2+n$ different energies (i.e., $2+n$ thresholds) are required for material decomposition. Such an imaging technique is called K-edge imaging and may be enabled by photon-counting detectors (Roessl and Proksa 2007).

The use of two or more thresholds also allows to further improve the image contrast by giving a higher weight to the lower energy X-ray photons. However, weighting some photons more than others also leads to a reduction of the SNR, like in energy-integrating detectors (Danielsson, Persson, and Sjölin 2020). Therefore, sets of weights that optimize the CNR for specific imaging tasks should be found.

### 10.3.3 Requirements on Photon-Counting Detectors

Given the exciting advantages of photon-counting detectors, why is this type of detector not widely implemented in medical X-ray imaging systems yet? An important reason is the very high X-ray fluence rate incident on the detector, such that pulse pile-up can occur. In case of pile-up, two or more X-ray photons are detected in the same pixel very shortly after one another, such that the resulting detector pulses are superimposed on each other. This distorts the measurement in two ways. Firstly, some X-ray photons may not be registered as a count. Consequently, the mean number of counts during a certain exposure time is no longer proportional to the incident X-ray fluence rate and the variance in the number of counts during that exposure time does not follow Poisson statistics anymore. More importantly, pile-up can cause a count to end up in the wrong energy bin. This worsens the spectral separation between energy bins and the material decomposition performance.

Persson et al. (2016) studied in detail the exact values of the X-ray fluence rate that can be expected from standard clinical CT protocols. They found a maximum fluence rate of about 350 Mcps/mm$^2$ provided that the patient is properly aligned. However, they also found that such high rates exclusively occur in projection lines that only pass through some skin, which are not the most relevant projection lines from an imaging perspective. The fluence rate in other projection lines can be considerably lower. Nevertheless, it means that the pixel size must be reduced to $0.5 \times 0.5$ mm$^2$ or smaller to mitigate the negative effects of pile-up even if the detector pulses last only tens of nanoseconds.

However, the smaller the pixel size, the more likely the escape of characteristic and Compton-scattered X-ray photons from a detector pixel. If these secondary X-rays leave the detector, they will only cause an incorrect energy measurement. However, they can also be absorbed in another pixel (X-ray crosstalk). In this case, not only the energy measurement is distorted, but also two or more counts are registered for a single incident X-ray photon. This phenomenon is referred to as count multiplicity. It can be seen as number-weighting of X-ray photons and leads to a reduction of the (zero-frequency) SNR (Michel et al. 2006). It also leads to pixel-to-pixel correlations and a non-white noise power spectrum (Xu et al. 2014). In addition, X-ray crosstalk increases the pile-up level in a pixel for a given incident rate of X-ray photons and has a small negative effect on the spatial resolution (Danielsson, Persson, and Sjölin 2020).

Furthermore, an X-ray photon-counting detector should have sufficient energy resolution in the diagnostic energy range to assign a large fraction of the X-ray photons that are fully absorbed in a single pixel to the correct energy bin. High density of the detection material is also preferred, because it allows to use a thin (few mm) detector to absorb X-ray photons in a dose-efficient way. Thin detectors are also beneficial from a spatial resolution point-of-view. Lastly, the detector should operate well at room temperature.

## 10.4 CdTe/CZT Detectors

The requirements of short pulse duration and good energy resolution point in the direction of semiconductor detectors. CdTe- and CdZnTe(CZT)-based detectors are particularly interesting, because they can be operated at room-temperature and have high density (5.8 g $cm^{-3}$). Such detectors directly convert an X-ray photon into a set of electron-hole pairs. The electrons and holes drift to opposite electrodes under the influence of an electric field (bias voltage 800–1000 V, detector thickness ~2 mm) on which they induce current pulses. Only pixilation of the anode is required. The detection layer can be monolithic. This allows for relatively easy fabrication of sub-mm pixels. See Figure 10.2(a) for a schematic overview of this detector concept. Recent progress in the growth/synthesis of these materials and in detector electronics has brought photon-counting medical X-ray imaging within reach (Flohr et al. 2020). Some medical device companies have recently built prototype photon-counting CT scanners, which are used to evaluate the theoretical benefits of photon-counting detectors in clinical practice.

Two Siemens prototypes are equipped with 1.6 mm thick CdTe detectors. The anodes consist of $0.225 \times 0.225$ $mm^2$ pixels, which are all connected to their own, in-house developed application-specific integrated circuit (ASIC) that has two thresholds. Pixels can be combined, such that the system is operated in high-resolution mode ($2 \times 2$ pixels) or in macro-mode ($4 \times 4$ pixels) (Yu et al. 2016). The latter yields an effective pixel size of $0.9 \times 0.9$ $mm^2$,

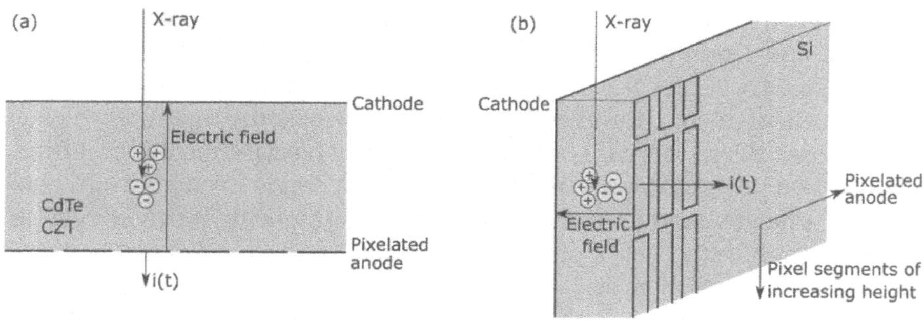


**FIGURE 10.2**
(a) Schematic cross-section of a face-on CdTe/CZT detector with pixelated anode contacts. The interaction of an X-ray photon results in ~200–250 electron-hole pairs per keV of deposited energy, but only three are shown here. The electrons and holes drift to opposite electrodes under the influence of an externally applied electric field. The current pulse *i(t)* induced on the anodes is fed into an ASIC. (b) Schematic overview of an edge-on silicon detector, which is being investigated as an alternative to face-on CdTe/CZT detectors. Edge-on means that the semiconductor wafer is put on its side, such that the electric field is perpendicular to the direction of the impinging X-rays. The pixelated anode can be divided in segments of increasing height. A current pulse *i(t)* will be induced in one of the pixel segments, depending on the depth-of-interaction of the X-ray photon.

comparable to the pixel size in energy-integrating CT detectors, whereas the high-resolution mode potentially allows for improved spatial resolution. However, spatial resolution is also determined by the focal spot size of the X-ray tube, for example. Reducing the spot size usually means that the tube current has to be reduced, as well. This may limit the applicability of improved spatial resolution in CT imaging (Flohr et al. 2020). Different threshold values can be assigned to the pixels within a group of $2 \times 2$ or $4 \times 4$ pixels, such that the effective number of energy bins is increased. However, this comes at the cost of some dose-efficiency (Yu et al. 2016). This is not an issue anymore in a more recent Siemens prototype that has $0.275 \times 0.322$ mm$^2$ pixels, all of which are equipped with four thresholds (Flohr et al. 2020).

Philips has built a prototype scanner based on a 2 mm thick CZT detector. The anode is subdivided into pixels of $0.5 \times 0.5$ mm$^2$, which are coupled to an in-house developed ASIC (ChromAIX2) with five thresholds (Steadman, Herrmann, and Livne 2017).

### 10.4.1 CdTe/CZT Detector Challenges

Challenges in the development of CdTe/CZT detectors can be split up in three categories. The first one is related to the small pixel size and applies to any semiconductor detector. The charge cloud resulting from an X-ray interaction in the CdTe/CZT layer grows along its way to the electrodes, due to diffusion and mutual repulsion of the charges. Consequently, if an X-ray photon deposits (part of) its energy in the proximity of a virtual pixel border in the detection layer, it is likely that a current pulse will be induced in more than one pixel. This phenomenon is referred to as charge sharing. It leads to a broadening of the low-energy side of the full-energy peak in a gamma spectrum and therefore to a degradation of the FWHM energy resolution of the detector. The other consequences of charge sharing are the same as those of X-ray crosstalk described in Section 10.3.3. Charge sharing becomes more severe as the detector thickness increases and/or the pixel size decreases. Hence, not only X-ray crosstalk, but also charge sharing discourages further reduction of the pixel size for rate capability reasons. In this respect, it is interesting to note that Philips previously used a 3 mm thick CdTe detector with $0.3 \times 0.3$ mm$^2$ pixels (Steadman et al. 2011). However, a 2 mm thick CZT detector with a pixel size of $0.5 \times 0.5$ mm$^2$ was later selected, because of better spectral performance (Steadman, Herrmann, and Livne 2017). For example, FWHM energy resolutions of 18% and 8% at 60 keV were reported for the $0.3 \times 0.3$ mm$^2$ and $0.5 \times 0.5$ mm$^2$ pixels, respectively.

The second challenge is related to the intrinsically weak raw signal from a semiconductor detector. In the order of $10^2$ electron-hole pairs are created per keV of deposited X-ray energy in CdTe and CZT. Therefore, the induced current pulses are one-by-one integrated and amplified using a charge-sensitive amplifier. The next step in the pulse processing chain is a pulse-shaping circuit, which outputs (semi-)Gaussian pulses (see Figure 10.4) (Ballabriga et al. 2020).

In the case of the Philips prototype, for example, these pulses have an FWHM of 20 ns (Steadman et al. 2011, Steadman, Herrmann, and Livne 2017). A number of pulse-height discriminators and associated counters are used to obtain energy-resolved photon-counting data. Each detector pixel must be one-to-one coupled to such an elaborate read-out channel in an ASIC. Ideally, there is a direct physical interconnection between the pixel and its read-out channel and the pixel pitch and the pitch of the read-out channels are the same (hybrid pixel technology). This offers low noise and high bandwidth for a given power consumption. However, the costs of the interconnection process are high for relatively small volumes (Ballabriga et al. 2020).

Lastly, there exist the issues of relatively low hole mobility (about ten times lower than the electron mobility) and charge trapping in CdTe and CZT. Especially holes are likely to get trapped in defects and impurities in the CdTe/CZT layer while drifting to the cathode. Frequently-mentioned defects in literature include (sub-)grain boundary networks and Te inclusions (Roy et al. 2018, Roy et al. 2019). Low hole mobility and trapping affect the detector pulse characteristics. If defects are not uniformly distributed throughout the detector, the pulse characteristics will be pixel-dependent and pixel-specific calibrations will be needed. Low hole mobility and trapping are also the reasons why a net positive charge may be present in the CdTe/CZT layer when the next X-ray photon arrives. This is called polarization and it distorts the electric field in the detector, which also affects the detector pulse characteristics. Polarization becomes more severe at higher count rates, when a larger net positive charge builds up in the CdTe/CZT layer. Hence, the pulse characteristics of a pixel may depend on the irradiation history of that pixel. As a consequence, the calibrations may not hold anymore during CT imaging, which deteriorates the measurement of counts and energy. This may lead to ring artifacts and less accurate material decomposition, for example (Flohr et al. 2020, Sjölin and Danielsson 2016). Therefore, stable and reliable detector performance requires CdTe/CZT with a very low density of trapping centers, uniformly distributed throughout the layer. Cost-effective production of large-area detectors of reasonable quality remains a challenge, despite substantial improvements over the past decades. However, the intrinsically, poor thermophysical properties of CdTe/CZT near and below the melting point, which are responsible for (some of) the earlier-mentioned defects, may limit further improvements (Roy et al. 2018, Roy et al. 2019). In addition, it is unclear how continuous exposure to high X-ray fluence rates affects CdTe/CZT detector performance in the long term (Taguchi and Iwanczyk 2013).

A semiconductor detector can be equipped with Schottky (blocking) or Ohmic (conducting) electrode contacts. Considerable polarization has been observed at relatively low X-ray fluence rates in detectors with Schottky contacts (Herrmann et al. 2020). It has therefore been proposed to use Ohmic cathode contacts instead, because such contacts inject electrons into the CdTe/CZT layer when there is a net positive charge, thereby combatting polarization. However, Herrmann et al. (2020) observed that this results in the variance of the number of observed counts exceeding the mean of the number of observed counts in a given time interval, which they consider unacceptable for X-ray imaging. They explain how the injection of electrons leads to a baseline shift that is not constant in time, but contains higher frequency components, regardless of the exact injection mechanism. Such fast changes in the baseline can lead to counting threshold crossings, thereby disturbing the Poisson statistics of the number of observed counts. Using a baseline restorer does not solve the problem, because the restorer can only eliminate slow baseline changes. Otherwise, it would also eliminate the X-ray photon-induced pulses.

## 10.5 Silicon Detectors

Another type of room-temperature semiconductor detector has been proposed and developed by researchers from KTH Royal Institute of Technology, Stockholm, Sweden (Bornefalk and Danielsson 2010). This detector is based on silicon, a very mature semiconductor material that can be produced in a cost-effective way thanks to its widespread use in the electronics industry. The density of trapping centers in Si must be considerably lower than in CdTe and CZT, given that the mobility-lifetime product is three to four orders of magnitude better, while the mobilities are in the same order of magnitude (Ballabriga et al. 2020). This means hole trapping and polarization effects are of much less concern.

On the other hand, the density (2.3 g $cm^{-3}$) and atomic number (14) of silicon are rather low. This means that a few centimeters (i.e., 3–6 cm) of material are needed to absorb in a dose-efficient way X-ray photons in the diagnostic energy range, which could have negative consequences for the spatial resolution of the detector. The required thickness is achieved by placing silicon wafers with a thickness of 0.5 mm on their sides. As shown in Figure 10.2(b), the electrode configuration is such that charge carriers drift perpendicular to the direction of the incident X-ray photons. In this way, charge carriers do not need to travel several centimeters through the material. Furthermore, multiple anode contacts with a constant width of 0.4 mm and an exponentially increasing length can be placed along the depth direction of a pixel. This geometry dramatically increases the rate capability of the detector and practically reduces charge sharing to only one dimension, but it also increases the power consumption per unit detector area to a level much higher than that of other photon-counting and conventional energy-integrating detectors (Ballabriga et al. 2020). The fact that hybrid pixel detector technology cannot be used and relatively long interconnections between anode and ASIC are needed further increases the power consumption. The associated additional demand for active cooling increases the operating costs (Sundberg et al. 2020). The relatively low density and atomic number of silicon also imply that Compton scattering is the dominant X-ray interaction mechanism in the detector. The initial X-ray energy deposition is therefore smaller and the first counting threshold has to be set lower than in CdTe/CZT detectors for dose-efficient X-ray detection. This puts heavier requirements on the noise performance of the ASIC. X-ray crosstalk is also more of a concern. In order to reduce it, 20 μm thick tungsten foils can be placed between the wafers (Bornefalk and Danielsson 2010).

Recently, a prototype photon-counting CT scanner based on this technology was built (da Silva et al. 2019). The technology seems to be rapidly gaining interest in the CT community. All in all, two different detector concepts are under investigation (Figure 10.2), both of which have their advantages and challenges. Thus, it currently remains unclear what the best choice is, which leaves room for the development of alternative detector concepts.

## 10.6 Scintillation Detectors with Photon-Counting Capability

Scintillation detectors could be one such alternative detector concept for photon-counting X-ray imaging. These detectors are based on the principle of indirect conversion: an energy deposition by an X-ray photon in a scintillator leads to a pulse of optical photons, which is in turn converted into a current pulse by a light sensor (see Figure 10.3(a)). Scintillation detectors are widely applied in medical X-ray imaging systems operating in energy-integrating mode. This could be the reason why scintillation detectors are sometimes presented as synonymous to energy-integrating detectors in the medical X-ray photon-counting literature. However, the operating mode is mainly determined by the way the pulses from the detector are processed. There is no fundamental reason why scintillation detectors cannot be operated in photon-counting mode. In fact, photon-counting scintillation detectors are widely employed in nuclear medicine imaging systems, such as single-photon emission computed tomography (SPECT) and positron emission tomography (PET).

### 10.6.1 Choice of Light Sensor

The light sensor in an energy-integrating scintillation detector for medical X-ray imaging is usually a photodiode, which generates one electron-hole pair per detected optical photon. Since bright scintillators generate in the order of tens of photons per keV of deposited X-ray energy, the resulting current pulse contains tens of charge carriers per keV. Note that this is about an order of magnitude smaller than the amount of charge produced per keV in

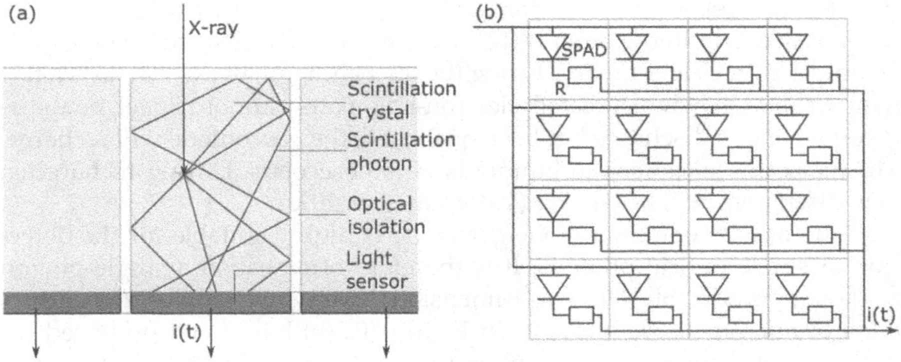


**FIGURE 10.3**
(a) Schematic cross-section of a pixelated scintillation detector for photon-counting X-ray imaging. Each scintillation crystal is one-to-one coupled to a light sensor, such as a silicon photomultiplier (SiPM), which converts the pulse of optical photons generated by the interaction of an X-ray photon in the scintillator into a current pulse *i(t)*. Only five optical photons are shown here, although typically in the order of $10^2$–$10^3$ scintillation photons are generated for medical X-ray energies. Reflective optical isolation is needed to guide the photons toward the light sensor and prevent light sharing with nearby pixels. (b) Schematic top view of an SiPM, i.e., a two-dimensional array of single-photon avalanche diodes (SPADs). The SPADs on a single SiPM are connected in parallel and each one is equipped with a quenching resistor (*R*). Only a $4 \times 4$ array of SPADs is shown here. Practical SiPMs typically consist of $10^2$–$10^5$ SPADs. Figure from van der Sar, Brunner, and Schaart (2021).

Si, CdTe, and CZT. In scintillation detectors based on conventional photodiodes, it is therefore challenging to get the signal from a single X-ray photon above the noise level of the electronics, as is required to operate the detector in photon-counting mode.

Fortunately, there exist light sensors with internal gain. These multiply the initial number of charge carriers by several orders of magnitude. For example, a photomultiplier tube (PMT) multiplies photoelectrons emitted from a photocathode via a series of dynodes in a vacuum tube. This yields a gain in the order of $10^6$–$10^8$. However, PMTs are bulky and a reduction of their size to the sub-mm level, as required to handle the high incident X-ray fluence rate, is challenging.

Another way to achieve internal gain is to increase the bias voltage applied to a photodiode, such that avalanche multiplication of the charge carriers takes place. The gain of such an avalanche photodiode (APD) can be high enough (typically in the order of $10^2$–$10^3$) to enable X-ray photon-counting with scintillators. In addition, the dimensions of APDs can be miniaturized to the sub-mm scale. A drawback is that the gain strongly depends on bias voltage and temperature, so accurate stabilization of both is required for reliable detector operation.

If the bias voltage is further increased, such that it exceeds the breakdown voltage of the photodiode, a single electron-hole pair is multiplied without limit (discharge). A series resistor can be used to quench the discharge, such that a stable gain in the order of $10^6$ is achieved. In other words, a diode can be operated in Geiger mode. During the quenching phase, which lasts a few nanoseconds, newly generated electron-hole pairs cannot trigger new discharges. Once a discharge has been quenched, the photodiode will recharge. This typically takes tens to hundreds of nanoseconds. During recharging, new discharges will be weaker (Van Dam et al. 2010).

A photodiode operated in Geiger mode is highly suitable for the detection of single optical photons. It is therefore often called a single-photon avalanche diode (SPAD). Two-dimensional arrays of small-sized SPADs (tens of micrometers) as shown in Figure 10.3(b) have been proposed for detecting the pulses of optical photons from scintillators. Such an array is called a SiPM. SiPMs can be miniaturized to the sub-mm level and provide high internal gain, with limited dependency on bias voltage and temperature. SiPMs can therefore be considered a suitable light sensor for X-ray photon-counting scintillation detectors.

### 10.6.2 Relevant Scintillator Properties

Photon-counting X-ray imaging puts heavy requirements on both the scintillator and the SiPM. Regarding the scintillator, requirements such as high density and room-temperature operation are relatively easy to fulfill if one uses inorganic scintillation crystals. However, a short pulse duration, sub-mm pixel size, and reasonable energy resolution may be more difficult to achieve. The scintillator properties that determine these performance parameters are described here.

Inorganic scintillators are transparent insulators, in which the energy deposition by an X-ray photon leads to the creation of hot electrons (photoelectrons, Auger electrons, and Compton electrons), which lose their energy by exciting other electrons from the valence band to the conduction band. The excited electrons freely migrate through the scintillator, but are quickly trapped in luminescence centers. There, they recombine under emission of an optical photon. The lifetime of an electron in a luminescence center determines the decay time constant of the scintillator, which characterizes the exponentially decaying pulse of optical photons following the absorption of an X-ray photon. Hence, it takes 2.3 decay time constants before the pulse intensity has decayed to 10% of the maximum intensity, for example. For completion, it is to be noted that some scintillators have multiple decay components, the decay time constants of which may differ substantially. For high-count rate applications such as photon-counting X-ray imaging, a short decay time constant is crucial, whereas the presence of additional slow decay components is undesirable, because it leads to a rate-dependent shift of the baseline signal.

The energy resolution $R$ of a scintillation detector is mainly determined by two components (Dorenbos et al. 1995). The first one is related to the statistics of optical photon detection ($R_{stat}$), whereas the second one is called the scintillator's intrinsic resolution ($R_{intr}$). $R$ can be obtained by Pythagorean addition: $R^2 = R_{stat}{}^2 + R_{intr}{}^2$.

$R_{stat}$ will in general improve if more optical photons are generated and detected. This means a high light yield [photons/keV] and good light collection efficiency contribute to a good energy resolution. However, the optical photon emission is isotropic and the light sensor, in this case the SiPM, is usually optically coupled to only one side of a scintillation crystal. High light collection efficiency therefore requires that the optical photons are guided toward the light sensor in an efficient way. As shown in Figure 10.3(a), this is traditionally achieved by inserting reflectors in between, and on top of, the individual scintillation crystals of a 2D pixel array. The resulting optical isolation of each pixel also prevents light sharing among pixels (cf. charge sharing in semiconductor detectors). However, the X-ray insensitive area occupied by the reflectors can be relatively large for sub-mm pixels, such that the geometric detection efficiency, and therefore the dose efficiency, are reduced. Hence, thin (tens of micrometers) reflectors, or more innovative optical isolation techniques, are required for scintillator-based X-ray photon-counting detectors.

$R_{intr}$ is mainly determined by the non-proportionality of the scintillator, which refers to the light yield being a function of energy. A non-proportionality factor at the energy-of-interest should be multiplied by the reported light yield at 662 keV to obtain the light yield at the energy-of-interest. The stronger the deviation from ideal (i.e., proportional) behavior, the worse this component of the energy resolution. This originates from the fact that different incident X-ray photons of the same energy yield different combinations of secondary electron energies, which will give rise to different numbers of optical photons if the light yield is a function of energy. This increases the variance in the number of detected optical photons, negatively affecting the energy resolution. Other factors that contribute to the energy resolution are an inhomogeneous distribution of luminescence centers throughout the scintillation crystal and excess variance in the light collection efficiency for detected X-rays of a given energy. However, these two contributions should be limited for commercially grown crystals of small dimensions, as required for scintillator-based X-ray photon-counting detectors.

### 10.6.3 Relevant SiPM Properties

Not only the scintillator, but also the SiPM affects the pulse duration and the energy resolution. Each SPAD on the SiPM generates a current pulse in response to the detection of an optical photon. This single-SPAD response can be described as a short spike of a few nanoseconds corresponding to the discharge and quenching phases, followed by an exponential decay corresponding to the recharge phase (Marano et al. 2014).

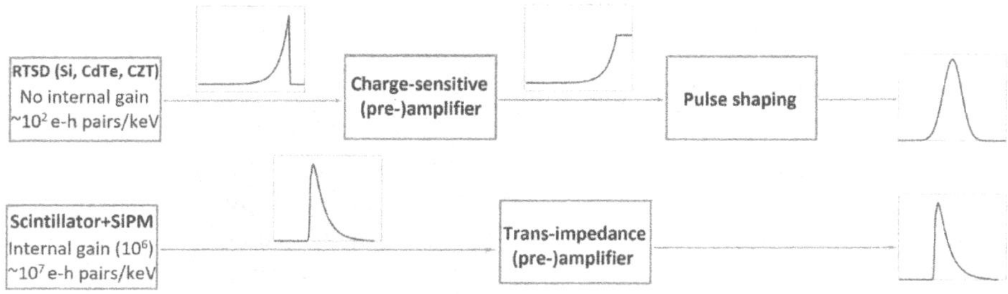


**FIGURE 10.4**
The top row shows the typical shape of the raw current pulse from a room-temperature semiconductor detector (RTSD). This is a relatively weak pulse, because only ~$10^2$ electron-hole (e-h) pairs are generated per keV of deposited X-ray energy. The raw pulse must therefore be processed by a charge-sensitive amplifier, which integrates and amplifies the pulse, and a pulse shaping circuit, which typically outputs (semi-)Gaussian shaped voltage pulses. The bottom row shows the typical shape of the raw current pulse from an SiPM-based scintillation detector. Due to the internal gain of the SiPM, ~$10^7$ e-h pairs are generated per keV of deposited energy. Such a relatively strong current pulse can simply be converted into a voltage pulse using a trans-impedance amplifier, which maintains the original pulse shape.

The exponential decay is characterized by the recharge time constant. Since all SPADs on a single SiPM are connected in parallel (see Figure 10.3(b)), the detector pulse can be approximated by a convolution of the scintillator's light pulse and the single-SPAD response. Hence, the recharge time constant will have a substantial effect on the pulse duration, if it is comparable to or larger than the decay time constant. The typical shape of a pulse from an SiPM-based scintillation detector is shown in Figure 10.4. This figure also shows that due to the high internal gain of the SiPM (~$10^6$ for a bias voltage of only tens of volts), relatively simple front-end electronics can be used, which maintain the original pulse shape, in contrast to the more complex front-end electronics required for room-temperature semiconductor detectors.

The SiPM also affects $R_{stat}$. For example, the photodetection efficiency (PDE) of an SiPM describes which fraction of the optical photons that have reached the SiPM are detected. The PDE, and therefore the energy resolution, can be improved by having (1) a high fill factor, which is easiest to obtain if the SPADs are large so that the dead areas between SPADs remain relatively small; (2) a good match between the scintillator's emission spectrum and the spectral sensitivity of the SPADs; and (3) a high overvoltage, which equals the difference between applied bias voltage and breakdown voltage (Zappalà et al. 2016). It is to be noted, though, that the first and the third option can also have a negative effect on $R_{stat}$ as described below.

Large SPADs reduce the number of SPADs that fit on an SiPM. This is especially relevant for the sub-mm pixels required for photon-counting X-ray imaging. The smaller the number of SPADs on an SiPM, the more likely that an optical photon from the scintillator is detected by a SPAD that is still recharging, such that the new discharge will be weaker. Consequently, the signal generated by the SPAD seems to correspond to "a fraction of an optical photon." The response of the SiPM is no longer proportional to the number of detected optical photons (Van Dam et al. 2010) and the SiPM is said to saturate. This has a negative effect on $R_{stat}$. The effect can be mitigated by reducing the SPAD size and the value of the recharge time constant. Interestingly, one way to reduce the value of the recharge time constant is by reducing the SPAD size (Acerbi et al. 2017).

A high overvoltage increases the PDE, but also the noise in SiPMs. Dark counts are the main source of uncorrelated noise. These are discharges due to thermally-induced electron hole-pairs. Typical dark-count rates of modern SiPMs at room-temperature are low ($10^4$–$10^5$ $s^{-1}$ $mm^{-2}$) compared to the X-ray fluence rate in medical imaging, such that the X-ray induced pulses are hardly affected by this noise source. The main correlated noise sources are afterpulsing and optical crosstalk. The former refers to discharges initiated by charge carriers that were trapped during an avalanche, but escaped some time later. The afterpulsing probability in modern SiPMs has been greatly reduced. Optical crosstalk is therefore the most important noise source for medical X-ray photon-counting applications. It is caused by the multitude of infrared photons that are emitted from a discharging SPAD. These optical crosstalk photons have a small chance to trigger a discharge in a nearby SPAD on the same SiPM. This process has been mathematically described by Vinogradov (2012). It worsens $R_{stat}$, because it more strongly increases the variance in the number of detected optical photons than the mean. In addition, it contributes to SiPM saturation.

### 10.6.4 Suitable Scintillators and SiPMs

A main challenge for SiPM-based X-ray photon-counting detectors is to find or develop scintillators and SiPMs that meet the requirements discussed in Sections 10.6.2 and 10.6.3, respectively. Regarding commercially available scintillators, cerium-doped inorganic scintillation crystals combine short decay time constants (10–100 ns) with high light yields (10–100 photons/keV). For example, $LaBr_3$:Ce, $LuAlO_3$:Ce (LuAP:Ce), and $YAlO_3$:Ce (YAP:Ce) appear interesting candidates, because they are among the fastest and most proportional cerium-doped scintillators. Specific challenges for these materials are a hygroscopic nature ($LaBr_3$:Ce) or slow decay components (LuAP:Ce and, to a lesser extent, YAP:Ce). A hygroscopic material needs a special sealing before it can be exposed to moisture in air. However, this has not prevented $LaBr_3$:Ce from being used in numerous commercial applications. Likewise, another hygroscopic scintillator, NaI:Tl, has been widely applied in nuclear medicine imaging equipment. The slow components in LuAP:Ce and YAP:Ce can be reduced (but, so far, not fully eliminated) by crystal engineering. Thus, it may be necessary to develop signal processing methods that mitigate the associated rate-dependent baseline shift.

All of the earlier-mentioned scintillators emit photons in the blue/near-ultraviolet part of the electromagnetic spectrum. This means that the SiPM should have a high PDE for those wavelengths. A variety of SiPMs have been developed for PET detectors by different manufacturers, which satisfy this requirement with a peak PDE around 420 nm. These commercially available SiPMs typically have a SPAD pitch of several tens of μm and recharge time constants in the order of tens of ns. With such SiPMs, the recharge time constant may dominate the pulse duration. Moreover, saturation may negatively affect the energy resolution of a finely pixelated detector. Fortunately, SiPMs with a smaller SPAD pitch, accompanied by a substantially shorter recharge time constant, are currently under development.

We recently developed and validated a pulse shape model that allows us to calculate the expected FWHM energy resolution, pulse duration, and associated rate capability, of SiPM-based scintillation detectors for photon-counting CT (van der Sar, Brunner, and Schaart 2021). The model takes into account the relevant scintillator and SiPM properties discussed in Sections 10.6.2 and 10.6.3, respectively. The pulse duration is defined as the moment in time at which the normalized cumulative signal of a detector output pulse reaches a value of 0.95 ($t_{95}$), as illustrated in Figure 10.5. Given the exponential distribution of X-ray photon inter-arrival times, a fluence

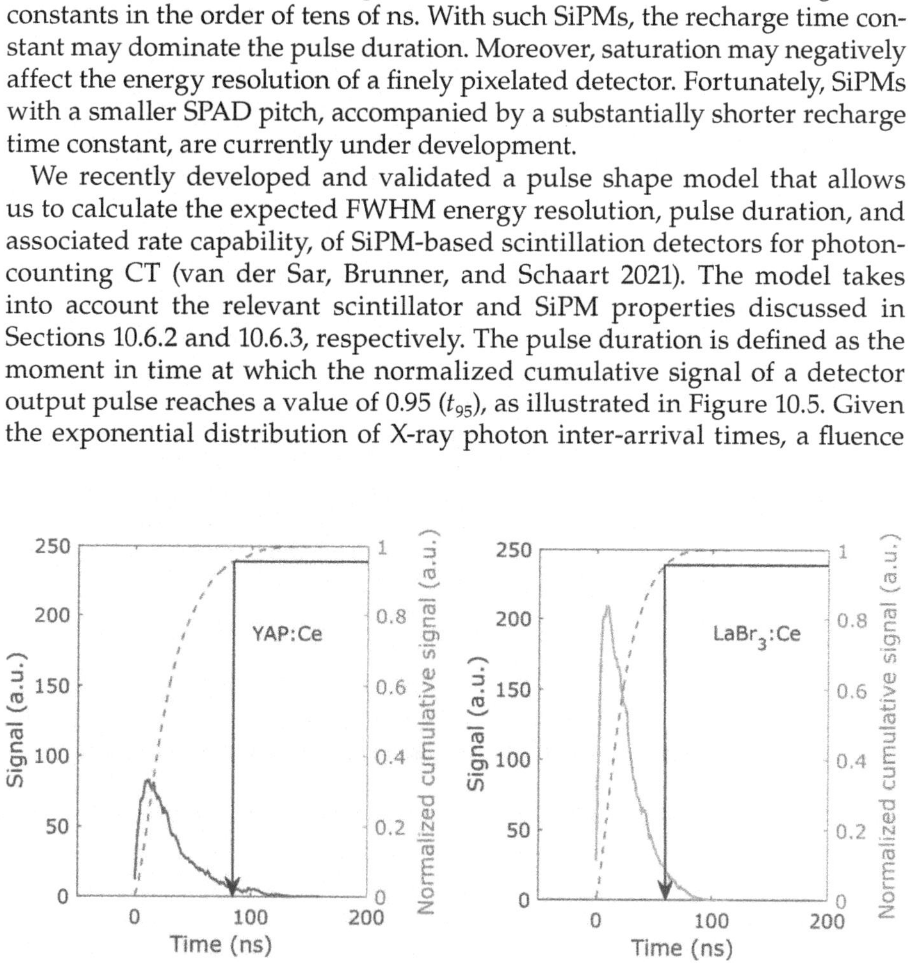


**FIGURE 10.5**
Two examples of pulse shapes from SiPM-based scintillation detectors simulated with the model presented in van der Sar, Brunner, and Schaart (2021). The pulse duration ($t_{95}$) is defined as the moment in time at which the normalized cumulative signal reaches a value of 0.95. The left pulse was simulated for a detector pixel with dimensions of 0.33 × 0.33 mm$^2$, consisting of a YAP:Ce scintillation crystal (Table 10.1) coupled to an SiPM with a SPAD pitch of 15 μm (Table 10.2). The right pulse was simulated for a detector pixel with dimensions of 0.4 × 0.4 mm$^2$, consisting of a $LaBr_3$:Ce scintillation crystal (Table 10.1) coupled to an SiPM with a SPAD pitch of 15 μm (Table 10.2). The simulated energy deposition equaled 60 keV and the light collection efficiency was set to a value of 1.0 in both cases.

**TABLE 10.1**

Properties of Two Potential Scintillators for Photon-Counting X-ray Detectors

| | YAP:Ce ($YAlO_3$:Ce) | $LaBr_3$:Ce |
|---|---|---|
| Density (g cm$^{-3}$) | 5.4 | 5.1 |
| Required thickness (mm) | 7.0 | 3.5 |
| Decay time constant (ns) | 25 | 16 |
| Light yield (photons/keV) | 25 | 63 |
| Non-proportionality factor @ 60 keV | 1.00 | 0.98 |
| Intrinsic energy resolution @ 60 keV (%) | 4.0 | 8.0 |

YAP:Ce data were obtained from Crytur, $LaBr_3$:Ce data from Saint Gobain Crystals. However, the non-proportionality factors and intrinsic resolutions were obtained from Moszyński et al. (2016) and Moszyński et al. (1998) for YAP:Ce, and from Chewpraditkul and Moszyński (2011) and Swiderski et al. (2009) for $LaBr_3$:Ce. The required thickness corresponds to an X-ray detection efficiency of almost 100% at 50 keV, about 90% at 100 keV, and about 55% at 150 keV.

rate $r_{50,pix}$ was calculated at which an X-ray photon has a 50% chance to arrive within $t_{95}$ from the previous X-ray photon on the same pixel: $r_{50,pix} = \ln(2)/t_{95}$. In that case, the resulting pulse can be considered to be affected by pile-up. $r_{50,pix}$ was converted into a rate capability per mm$^2$ $r_{50}$ by dividing $r_{50,pix}$ by the pixel size squared.

In van der Sar, Brunner, and Schaart (2021) we calculate the expected performance of detectors based on LuAP:Ce and $LaBr_3$:Ce as well as on LYSO:Ce, which is commonly employed in PET detectors. Here, we show the results for $LaBr_3$:Ce and present new data on YAP:Ce. The properties of these two scintillators are shown in Table 10.1. We also consider two SiPM technologies, i.e., a typically commercially available technology and a prototype SiPM with smaller SPAD pitch. Their properties are shown in Table 10.2. The detector performance is evaluated for 60 keV X-ray photons. The results are compared with the performance of the Philips CdTe/CZT detectors that were mentioned in Sections 10.4 and 10.4.1. These detectors produce Gaussian-shaped pulses with an FWHM of 20 ns, corresponding to $t_{95} = 34$ ns ($4\sigma$).

Figure 10.6 shows the outcomes of the model calculations for the combination of YAP:Ce and SiPMs with a SPAD pitch of 30 μm. As shown in Figure 10.6(a), pixel sizes as small as 200 μm are needed to achieve a similar rate capability as a CdTe/CZT detector with 500 μm pixels, for example. This is primarily due to the long recharge time constant of 55 ns of these SiPMs, which causes $t_{95}$ to be just below 200 ns. Also note that the energy resolution severely degrades for such small pixels due to SiPM saturation (Figure 10.6(b)).

Figure 10.7 shows the results of the model computations for the same YAP:Ce scintillator, when combined with the SiPMs with a SPAD pitch of

**TABLE 10.2**

Properties of a Commercially Available SiPM Technology with a SPAD Pitch of 30 μm and a Prototype SiPM with a Reduced SPAD Pitch of 15 μm, Both from Broadcom Inc (private communications)

| | SPAD pitch = 30 μm | | SPAD pitch = 15 μm | |
|---|---|---|---|---|
| | **Overvoltage 3.0 V** | **Overvoltage 7.0 V** | **Overvoltage 3.0 V** | **Overvoltage 7.0 V** |
| Recharge time constant (ns) | **55** | 50 | 9.0 | **7.0** |
| Photodetection efficiency | **0.41** (0.44) | 0.48 (0.55) | 0.21 (0.23) | **0.28** (0.30) |
| Optical crosstalk parameter λ | **0.1235** | 0.5753 | 0.0128 | **0.1235** |

The values in bold were used for the model calculations. The values for the photodetection efficiency take into account the emission spectra of YAP:Ce and $LaBr_3$:Ce, whereas the values in between brackets are the peak photodetection efficiencies at 420 nm. See Vinogradov (2012) for the exact meaning of the λ parameter. Data reproduced from van der Sar, Brunner, and Schaart (2021).

15 μm. The rate capability has considerably improved and the negative effect of SiPM saturation has been greatly reduced. The value of $t_{95}$ is about 80 ns, due to the much shorter recharge time constant of these SiPMs. Figure 10.7(a) shows that a pixel size of 333 μm would therefore provide

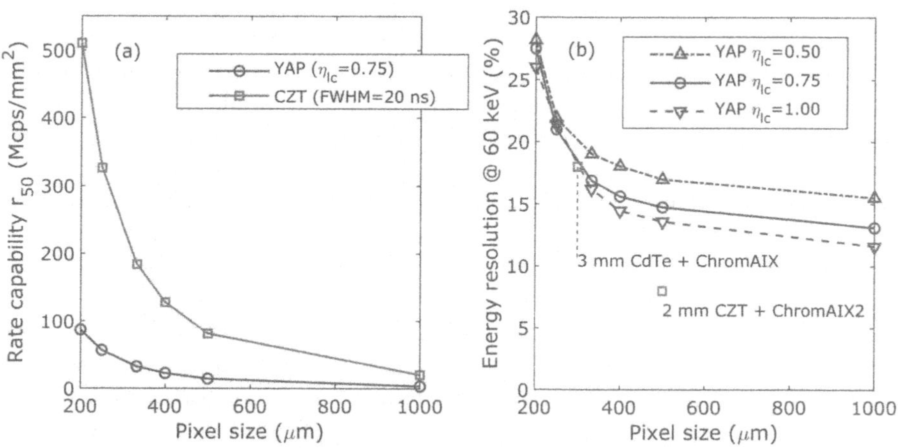


**FIGURE 10.6**

(a) Rate capability $r_{50}$ and (b) energy resolution at 60 keV, as a function of pixel size and light collection efficiency ($\eta_{lc}$), for the YAP:Ce scintillator coupled to the SiPMs with a SPAD pitch of 30 μm. The pulse duration, and therefore $r_{50}$, hardly depends on $\eta_{lc}$, so it is only shown for a value of 0.75. For comparison, data points reported for CdTe- and CZT-based photon-counting detectors are also shown (Steadman et al. 2011; Steadman, Herrmann, and Livne 2017).

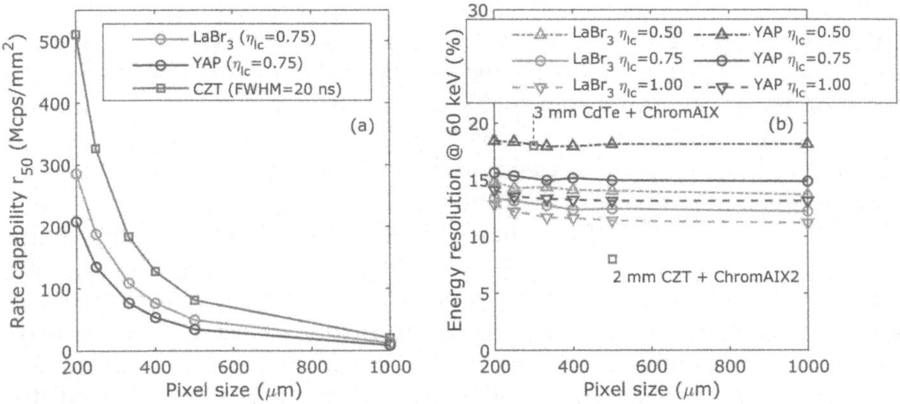


**FIGURE 10.7**
(a) Rate capability $r_{50}$ and (b) energy resolution at 60 keV, as a function of pixel size and light collection efficiency ($\eta_{lc}$), for the YAP:Ce and $LaBr_3$:Ce scintillators coupled to the SiPMs with a SPAD pitch of 15 μm. The pulse duration, and therefore $r_{50}$, hardly depends on $\eta_{lc}$, so it is only shown for a value of 0.75. Data for $LaBr_3$:Ce are the same as in van der Sar, Brunner, and Schaart (2021). For comparison, data points reported for CdTe- and CZT-based photon-counting detectors are also shown (Steadman et al. 2011), (Steadman, Herrmann, and Livne 2017).

similar rate capability as a CdTe/CZT detector with 500 μm pixels, for example. As shown in Figure 10.7(b), the values of the achievable energy resolution lie in between the two reference values for the CdTe/CZT detectors.

Figure 10.7 furthermore shows that even better rate capability and energy resolution can be achieved with $LaBr_3$:Ce in combination with the SiPMs with a SPAD pitch of 15 μm. Due to the shorter decay time constant of $LaBr_3$:Ce, $t_{95}$ is less than 60 ns. As can be appreciated from Figure 10.7(a), a pixel size of 400 μm would be sufficient to achieve similar rate capability as a CdTe/CZT detector with 500 μm pixels, for example. In addition, Figure 10.7(b) shows that the achievable energy resolution may be as good as 12% at 60 keV. Examples of a simulated detector pulse from a YAP:Ce-based detector with a pixel size of 333 μm and from a $LaBr_3$:Ce-based detector with a pixel size of 400 μm were shown in Figure 10.5.

## 10.7 Discussion and Conclusion

The results of our preliminary investigations show that SiPM-based scintillation detectors may have potential as an alternative for CdTe/CZT-based detectors for photon-counting X-ray imaging, provided that very fast scintillators with state-of-the art energy resolution, such as YAP:Ce and

$LaBr_3$:Ce, and SiPMs with a small SPAD pitch (e.g., 15 μm) and short recharge time constants (e.g., 7 ns) are used. The required pixel size is 400 μm or smaller, depending on the desired rate capability and the acceptable degradation of the spectral response due to secondary X-rays escaping a pixel. Such a small pixel size also means that innovative ways to realize dose-efficient optical isolation between pixels need to be investigated.

An important advantage of scintillation detectors is that they are based on light transport rather than charge transport, so charge sharing, charge trapping, polarization, and associated issues do not play a role. Hence, stable and reliable performance for long periods of time may be easier to achieve, as illustrated by the widespread use of scintillation detectors in other medical imaging modalities. Moreover, state-of-the-art PET scanners are already equipped with SiPM-based scintillation detectors. It is acknowledged that PET puts less stringent requirements on pulse duration, pixel size, and energy resolution than X-ray photon-counting. Yet, this example shows that high-performance medical imaging systems based on SiPM-based scintillation detectors can be produced on a large scale in a cost-effective way.

In conclusion, fast scintillators with state-of-the-art energy resolution coupled to SiPMs with a reduced SPAD pitch of, e.g., 15 μm may have potential for photon-counting X-ray imaging. The present result and the advantages of this detector concept warrant further research, so as to bring a cost-effective SiPM-based X-ray photon-counting detector showing stable and reliable performance within reach.

## References

Acerbi, Fabio, Alberto Gola, Veronica Regazzoni, Giovanni Paternoster, Giacomo Borghi, Nicola Zorzi, and Claudio Piemonte. 2017. "High efficiency, ultra-high-density silicon photomultipliers." *IEEE Journal of Selected Topics in Quantum Electronics* 24 (2):1–8. doi: 10.1109/JSTQE.2017.2748927.

Alvarez, Robert E, and Albert Macovski. 1976. "Energy-selective reconstructions in x-ray computerised tomography." *Physics in Medicine & Biology* 21 (5):733. doi: 10.1088/0031-9155/21/5/002.

Ballabriga, R, J Alozy, FN Bandi, M Campbell, N Egidos, JM Fernandez-Tenllado, EHM Heijne, I Kremastiotis, X Llopart, and BJ Madsen. 2020. "Photon counting detectors for X-ray imaging with emphasis on CT." *IEEE Transactions on Radiation and Plasma Medical Sciences*. doi: 10.1109/TRPMS.2020.3002949.

Bornefalk, Hans, and Mats Danielsson. 2010. "Photon-counting spectral computed tomography using silicon strip detectors: a feasibility study." *Physics in Medicine & Biology* 55 (7):1999. doi: 10.1088/0031-9155/55/7/014.

Chewpraditkul, Weerapong, and Marek Moszyński. 2011. "Scintillation properties of $Lu_3Al_5O_{12}$, $Lu_2SiO_5$ and $LaBr_3$ crystals activated with cerium." *Physics Procedia* 22:218–226. doi: 10.1016/j.phpro.2011.11.035.

da Silva, Joakim, Fredrik Grönberg, Björn Cederström, Mats Persson, Martin Sjölin, Zlatan Alagic, Robert Bujila, and Mats Danielsson. 2019. "Resolution characterization of a silicon-based, photon-counting computed tomography prototype capable of patient scanning." *Journal of Medical Imaging* 6 (4):043502. doi: 10.1117/1.JMI.6.4.043502.

Danielsson, Mats, Mats Persson, and Martin Sjölin. 2020. "Photon-counting x-ray detectors for CT." *Physics in Medicine & Biology* in press. doi: 10.1088/1361-6560/abc5a5.

Dorenbos, P, JTM de Haas, and CWE Van Eijk. 1995. "Non-proportionality in the scintillation response and the energy resolution obtainable with scintillation crystals." *IEEE Transactions on Nuclear Science* 42 (6):2190–2202. doi: 10.1109/23.489415.

Flohr, Thomas, Martin Petersilka, Andre Henning, Stefan Ulzheimer, Jiri Ferda, and Bernhard Schmidt. 2020. "Photon-counting CT review." *Physica Medica* 79:126–136. doi: 10.1016/j.ejmp.2020.10.030.

Herrmann, Christoph, Ira Blevis, Roger Steadman, and Amir Livne. 2020. "Issues of ohmic contacts in human medical photon-counting CT detectors." *IEEE Transactions on Radiation and Plasma Medical Sciences*. doi: 10.1109/TRPMS.2020.3026832.

Marano, Davide, Massimiliano Belluso, Giovanni Bonanno, Sergio Billotta, Alessandro Grillo, Salvatore Garozzo, and Giuseppe Romeo. 2014. "Accurate analytical single-photoelectron response of silicon photomultipliers." *IEEE Sensors Journal* 14 (8):2749–2754. doi: 10.1109/JSEN.2014.2316363.

McCollough, Cynthia H, Shuai Leng, Lifeng Yu, and Joel G Fletcher. 2015. "Dual-and multi-energy CT: principles, technical approaches, and clinical applications." *Radiology* 276 (3):637–653. doi: 10.1148/radiol.2015142631.

Michel, T, G Anton, M Böhnel, J Durst, M Firsching, A Korn, B Kreisler, A Loehr, F Nachtrab, and D Niederlöhner. 2006. "A fundamental method to determine the signal-to-noise ratio (SNR) and detective quantum efficiency (DQE) for a photon counting pixel detector." *Nuclear Instruments and Methods in Physics Research Section A: Accelerators, Spectrometers, Detectors and Associated Equipment* 568 (2): 799–802. doi: 10.1016/j.nima.2006.08.115.

Moszyński, M, M Kapusta, D Wolski, W Klamra, and B Cederwall. 1998. "Properties of the YAP:Ce scintillator." *Nuclear Instruments and Methods in Physics Research Section A* 404 (1):157–165. doi: 10.1016/S0168-9002(97)01115-7.

Moszyński, M, A Syntfeld-Każuch, L Swiderski, M Grodzicka, J Iwanowska, P Sibczyński, and T Szczęśniak. 2016. "Energy resolution of scintillation detectors." *Nuclear Instruments and Methods in Physics Research Section A* 805:25–35. doi: 10.1016/j.nima.2015.07.059.

Persson, Mats, Robert Bujila, Patrik Nowik, Henrik Andersson, Love Kull, Jonas Andersson, Hans Bornefalk, and Mats Danielsson. 2016. "Upper limits of the photon fluence rate on CT detectors: case study on a commercial scanner." *Medical Physics* 43 (7):4398–4411. doi: 10.1118/1.4954008.

Prince, Jerry L, and Jonathan M Links. 2015. *Medical Imaging Signals and Systems*. Upper Saddle River, NJ: Pearson Prentice Hall.

Roessl, E, and R Proksa. 2007. "K-edge imaging in x-ray computed tomography using multi-bin photon counting detectors." *Physics in Medicine & Biology* 52 (15):4679. doi: 10.1088/0031-9155/52/15/020.

Roy, UN, GS Camarda, Y Cui, R Gul, A Hossain, G Yang, J Zazvorka, V Dedic, J Franc, and RB James. 2019. "Role of selenium addition to CdZnTe matrix for room-temperature radiation detector applications." *Scientific Reports* 9 (1):1–7. doi: 10.1038/s41598-018-38188-w.

Roy, UN, GS Camarda, Y Cui, R Gul, G Yang, and RB James. 2018. "Charge-transport properties of as-grown $Cd_{1-x}Zn_xTe_{1-y}Se_y$ by the traveling heater method." *AIP Advances* 8 (12):125015. doi: 10.1063/1.5064373.

Sjölin, Martin, and Mats Danielsson. 2016. "Relative calibration of energy thresholds on multi-bin spectral x-ray detectors." *Nuclear Instruments and Methods in Physics Research Section A: Accelerators, Spectrometers, Detectors and Associated Equipment* 840:1–4. doi: 10.1016/J.NIMA.2016.09.045.

Steadman, Roger, Christoph Herrmann, and Amir Livne. 2017. "ChromAIX2: a large area, high count-rate energy-resolving photon counting ASIC for a spectral CT prototype." *Nuclear Instruments and Methods in Physics Research Section A: Accelerators, Spectrometers, Detectors and Associated Equipment* 862:18–24. doi: 10.1016/j.nima.2017.05.010.

Steadman, Roger, Christoph Herrmann, Oliver Mülhens, and Dale G Maeding. 2011. "ChromAIX: fast photon-counting ASIC for spectral computed tomography." *Nuclear Instruments and Methods in Physics Research A* 648:S211–S215. doi: 10.1016/j.nima.2010.11.149.

Sundberg, Christel, Mats U Persson, Martin Sjölin, J Jacob Wikner, and Mats Danielsson. 2020. "Silicon photon-counting detector for full-field CT using an ASIC with adjustable shaping time." *Journal of Medical Imaging* 7 (5):053503. doi: 10.1117/1.JMI.7.5.053503.

Swank, Robert K. 1973. "Absorption and noise in x-ray phosphors." *Journal of Applied Physics* 44 (9):4199–4203. doi: 10.1063/1.1662918.

Swiderski, Lukasz, Marek Moszyński, Antoni Nassalski, Agnieszka Syntfeld-Kazuch, Tomasz Szczesniak, Kei Kamada, Kousuke Tsutsumi, Yoshiyuki Usuki, Takayuki Yanagida, and Akira Yoshikawa. 2009. "Scintillation properties of praseodymium doped LuAG scintillator compared to cerium doped LuAG, LSO and $LaBr_3$." *IEEE Transactions on Nuclear Science* 56 (4):2499–2505. doi: 10.1109/TNS.2009.2025040.

Taguchi, Katsuyuki, and Jan S Iwanczyk. 2013. "Vision 20/20: single photon counting x-ray detectors in medical imaging." *Medical Physics* 40 (10):100901.

Tao, Shengzhen, Kishore Rajendran, Cynthia H McCollough, and Shuai Leng. 2019. "Feasibility of multi-contrast imaging on dual-source photon counting detector (PCD) CT: an initial phantom study." *Medical Physics* 46 (9):4105–4115. doi: 10.1002/mp.13668.

Van Dam, Herman T, Stefan Seifert, Ruud Vinke, Peter Dendooven, Herbert Löhner, Freek J Beekman, and Dennis R Schaart. 2010. "A comprehensive model of the response of silicon photomultipliers." *IEEE Transactions on Nuclear Science* 57 (4): 2254–2266. doi: 10.1109/TNS.2010.2053048.

van der Sar, Stefan J, Stefan E Brunner, and Dennis R Schaart. 2021. "Silicon photomultiplier-based scintillation detectors for photon-counting CT: a feasibility study." *Medical Physics* in press. doi: 10.1002/mp.14886

Vinogradov, S. 2012. "Analytical models of probability distribution and excess noise factor of solid state photomultiplier signals with crosstalk." *Nuclear Instruments and Methods in Physics Research Section A: Accelerators, Spectrometers, Detectors and Associated Equipment* 695:247–251. doi: 10.1016/j.nima.2011.11.086.

Xu, J, W Zbijewski, G Gang, JW Stayman, K Taguchi, M Lundqvist, E Fredenberg, JA Carrino, and JH Siewerdsen. 2014. "Cascaded systems analysis of photon counting detectors." *Medical Physics* 41 (10):101907. doi: 10.1118/1.4894733.

Yu, Zhicong, Shuai Leng, Steven M Jorgensen, Zhoubo Li, Ralf Gutjahr, Baiyu Chen, Ahmed F Halaweish, Steffen Kappler, Lifeng Yu, and Erik L Ritman. 2016. "Evaluation of conventional imaging performance in a research whole-body CT system with a photon-counting detector array." *Physics in Medicine & Biology* 61 (4):1572. doi: 10.1088/0031-9155/61/4/1572.

Zappalà, Gaetano, Fabio Acerbi, Alessandro Ferri, A Gola, Giovanni Paternoster, Veronica Regazzoni, Nicola Zorzi, and Claudio Piemonte. 2016. "Study of the photo-detection efficiency of FBK high-density silicon photomultipliers." *Journal of Instrumentation* 11 (11):P11010. doi: 10.1088/1748-0221/11/11/P11010.

# *Index*

Note: Locators in *italics* represent figures and **bold** indicate tables in the text.

## D